Andreas Sudmann, Anna Echterhölter, Markus Ramsauer, Fabian Retkowski,
Jens Schröter, Alexander Waibel (eds.)
Beyond Quantity

KI-Kritik / AI Critique | Volume 6

Die E-Book-Ausgabe erscheint im Rahmen der »Open Library Medienwissenschaft 2023« im Open Access. Der Titel wurde dafür von deren Fachbeirat ausgewählt und ausgezeichnet. Die Open-Access-Bereitstellung erfolgt mit Mitteln der »Open Library Community Medienwissenschaft 2023«.

Die Formierung des Konsortiums wurde unterstützt durch das BMBF (Förderkennzeichen 16TOA002).

Die Open Library Community Medienwissenschaft 2023 ist ein Netzwerk wissenschaftlicher Bibliotheken zur Förderung von Open Access in den Sozial- und Geisteswissenschaften:

Vollsponsoren: Technische Universität Berlin / Universitätsbibliothek | Universitätsbibliothek der Humboldt-Universität zu Berlin | Staatsbibliothek zu Berlin – Preußischer Kulturbesitz | Universitätsbibliothek Bielefeld | Universitätsbibliothek Bochum | Universitäts- und Landesbibliothek Bonn | Technische Universität Braunschweig | Universitätsbibliothek Chemnitz | Universitäts- und Landesbibliothek Darmstadt | Sächsische Landesbibliothek, Staats- und Universitätsbibliothek Dresden (SLUB Dresden) | Universitätsbibliothek Duisburg-Essen | Universitäts- und Landesbibliothek Düsseldorf | Goethe-Universität Frankfurt am Main / Universitätsbibliothek | Universitätsbibliothek Freiberg | Albert-Ludwigs-Universität Freiburg / Universitätsbibliothek | Niedersächsische Staats- und Universitätsbibliothek Göttingen | Universitätsbibliothek der FernUniversität in Hagen | Staats- und Universitätsbibliothek Hamburg | Gottfried Wilhelm Leibniz Bibliothek - Niedersächsische Landesbibliothek | Technische Informationsbibliothek (TIB) Hannover | Karlsruher Institut für Technologie (KIT) | Universitätsbibliothek Kassel | Universität zu Köln, Universitäts- und Stadtbibliothek | Universitätsbibliothek Leipzig | Universitätsbibliothek Mannheim | Universitätsbibliothek Marburg | Ludwig-Maximilians-Universität München / Universitätsbibliothek | FH Münster | Bibliotheks- und Informationssystem (BIS) der Carl von Ossietzky Universität | Oldenburg | Universitätsbibliothek Siegen | Universitätsbibliothek Vechta | Universitätsbibliothek der Bauhaus-Universität Weimar | Zentralbibliothek Zürich | Zürcher Hochschule der Künste

Sponsoring Light: Universität der Künste Berlin, Universitätsbibliothek | Freie Universität Berlin | Hochschulbibliothek der Fachhochschule Bielefeld | Hochschule für Bildende Künste Braunschweig | Fachhochschule Dortmund, Hochschulbibliothek | Hochschule für Technik und Wirtschaft Dresden - Bibliothek | Hochschule Hannover - Bibliothek | Hochschule für Technik, Wirtschaft und Kultur Leipzig | Hochschule Mittweida, Hochschulbibliothek | Landesbibliothek Oldenburg | Akademie der bildenden Künste Wien, Universitätsbibliothek | Jade Hochschule Wilhelmshaven/Oldenburg/Elsfleth | ZHAW Zürcher Hochschule für Angewandte Wissenschaften, Hochschulbibliothek

Mikrosponsoring: Ostbayerische Technische Hochschule Amberg-Weiden | Deutsches Zentrum für Integrations- und Migrationsforschung (DeZIM) e.V. | Max Weber Stiftung – Deutsche Geisteswissenschaftliche Institute im Ausland | Evangelische Hochschule Dresden | Hochschule für Bildende Künste Dresden | Hochschule für Musik Carl Maria Weber Dresden Bibliothek | Filmmuseum Düsseldorf | Universitätsbibliothek Eichstätt-Ingolstadt | Bibliothek der Pädagogischen Hochschule Freiburg | Berufsakademie Sachsen | Bibliothek der Hochschule für Musik und Theater Hamburg | Hochschule Hamm-Lippstadt | Bibliothek der Hochschule für Musik, Theater und Medien Hannover | HS Fresenius gemGmbH | ZKM Zentrum für Kunst und Medien Karlsruhe | Hochschule für Grafik und Buchkunst Leipzig | Hochschule für Musik und Theater »Felix Mendelssohn Bartholdy« Leipzig, Bibliothek | Filmuniversität Babelsberg KONRAD WOLF - Universitätsbibliothek | Universitätsbibliothek Regensburg | THWS Technische Hochschule Würzburg-Schweinfurt | Hochschule Zittau/Görlitz, Hochschulbibliothek | Westsächsische Hochschule Zwickau | Palucca Hochschule für Tanz Dresden

Andreas Sudmann, Anna Echterhölter, Markus Ramsauer, Fabian Retkowski, Jens Schröter, Alexander Waibel (eds.)

Beyond Quantity

Research with Subsymbolic AI

[transcript]

In cooperation with Julia Herbach, Liana Popa, and Jeffrey Röchling.

Funded by the Volkswagen Foundation

Bibliographic information published by the Deutsche Nationalbibliothek
The Deutsche Nationalbibliothek lists this publication in the Deutsche Nationalbibliografie; detailed bibliographic data are available in the Internet at http://dnb.d-nb.de

First published in 2023 by transcript Verlag, Bielefeld

Cover layout: Maria Arndt, Bielefeld
Cover illustration: Julia Herbach, Cologne
Printed by: Majuskel Medienproduktion GmbH, Wetzlar
https://doi.org/10.14361/9783839467664
Print-ISBN: 978-3-8376-6766-0
PDF-ISBN: 978-3-8394-6766-4
EPUB-ISBN: 978-3-7328-6766-0
ISSN of series: 2698-7546
eISSN of series: 2703-0555

Printed on permanent acid-free text paper.

Contents

Acknowledgements

First of all, we would like to thank our authors for the wonderful contributions and discussions as well as our student assistants very much for their support in the realization of this book: Julia Herbach for her assistance with the organization of the "Beyond Quantity" conference, the coordination and editing of the book's contributions, as well as the book's cover design; Liana Popa for the coordination, and especially the extensive linguistic editing of all contributions to this book; Jeffrey Röchling for his assistance during the review process and the editing of the book's contributions. We also very much appreciate the work of many of our colleagues for providing reviews and/or feedback for this collection. Further thanks go to our publisher transcript, in particular to Johanna Mittelgöker, as editor of this publication, to Dagmar Buchwald, as well as to Anna Tuschling and Bernhard Dotzler, the editors of the series "AI Critique".

This book publication is part of the research group "How is AI Changing Science? Research in the Era of Learning Algorithms" (HiAICS), which has been generously funded by the Volkswagen Foundation since 2022 (planning grant since 2019).

Finally, we would like to thank the Sorbonne Center of Artificial Intelligence, especially Gérard Biau, Xavier Fresquet and Nora Roger, for being such wonderful hosts to the opening conference of the HiAICS research group.

Introduction

Andreas Sudmann, Anna Echterhölter, Markus Ramsauer, Fabian Retkowski, Jens Schröter, Alexander Waibel

For more than ten years now, we have witnessed an AI boom affecting basically all areas of culture and society, including the scientific field. This book explores the potentially profound transformation in academic research.

Such a focus is not only aiming at the question of AI's impact, be it as a technology, a component of a larger infrastructure, or a tool. It is also about exploring what AI as a concept actually means, which different techniques and approaches it addresses, to what extent it might be important to continue the long tradition of problematizing it, and last but not least, how a particular understanding of AI might be transformed by the practices and conditions of its scientific situatedness and application (Suchman 2006).

As a point of departure for the following considerations, we engage with the history of AI as a contest between two fundamental approaches: the symbolic and the subsymbolic (see also Dreyfus/Dreyfus 1988: 15–43). The former, also known as GOFAI (Good Old-Fashioned Artificial Intelligence), processes knowledge and tasks based on logical or rule-based procedures. Knowledge is explicitly represented, often hard-coded and manually entered into the system by experts. The latter is characterized by the fact that corresponding procedures seek to find patterns and correlations in data automatically. This approach involves statistical and neural models to learn from data without relying on explicitly defined rules. Knowledge representation operates in an implicit manner. For example, knowledge can be implicitly encoded in the weights of a neural network. While this allows these systems to process sizable amounts of complex, unstructured data, it is also responsible for their black-box nature.

Subsymbolic AI and its scientific impact are the focus of this book. More specifically, the contributions from various fields shed light on artificial neural networks (ANNs) as the currently dominant and discourse-determining forms of AI, which are broadly inspired by the neuroinformatic model of the brain (mostly related to humans, but also with regard to animals).[1] In fact, prima facie it seems as if the subsymbolic approach of ANNs (mostly even only based on backpropagation) has become synonymous with AI as it has largely supplanted symbolic AI and even other machine learning methods, including symbolic learning, statistical learning, and Hidden Markov models, among others.

However, such a thesis must be differentiated more precisely in at least two respects: on the one hand with regards to the long tradition of hybrid connections of symbolic and subsymbolic methods; on the other hand that this distinction becomes blurred when traditionally symbolic problems (e.g., language processing) are increasingly handled on a neural substrate as well (e.g., machine translation, parsing, large language models (LLMs)). The term "neurosymbolic AI" is pertinent in this context, as it refers to hybrid systems that integrate neural models with symbolic AI. In his Robert S. Engelmore Memorial Lecture at AAAI 2020, Henry Kautz (2022) provided a taxonomy of neurosymbolic AI systems. One of the categories he introduced was "symbolic Neuro symbolic" (ibid.: 118), which also directly applies to LLMs. Systems in this category have their inputs and outputs presented as a symbolic form and even natural language with its discrete tokens counts towards that. Although these systems are not widely regarded as neuro-symbolic, it does make the term more ambiguous. The "Neuro[Symbolic]" (ibid.: 119) category may be of greater interest and relevancy, as it embeds symbolic reasoning as part of the neural engine. A new development with contemporary LLMs, such as ChatGPT or Toolformer, is the ability to interact with plugins. One such plugin can be a symbolic reasoning engine like WolframAlpha.

However, the relations and connections between symbolic AI and the subsymbolic AI of ANNs are no impediment for focussing on the latter for the purposes of this book – on the contrary. Above all, given their specific history, the current relevance of ANNs is quite remarkable. Their technical foundations were already developed as early as the 1940s and 1950s (Sudmann 2018a; Sudmann 2018b), but more complex, foundational architectures emerged in the 1980s and 1990s as they enabled ANNs to operate on real-world problems that required context, shift invariance, or sequential processing (Waibel et al. 1987; Waibel et al. 1989; Hochreiter/Schmidhuber 1997; LeCun et al. 1998). Nevertheless, until the 2000s, ANNs were largely ignored by practitioners and struggled to find broad adoption: Because the computing of the day would

1 In addition to neural networks, other approaches or algorithms can also be subsumed as subsymbolic AI, for example k-nearest neighbors.

only permit training of small networks, simpler statistical methods could already deliver competitive performance. It would take 20 more years until computational resources and data had scaled sufficiently for ANNs to show their true potential: From networks with a dozen or hundred connections and a single hidden layer, we now see networks with 175 billion parameters (GPT-3) and dozens or hundreds of layers. And ANNs could now deliver (with the same or similar algorithms as in the 1980s) impressive performance advances over classical methods. In speech recognition, error rate reductions of 30% or more were observed on published benchmarks. In vision, significant improvements could be obtained over standardized object classification benchmarks (see ImageNet, Krizhevsky/Sutskever/Hinton 2012). And even in machine translation, performance leapt forward through the adoption of large recurrent neural encoder-decoder networks (Luong/Manning 2015). In many domains, e.g., speech (Nguyen/Stueker/Waibel 2020), vision, machine translation, performance now exceeds human capabilities over certain defined benchmarks.

Another decisive part was played by big tech. The AI renaissance was accelerated as soon as the information industry became aware of the economic potential of ANNs. This resulted in a concerted move to massively expand AI research activities, invest in computing resources, and to acquire and merge promising AI start-ups (like DeepMind, and others).

The widely broadcasted 2016 victory of the AlphaGo program over Go master Lee Sedol had a further reinforcing effect with regard to the perception of the lingering capabilities of AI. This media event significantly shaped public perception. Subsequently, experts in various scientific fields were alerted, and increasingly interested in AI and, ultimately, began to integrate the new technology into various methodological toolkits. Somewhat unexpectedly, the release of ChatGPT in 2022 proved to be another game-changer. AI could finally be experienced and utilized by a wider circle of users, an encounter that swept public perception, and made it impossible to overlook the ramifications of this new technology for the most basic practices of mainstream science, its quotation standards, and academic exams. Besides questions of authorship and reliability, one important provocation may lie with the political and moralistic overtones of these chatty AIs. Furthermore, as language models, they merely predict text based on massive amounts of past textual data und thus ethical standards or factual correctness can not be assured as of yet. Even if a majority-driven form of reinforcement learning from human feedback decides about the biases of such machines, a "mathematisation of ethics" and a quantitative vote for majority morals is at hand (Simanowski 2023: 73). Still, large self-supervised models like LLMs can digest virtually all of humanity's textual data and thus generate predictions with surprising accuracy and relevance, resulting in a powerful illusion of human-like intelligence and clarity.

Nevertheless, it is necessary to unravel this rather event-centered and also person-centered historiography in more detail. For example, backpropagation as a central learning algorithm of ANNs was already developed in the 1970s and 1980s, some elements of it even as early as the 1960s.[2] Accordingly, it is difficult to attribute the development of this algorithm to just one person or one group of people at a specific time. Moreover, it has been and continues to be the case that the development of AI in many areas is based on close cooperation between industry and science, but also the military. Not least for this reason, AI research has always been, to a considerable extent, applied research.

To understand these transdisciplinary effects of the new technology, we must examine the level of data practices and scientific methodologies. Several recent publications have been addressing the impact of new AI technologies on scientific practices (Athey 2018; Fecher et al. 2023; Gethmann et al. 2022; Okerlund et al. 2022). At the same time, it seems evident that we are witnessing the effects of a much longer history of data, statistics, formalization, modeling, and simulation. Since the early days of AI, attempts were made to put 'intelligent' systems to use in various academic settings[3], but the corresponding reflections, if they had their place in the sciences at all, remained, in most cases, either necessarily speculative or their lasting contribution to the development of a research field ultimately proved to be extremely limited. There were, for example, early attempts to use AI systems for specified scientific tasks such as proving theorems (see Feigenbaum/Feldman 1963; Dick 2011), but corresponding implementations of the systems were typically very far from actually advancing research in the respective areas of knowledge.

With the successive establishment of so-called expert systems starting in the 1970s, the application-oriented perspective of AI finally gained some relevance, but this upswing ultimately did not last either. It is quite telling that Pamela McCorduck's relevant study on the history of AI – *Machines who Think* –

2 For a technical history of backpropagation related to ANNs, see for example Schmidhuber (2022).

3 For discussing AI in the context of psychology, see for example Hunt (1968: 135–168); for organic chemistry, see Feigenbaum (1968: 23–27).

contains a separate chapter titled "Applied Artificial Intelligence", which introduces two of these early expert systems and their respective application contexts. But remarkably enough, this chapter begins by pointing out how AI is derided and mocked in terms of its supposed potential on a regular basis (McCorduck 1979: 272f.).

The latter has not fundamentally changed today, even in light of the considerable achievements of large language models like ChatGPT. There continues to be a pronounced interest as well as a certain pleasure to expose the shortcomings of even the most advanced AI systems. Nevertheless, there is a significant shift in this respect: Currently, AI is no longer a speculative concept at its core; the relevant point of reference for (critical) reflections now is the concrete implementation of corresponding systems, not only with respect to areas of academic knowledge but all areas of culture and society.

There is little doubt about the fundamental importance of AI in all spheres of social life, given the prevailing assessments in public discourse. Furthermore, there seems to be no sign of an imminent end to today's AI boom. Following many booms and busts of previous AI excitement and promised revolutions, AI has now found its firm and sustainable footing. This is especially true for applications of AI in various fields of science, as countless research examples demonstrate (for an exemplary overview of AI research projects in Europe, see "How is Artificial Intelligence Changing Science?" 2023).

Unsupervised and self-supervised algorithms and the increasing use of simulations and data augmentation have advanced practical AI applications to astonishing performance levels and opened new applications. Sharing of open-source code, tools and large pretrained models now also accelerate progress by leapfrogging from one accomplishment to another at unprecedented speed. Google DeepMind, for example, has released a series of specialized models that aim to assist researchers in their respective fields, including AlphaFold (Jumper et al. 2021) which is able to predict 3D structures of proteins more accurately than previous models and, more importantly, is in many cases accurate enough to replace real-life experiments. AlphaFold is arguably the organization's biggest success so far and is now deeply ingrained as a tool in medicine and life sciences (Varadi/Velankar, 2022). More recently, Google DeepMind published AlphaTensor (Fawzi et al. 2022) and AlphaDev (Mankowitz et al. 2023), both of which have been used in the research area of computer science to optimize algorithms and low-level code such as matrix multiplication and sorting algorithms. In the case of AlphaTensor, the model was able to find an algorithm to reduce the number of multiplications necessary for certain types of matrix multiplications. On the related blog posting, Google DeepMind's headline "optimising the world's code one algorithm at a time" (Mankowitz/Michi 2023), aptly describes its current approach.

At present, the contributions of AI to scientific challenges are not always as spectacular as in the case of AlphaFold; often enough, standard AI technologies are used as elements of methods or in everyday applications (although usually at much better performance). However, it is remarkable how diversely and broadly AI is now being applied in various fields of research. In sports science, ML-based pattern recognition is increasingly used for the performance analysis of athletes, players and teams (Araújo et al. 2021). In art history, a computer vision system has been able to identify connections between artworks by analyzing poses of human subjects in paintings (Jeniczek & Chum, 2019). ML has also, for a long time, been used in particle physics, due to the enormous datasets analyzed in this field. In 2012, one of the important discoveries, the Higgs boson, owed much to the application of machine learning (Radovic et al. 2018; Bourilkov 2019).

Even though the general AI boom has been felt in many scientific fields for years now, one should note that the application of AI in many disciplines is still in its infancy. In our view, it is therefore even more important and timely to recognize, reflect on, and historically document this transformation of the sciences by AI in statu nascendi. To address this challenge, our transdisciplinary research group, encompassing the disciplines of media studies, computer science, and the history of science, has started its work in 2019, respectively 2022, to investigate the ways in which research is conducted not only on/about AI but *with* AI, in various fields from the natural and social sciences to the humanities. In particular, we are interested in exploring how AI interacts with the established practices and methods of science, whether they are complemented, modified, and/or potentially replaced.

Three disciplines or domains of research are at the center of our inquiry: environmental sciences/climatology, social sciences/sociology, and film studies. Three additional fields – literary studies, medicine and economics – are investigated to broaden the range of disciplines to be studied, partly in order to capture the heterogeneous range of uses of AI more accurately and to better generalize our results across scientific disciplines. In a first programmatic paper, the research group has already discussed some key challenges and perspectives, as well as some general considerations (Echterhölter/Schröter/Sudmann 2021).

The question of the transformations of the sciences through AI requires description of precisely their different scales and dimensions, as well as the general heterogeneity of the aspects addressed by them. One way of illustrating the range of conceptualizations can be the marking of extreme positions and ways of thinking, thus allowing for a more nuanced perspective. For example, one might argue that as an advanced Artificial General Intelligence (AGI)[4] system evolves, it would also be capable of handling any (new) scientific problem. Another option would be to develop a system, however specialized, that is used for more or less specific scientific tasks or only within a certain domain or discipline. Both concepts can be imagined as systems of a "superintelligence", to pick up Bostrom's (2014) term, insofar as the abilities and skills of human systems are (or can be) clearly surpassed in both scenarios.

In the emphatic sense, AI stands for the possibility of a computer being able to gain its own insights, formulate questions and hypotheses at some point, and thus also complete all other steps along this path more or less autonomously. AI systems used for scientific (research) purposes can be further differentiated according to how human-like they have been designed and oriented. Explainable AI requirements make it at least likely for machine communication to remain connectable to human understanding and control.[5] This also applies to future machine-machine communication. AI processes can also be differentiated according to the extent to which they organize individual components/phases of scientific research processes autonomously or automatically, from formulating a research question to collecting data, analyzing and evaluating data, as well as presenting and disseminating research findings.

Furthermore, there is the fundamental question of which scientific problems seem a priori suitable to be addressed by AI at all. DeepMind (Hassabis 2022) has developed three criteria in this respect:

---

4 The notion denotes a hypothetical AI-system with cognitive, creative etc. capabilities comparable to or even exceeding those of humans. There are no realizations of such systems yet. Their development, if it will be possible at some point, is repeatedly discussed as a great danger.

5 In recent years, research on XAI systems has become increasingly important, and this applies in particular to scientific applications of AI. An overall very promising project in the German-speaking area was recently initiated with the Transregio "Constructing Explainability" of the Universities of Bielefeld and Paderborn.

1. massive combinatorial space,
2. clear objective function (metric) to optimise against,
3. either lots of data and/or an accurate and efficient simulator.

As can be seen from these criteria, the use of ML must be carefully considered, especially with regard to the significant resources and costs involved.

A relatively recent phenomenon to reflect within our study is the fact that more and more explicitly AI-driven tools or apps are either directly intended for scientific work or can be indirectly used for it. A plethora of commercial applications like SciSpace Copilot or Elicit has been launched, promising to automate certain research workflows and help with literature research or understanding literature. Language models like ChatGPT are actively used by researchers as assistance in the writing process of scientific documents, prompting repositories and journals like arXiv to define a 'use of generative AI language tools' policy for authors. Towards the end of last year, the domain-specific language model Galactica (Taylor et al. 2022) caused a stir among the research community. It is exclusively trained on scientific data like research papers, chemical formulas, and DNA sequences. The generated text of the model sounded convincingly scientific but triggered concerns that it could easily spread inaccuracies. At the same time, there is a class of models, such as Minerva (Lewkowycz et al. 2022) and AI Descartes (Cornelio et al. 2021) that are used in research itself and are intended to automate reasoning processes.

Beyond such specific applications, it seems important to us to explore the general tool character and principal potential of current data-driven, statistical AI systems in methodological terms. A few years ago, computer scientist Pedro Domingos described ML as the "scientific method on steroids" (Domingos 2015: 13). Such a description strikes us as highly questionable as it conceptualizes ML per se as a scientific method. In addition, the metaphor "on steroids" suggests that ML allows an almost illegal and unhealthy form of performance enhancement in this respect. Nevertheless, it is obvious that the performance level of learning algorithms significantly increases when corresponding systems are trained with more and more data and computational power. The present publication, therefore, is also motivated by an interest in discussing AI through the lens of the ways in which learning algorithms potentially reconfigure the epistemic relationship of qualities and quantities. More specifically, we would like to shift the perspective on this relationship by highlighting epistemic aspects *beyond quantity* and thus also illuminate perspectives beyond the dominant relation of AI and big data. Two aspects are particularly important to us in this respect:

Firstly, current approaches to AI, i.e., subsymbolic AI in the form of ANNs, are not merely capable of extracting information from large amounts of data and making it productive, but they can solve problems that can also be described as qualitative. They involve dealing with qualitative questions of content, aesthetics, style, e.g., in the field of natural language processing or computer vision, in ways that were unimaginable until recently.[6]

Secondly, current approaches in AI research are increasingly focused on reducing or avoiding dependence on large amounts of labeled data, e.g., through strategies of self-supervised learning, zero- or one-shot learning, transfer learning, or even the use of synthetic data or simulation data. Contemporary LLMs are, for example, the result of causal language modeling which is a type of self-supervised learning in the course of which the model is tasked with predicting the next token in a sequence while requiring no additional labels.

As a result of our project's opening conference hosted by the Sorbonne Center for Artificial Intelligence (SCAI), we present first explorations on the subject of AI in the natural sciences and humanities at a point in time where qualitative problems seem to come into reach to be handled by machines. At the same time, these discussions of European scientific applications tie in with concerns that lie beyond this subject area and concern general preconditions of digital humanities (DH) or also of STS. Certain problems of transformative processes in the scientific field, which are closely related to AI, emerge mutatis mutandis in other constellations as well.

It is important to keep in mind in this context that research on scientific practices in AI has been conducted in a wide variety of disciplines and analytical perspectives, such as science and technology studies, sociology, infrastructure studies, cultural anthropology, philosophy of science, and data science (Baurmann/Mans 1984; Carley 1996; Groß/Jordan 2023; Krämer 1994; Ligo et al. 2021; Manhart 1995), but also specifically in the disciplines involved in this project: media studies (MS), history of science (HS) and computer science (CS) itself.[7]

---

6 For the details, see our (the editor's) contribution to this book as well as the essay by Schröter and Sudmann, also published here.

7 For media studies, see for example Engemann/Sudmann 2018; Ernst et al. 2019; Mackenzie 2017; Mann/Matzner 2019; Pasquinelli 2017, 2023; Sudmann 2023. For the

Current investigations into AI research stem from various disciplines involved with the reflection of the sciences. The philosophy of science has deployed its specific expertise for problems of AI (about cognition, consciousness, etc.) and proliferates in the field of AI ethics in particular, a field in which, among other things, various critical perspectives on AI and its research are normatively negotiated (algorithmic biases, surveillance, opacity of technology, etc.; for an overview, see Coeckelbergh 2020; Dimock 2020; Mann/Matzner 2019). Moreover, AI clearly resonates with and functions as a catalyst for the research perspectives of the digital humanities (Jannidis 2013; Manovich 2017; Flückiger 2021).

To further enhance these discussions for the field of AI-based methods in the sciences, a more thorough investigation of scientific practices and infrastructures seems in order (Star 1999; Schabacher 2022). To keep track of current developments an integrated dialogue with computer science is of the essence. In addition to that, it seems highly desirable to observe, document and reflect the current shifts in scientific practices through AI-based methods. To capture these developments up close, a media ethnography of selected AI research projects is the most viable option and will be conducted as the research project unfolds (Drackle 2014; Dippel 2017; Schüttpelz/Gießmann 2015; Bareither 2019). The integrated approach to scientific practices will further draw on the strengths of media archaeology to situate technically mediated knowledge production in larger frameworks. To this end, we emphasize the technological aspect as well as the social embeddedness of the emerging technology (Dotzler 2006; Schröter 2020; Ernst/Schröter 2020). Historical depth is provided for these findings on scientific practices by recent results from the history of data use in various disciplines (Aronova/von Oertzen/Sepkoski 2017; Schlicht/Ledebur/Echterhölter 2021). In this newly developing field within the history of science, separate instances in data journeys are consulted (Leonelli/Tempini 2020), the emergence of specific algorithms are traced (Evans/Johns 2023) or models are investigated in and of themselves. One of the best researched cases may be weather models, which took a stunning trajectory from decentralized weather observers to dynamic climate models and eventually, their integration into the vast machines of computer simulations (Coen 2018; Edwards 2000; Edwards 2010; Gramelsberger 2010).

history of science, see Seising 2021; Cave/Dihal/Dillon 2020; Evans/Johns 2023. For computer science, see Vaswani et al. 2017; Devlin et al. 2019; Brown et al. 2020; Rombach et al. 2022; Kirillov et al. 2023.

A new technical option for the sciences and humanities calls for a critical reflection of emerging forms (such as databases, algorithms, frameworks, interfaces, etc.) related to the production of knowledge. An engagement with possible transformations of scientific practices demands a methodological approach which refrains from creating prima facie distinctions between internal and external factors shaping these transformations, namely approaches from media ethnography, media archaeology, or the history of quantification. Establishing an account of what factors are important for the origin, the implementation (or non-implementation), and not least the retention of AI technology can possibly serve as a gateway for criticizing these very conditions in which the scientific endeavor takes place.

Various contemporary debates on AI technologies revolve around their social and cultural effects. Problems of algorithmic biases, data privacy, or opacity of infrastructures are commonly placed in the normative framework of AI ethics. Critical discussions of the high hopes invested in AI, as well as its present limitations, also continue to play a crucial role in ongoing debates (Broussard 2018). There is still little knowledge, however, about the relationship between the assumed problematic aspects of AI and the ways in which AI affects research practices, methodologies, and outcomes across different sciences. Adequate assessment of the impact of AI on science, including reference to its socio-political implications, is therefore a major research desideratum.

As has been pointed out here, research on the research of AI is confronted with significant challenges. The transdisciplinary view on the problems of AI in science requires distinctive expertise in very heterogeneous fields. However, there is no such thing as universal competence. Therefore, the research group hosting these discussions is all the more dependent on the dialogue and support of scholars from different disciplines and has benefited considerably from their civic engagement across the disciplines. The main focus of this publication is to explore different ways of thinking about the uses of AI in a broad set of scientific fields. At the same time, and in relation to selected disciplines, we want to exemplarily demonstrate the application of AI in specific academic contexts.

List of contributions

In their joint paper, the **members of the project "How is Artificial Intelligence Changing Science?"** discuss nine preliminary theses regarding the possible effects of the use of different AI technologies in the sciences. I) It is questioned if the widespread rhetoric of an "AI revolution" is helpful to describe the shifts that occur with the introduction of AI technologies in the sciences. II) It is emphasized that AI technologies can only be understood by understanding their embeddedness in infrastructures and social contexts. III) It is stated that AI systems can process fuzziness and uncertainty in a new way. In IV) the conflict between the big tech industry and academia in the development of AI is being highlighted. Thesis V) elaborates on how the fast introduction of AI technologies causes an expert crisis. In VI) it is discussed that many disciplines split into a computational and a non-computational branch. VII) points to the connection of AI technologies with data extraction and data colonialism. In VIII) the thesis is formulated that the introduction of AI will alter the labor landscape profoundly. IX) asks how the self-improvement and the self-evaluation of AI have to be conceptualized.

Mathematics struck gold when employing infinitesimal quantities to solve practical problems towards the end of the 17th century. In a further decolonial reflection on the inherent problems of AI and pattern recognition and discrimination, **Clemens Apprich** investigates this calculus in historical and present debates about incalculability. The calculus, which still performs reliable approximations within the schemes of artificial neural networks, should not be tamed into absolute congruence. On the contrary, it might be the imperfections and approximations which may help us to cultivate procedural and plural approaches. In this sense, the immanence residing at the end of all approximations (which would be tantamount with mastering the visual and compositional realms of quality by new AIs) does not appear fully desirable. Apprich acknowledges the insurmountable incongruity of the mathematical setup of AI and suggests strategic uses, such as Ramon Amaro's for possible "black totalities"'.

Matteo Pasquinelli's approach to AI is informed by the joint traditions of materialistic epistemology and media theory. He argues strictly against "folk AI", a perception of this new technology which all too readily accepts a new and contextless entity and its miraculous abilities. Instead, a much longer history of mediated thought and neoliberal entanglement of AI is in order. Without a shadow of a doubt automated and mechanical ways of reasoning have been part and parcel of the scientific endeavor long before artificial neural networks. The paper revisits Rosenblatt's 1957 strategy to facilitate pattern recognition via the modeling of the labor of perception and supervision. It integrates this historical analysis of AI with Peter Damerow's theory of mental representation, the dialectics of tools and knowing, as well as neo-Gramscian approaches towards formalization, and the Hessen-Grossmann thesis of the labor dependency of all science. AI's advanced algorithms are not unique. They are the latest result in a long history of confluences and attempts at "epistemic scaffolding".

Markus Ramsauer offers a genealogy of the development of Early Warning Systems and the potential enhancement in the detection of danger via the use of AI. Taking the trope of birds as sentinels for future catastrophic developments as leitmotif, it is argued that the discovery of latent danger often depends on the use of non-human sensors or kinds of intelligence; be they animalistic or machinistic. This offers a lens for thinking about the concept of 'artificial' and 'non-artificial' intelligence beyond the question if machines can pass as human.

Jean Gabriel-Ganascia's text discusses AI not only in terms of a tool for scientific practice but as a science itself. As such, the author claims, it evades classifications as 'theoretical science', 'science of nature' or 'science of culture'. The reason for this special status can be explained by the history of AI development of which the author provides a brief outline. As a second strand, the article explores the possibilities of 'epistemological ruptures' through the use of AI in the humanities as well as in the 'hard sciences'. Whereas for the former, these tools can assist in assessing individual cases, it contributes to an 'automation of induction' for the latter.

Gabriele Schabacher discusses in her essay the centrality of the notion of "pattern" for subsymbolic artificial intelligence. She asks what the power of patterns in contexts of cognition or application is, by distinguishing two ways of conceptualizing patterns, namely *template* and *correlation*. The reconstruction shows how these two forms are peculiarly blended in the horizon of AI technologies. The first example is the application domain of security research and how the blending of template and correlation works there. The focus will be on German pilot projects in Berlin and Mannheim that test the use of intelligent video analysis. Finally, Schabacher comments on the statistical creativity of AI image generators such as DALL-E, highlights four overarching aspects associated with the work of patterns of AI technologies, and describes their effects on scientific understanding, but also on culture and society in general.

The revolutionary potentials of AI in healthcare are covered in detail in the overview by **Urvi Sonawane** and **Matthieu Komorowski**. The usage of AI-based technologies is currently quite limited, the authors discover, despite its enormous potential. Responsible bottlenecks like technical, ethical, legal, and human aspects are examined and the need for a multidisciplinary approach involving regulatory bodies, clinicians, government, and patient committees is argued for.

In her position paper, **Isabelle Bloch** argues that a hybrid point of view of designing AI, considering both knowledge data representation and reasoning, offers opportunities towards explainability. This idea is illustrated on the example of medical image understanding, formulated as a spatial reasoning problem.

In his contribution, **Giaccomo Landeschi** shows how computer-based applications had a profound impact on the discipline of archaeology and how different methods and techniques, such as satellite remote sensing, geophysical prospections, and more recently, airborne laser scanning (LiDAR), have been employed for surveying purposes. Nowadays, artificial intelligence has also started to play an important role in the analysis of archaeological contexts. In the case of Sweden, approximately 70 per cent of its land comprises forests where a substantial number of archaeological sites remain hidden beneath the vegetation, undiscovered and unmapped. Landeschi explains how a team of scientists from Lund University recently undertook a project to showcase the potential of utilizing deep learning-based analysis and convolutional neural networks for automatically identifying a specific category of archaeological features called 'clearance cairns' in LiDAR-derived raster imagery.

Sabine Wirth's paper sheds light on the ways how the concept of the interface matters for a critical understanding of AI technologies in use. From a media and culture studies perspective she discusses how research on machine learning techniques can profit from a critical perspective on interfaces. Drawing on the emerging field of critical interface studies, Wirth describes two examples of popular apps that rely on machine learning, and she outlines potential lines of inquiry and critical questions that address the central role of interfaces as mediators of AI within the field of popular media culture. Ultimately, this allows her to ask how critical interface studies can inform research on AI in science by providing an additional analytical layer.

In their contribution, **Andreas Sudmann** and **Jens Schröter** shed light on the role of media related to how AI is used in and potentially transforms different fields of academic research. Furthermore, they draw attention to some important problems of applied AI which thus require critical reflection, especially from a media studies perspective.

Johannes Breuer poses the question of how AI is changing scientific practice in the realm of the social sciences. His contribution "Putting AI into social science" highlights the importance of tools for different stages of the scientific endeavor. The author discusses a variety of AI-driven research tools which are suitable for the social sciences, emphasizing their potentially transformative potential as well as ethical challenges that go hand in hand with this transformation. The chapter concludes with an invocation to focus on partnerships *with* AI, rather than on replacement *by* AI.

The paper by **Evangelos Pournaras** reviews the specific epistemological challenges and also the ethical and integrity risks related to generative AI and LLMs. In particular, Pournaras discusses emerging practices for research ethics, proposing ten recommendations that shape a response for a more responsible research conduct in the era of AI.

In his paper, **Fabian Retkowski** aims at concisely indicating the current state of the art in abstractive text summarization. The current paradigm shifts towards pre-trained encoder-decoder models and large autoregressive language models are outlined and the challenges of evaluating summarization systems and the potential of instruction-tuned models for zero-shot summarization are discussed in further detail. Additionally, the work gives a brief overview of how summarization systems are currently being integrated into commercial applications.

Sabina Leonelli maintains in her chapter that despite ever larger amounts of data and proclaimed bias-reducing algorithms, the employment of AI tools in scientific research is still heavily affected by the quality of the training data. The hardly traceable origin of data, combined with their often diverse nature and purpose, leads to what the author calls "in-practice opacity". Instead of focusing on quantitative modes of reproducibility as a panacea for making science transparent, the author calls for extended attention to questions about the quality and the funding of research data.

The use of data has been a key element of statistics, yet the dimensions of current data usages constitute a new situation. **Gérard Biau** is in a unique position to answer a set of questions about the changes affected by AI in this particular field of mathematics: He works at the Probability, Statistics, and Modeling Laboratory (LPSM), serves as director of the Sorbonne Center for Artificial Intelligence (SCAI), and was president of the *Société française de statistique*. Biau states that the impact of AI on mathematics is decisive. Some statistical tools, which have been stable for decades, are currently being revised. AIs start to make suggestions regarding results, or are instrumental in verifying the most advanced new proofs.

In his interview with **Sybille Krämer**, Jens Schröter poses nine questions which closely follow Krämer's writings over the decades. Her work has been, from the very beginning, revolving around questions that are of special relevance to understanding subsymbolic AI today. This starts with the question on the culturally shaped exteriority of the human mind, the relation of AI to the fundamental role of the analog and the digital, or the connection of AI to the field of digital humanities. Further fundamental points are discussed like the question if AI can be understood as a "cultural technique", especially when we observe the increasing role of computers in science. Finally, Krämer addresses questions of explainability and critique.

List of references

Araújo, Duarte/Couceiro, Micael/Seifert, Ludovic/Sarmento, Hugo/Davids, Keith (eds.) (2021): Artificial Intelligence in Sport Performance Analysis, London: Routledge.

Aronova, Elena/von Oertzen, Christine/Sepkoski, David (2017): "Introduction: Historicizing Big Data." In: Osiris 32/1, pp. 1–17.

Athey, Susan (2018): "The Impact of Machine Learning on Economics." In: Ajay Agrawal/Joshua Gans/Avi Goldfarb (eds.), The Economics of Artificial Intelligence: An Agenda, Chicago: University of Chicago Press, pp. 507–547.

Baurmann, Michael/Mans, Dieter (1984): "Künstliche Intelligenz in den Sozialwissenschaften." In: Analyse & Kritik 6/2, pp. 103–159.

Bostrom, Nick (2014): Superintelligence: Paths, Dangers, Strategies, Oxford: University of Oxford Press.

Bourilkov, Dimitri (2019): "Machine and Deep Learning Applications in Particle Physics". In: International Journal of Modern Physics A, 34/35, 1930019.

Broussard, M. (2018): Artificial Unintelligence: How Computers Misunderstand the World. Cambridge, MA: The MIT Press.

Brown, Tom/Mann, Benjamin/Ryder, Nick/Subbiah, Melanie/Kaplan, Jared D./Dhariwal, Prafulla/Neelakantan, Arvind/et al. (2020): "Language Models are Few-Shot Learners." In: Advances in Neural Information Processing Systems 33, pp. 1877–1901.

Carley, Kathleen M. (1996): "Artificial Intelligence within Sociology." In: Sociological Methods & Research 25/1, pp. 3–30.

Cave, Stephen/Dihal, Kanta/Dillon, Sarah (eds.) (2020): AI Narratives: A History of Imaginative Thinking about Intelligent Machines, Oxford: Oxford University Press.

Coeckelbergh, Mark (2020): AI Ethics. Cambridge, MA: The MIT Press.

Coen, Deborah (2018): Climate in Motion: Science, Empire, and the Problem of Scale, Chicago: The University of Chicago Press.

Cornelio, Cristina/Dash, Sanjeeb/Vernon, Austel/Josephson, Tyler/Goncalves, Joao/Clarkson, Kenneth/Megiddo, Nimrod/et al. (2021): AI Descartes: Combining Data and Theory for Derivable Scientific Discovery, arXiv Preprint (https://arxiv.org/pdf/2109.01634.pdf).

Devlin, Jacob/Chang, Ming-Wei/Lee, Kenton/Toutanova, Kristina (2019): "BERT: Pre-training of Deep Bidirectional Transformers for Language Understanding." In: Proceedings of the 2019 Conference of the North American Chapter of the Association for Computational Linguistics: Human Language Technologies, Volume 1 (Long and Short Papers), Minneapolis, MS, USA, pp. 4171–4186.

Dick, Stephanie (2011): "AfterMath: The Work of Proof in the Age of Human-Machine Collaboration." In: Isis 102/3, pp. 494–505.

Dimock, Wai Chee (2020): "AI and the Humanities." In: PMLA: Publications of the Modern Language Association of America 135/3, pp. 449–454.

Dippel, Anne (2017): "Das Big Data Game: Zur Spielerischen Konstitution kollaborativer Wissensproduktion in der Hochenergiephysik am CERN." In: NTM – Zeitschrift Für Geschichte der Wissenschaften, Technik und Medizin 25/4, pp. 485–517.

Domingos, Pedro (2015): The Master Algorithm. How the Quest for the Ultimate Learning Machine Will Remake Our World, New York: Basic Books.

Dotzler, Bernhard J. (2006): Diskurs und Medium. Zur Archäologie der Computerkultur, München: Wilhelm Fink Verlag.

Dracklé, Dorle (2014): "Medienethnologie/Medienethnographie." In: Jens Schröter (ed.), Handbuch Medienwissenschaft, Stuttgart: Metzler, pp. 393–404.

Dreyfus, Hubert/Dreyfus, Stuart (1988): "Making a Mind Versus Modeling the Brain: Artificial Intelligence Back at a Branchpoint." In: Stephen R. Graubard (ed.), The Artificial Intelligence Debate. False Starts, Real Foundations, Cambridge, MA, and London: The MIT Press, pp. 15–43.

Echterhölter, Anna/Schröter, Jens/Sudmann, Andreas (2021): How Is Artificial Intelligence Changing Science? Research in the Era of Learning Algorithms, MediarXiv Preprint (https://doi.org/10.33767/osf.io/28pnx).

Edwards, Paul (2000): "The World in a Machine. Origins and Impacts of Early Computerized Global Systems Models." In: Agatha C. Hughes/Thomas Parke Hughes (eds.), Systems, Experts, and Computers, Cambridge, MA: The MIT Press, pp. 221–253.

Edwards, Paul (2010): A Vast Machine. Computer Models, Climate Data, and the Politics of Global Warming, Cambridge, MA: The MIT Press.

Engemann, Christoph/Sudmann, Andreas (eds.). (2018): Machine Learning – Medien, Infrastrukturen und Technologien der Künstlichen Intelligenz, Bielefeld: transcript.

Ernst, Christoph/Schröter, Jens (2020): Zukünftige Medien: Eine Einführung, Wiesbaden: Springer.

Evans, James/Johns, Adrian (eds.) (2023): Beyond Craft and Code: Human and Algorithmic Cultures, Past and Present, Osiris 38/1.

Fawzi, Alhussein/Balog, Matej/Huang, Aja/Hubert, Thomas/Romera-Paredes, Bernardino/Barekatain, Mohammadamin/Novikov, Alexander/et al. (2022): "Discovering Faster Matrix Multiplication Algorithms with Reinforcement Learning." In: Nature 610/7930, pp. 47–53.

Fecher, Benedikt/Hebing, Marcel/Laufer, Melissa/Pohle, Jörg/Sofsky, Fabian (2023): Friend or Foe? Exploring the Implications of Large Language Models on the Science System, arXiv Preprint (https://doi.org/10.48550/arXiv.2306.09928).

Feigenbaum, Edward A. (1968): Artificial Intelligence: Themes in the Second Decade, Memo Number 67, AD680487, Stanford: Stanford University.

Feigenbaum, Edward A./Feldman, Julian (eds.) (1963): Computers and Thought, New York: McGraw-Hill.

Flückiger, Barbara (2017): "A Digital Humanities Approach to Film Colors." In: The Moving Image: The Journal of the Association of Moving Image Archivists 17/2, pp. 71–94.

Flückiger, Barbara (2020): "Digitale Werkzeuge zur ästhetischen Analyse von Filmfarben." In: Montage AV 29/1, pp. 157–172.

Flückiger, Barbara (2021): "Color and Expressive Movement. A Computer-assisted Analysis." In: Mediaesthetics – Zeitschrift für Poetologien Audiovisueller Bilder 4 (https://doi.org/10.17169/mae.2021.86).

Gethmann, Carl Friedrich/Buxmann, Peter/Distelrath, Julia/Humm, Bernhard G./Lingner, Stephan/Nitsch, Verena/Schmidt, Jan C./Spiecker genannt Döhmann, Indra (2022): Künstliche Intelligenz in der Forschung. Neue Möglichkeiten und Herausforderungen für die Wissenschaft, Berlin and Heidelberg: Springer.

Gramelsberger, Gabriele (2010): Computerexperimente. Zum Wandel der Wissenschaft im Zeitalter des Computers. Bielefeld: transcript.

Groß, Richard/Jordan, Rita (2023): KI-Realitäten. Modelle, Praktiken und Topologien maschinellen Lernens, Bielefeld: transcript.

Hassabis, Demis (2022): "Dr Demis Hassabis: Using AI to Accelerate Scientific Discovery." University of Oxford, uploaded to YouTube on August 3, 2022 (https://youtu.be/AU6HuhrC65k).

Hochreiter, Sepp/Schmidhuber, Jürgen (1997): "Long Short-Term Memory." In: Neural Computation 9/8, pp. 1735–1780.

How is Artificial Intelligence Changing Science? (2023): "AI Projects in Europe. Selected Disciplines.", last updated February 14, 2023 (https://howisaichangingscience.eu/ai-projects-in-europe/).

Hunt, Earl (1968): "Computer Simulation: Artificial Intelligence Studies and Their Relevance to Psychology." In: Annual Review of Psychology 19/1, pp. 135–168.

Jannidis, Fotis (2013): "Computerphilologie." In: Thomas Anz (ed.), Handbuch Literaturwissenschaft, Band 2, Methoden und Theorien, Stuttgart and Weimar: Metzler, pp. 27–40.

Jenicek, Tomas/Chum, Ondrej (2019): Linking Art through Human Poses, arXiv Preprint (http://arxiv.org/abs/1907.03537).

Jumper, John/Evans, Richard/Pritzel, Alexander/Green, Tim/Figurnov, Michael/Jumper Ronneberger, Olaf/Tunyasuvunakool, Kathrynet/et al. (2021): "Highly Accurate Protein Structure Prediction with AlphaFold." In: Nature 596/7873, pp. 583–589.

Kautz, Henry (2022): "The Third AI Summer: AAAI Robert S. Engelmore Memorial Lecture." In: AI Magazine 43/1, pp. 105–125.

Kirillov, Alexander/Mintun, Eric/Ravi, Nikhila/Mao, Hanzi/Rolland, Chloé/Gustafson, Laura/Xiao, Tete/et al. (2023): Segment Anything, arXiv Preprint (https://arxiv.org/abs/2304.02643).

Krämer, Sybille (ed.) (1994): Geist, Gehirn, künstliche Intelligenz: Zeitgenössische Modelle des Denkens. Ringvorlesung an der Freien Universität Berlin, Berlin and New York: de Gruyter.

Krizhevsky, Alex/Sutskever, Ilya/Hinton, Geoffrey E. (2012): "ImageNet Classification with Deep Convolutional Neural Networks." In: Advances in Neural Information Processing Systems 25, pp. 1097–1105.

LeCun, Yann/Bottou, Leon/Bengio, Yoshua/Haner, Patrick (1998): "Gradient-Based Learning Applied to Document Recognition". In: Proceedings of the IEEE 86, pp. 2278–2324.

Lewkowycz, Aitor/Andreassen, Anders/Dohan, David/Dyer, Ethan/Michalewski, Henryk/Ramasesh, Vinay V./Slone, Ambrose/et al. (2022): Solving Quantitative Reasoning Problems with Language Models, arXiv Preprint (https://arxiv.org/pdf/2206.14858.pdf).

Ligo, Alexandre K./Rand, Krista/Basset, Jason/Galaitsi, S.E./Trump, Benjamin D./Jayabalasingham, Bamini/Collins, Thomas/Linkov, Igor (2021): "Comparing the Emergence of Technical and Social Sciences Research in Artificial Intelligence." In: Frontiers in Computer Science 3, pp. 1–13.

Luong, Minh-Thang/Manning, Cristopher (2015): "Stanford Neural Machine Translation Systems for Spoken Language Domains". In: Proceedings of the 12th International Workshop on Spoken Language Translation: Evaluation Campaign, Da Nang, Vietnam, pp. 76–79.

Mackenzie, Adrian (2017): Machine Learners. Archaeology of a Data Practice, Cambridge, MA: The MIT Press.

Manhart, Klaus (1995): KI-Modelle in den Sozialwissenschaften. Logische Struktur und wissensbasierte Systeme von Balancetheorien, München: Oldenbourg Wissenschaftsverlag.

Mankowitz, Daniel J./Michi, Andrea (2023): "AlphaDev Discovers Faster Sorting Algorithms." June 7, 2023 (https://www.deepmind.com/blog/alphadev-discovers-faster-sorting-algorithms).

Mankowitz, Daniel J./Michi, Andrea/Zhernov, Anton/Gelmi, Marco/Selvi, Marco/Paduraru, Cosmin/Leurent, Edouard/et al. (2023): "Faster Sorting Algorithms Discovered Using Deep Reinforcement Learning." In: Nature 618/7964, pp. 257–263.

Mann, Monique/Matzner, Tobias (2019): "Challenging Algorithmic Profiling: The Limits of Data Protection and Anti-discrimination in Responding to Emergent Discrimination." In: Big Data & Society 6/2, pp. 1–11.

Manovich, Lev (2017): "Cultural Analytics, Social Computing and Digital Humanities." In: Mirko Tobias Schäfer/Karin van Es (eds.), The Datafied Society. Studying Culture Through Data, Amsterdam: Amsterdam University Press, pp. 55–68.

McCorduck, Pamela (1979): Machines Who Think. A Personal Inquiry into the History and Prospects of Artificial Intelligence, San Francisco: W.H. Freeman and Company.

Nguyen, Thai-Son/Stueker, Sebastian/Waibel, Alex (2020): Super-Human Performance in Online Low-latency Recognition of Conversational Speech, arXiv Preprint (https://doi.org/10.48550/arXiv.2010.03449).

Okerlund, Johanna/Klasky, Evan/Middha, Aditya/Kim, Sujin/Rosenfeld, Hannah/Kleinman, Molly/Parthasarathy, Shobita (2022): "What's in the Chatterbox? Large Language Models, Why They Matter, and What We Should Do About Them, Research Report Ford School of Public Policy.", April 2022 (https://stpp.fordschool.umich.edu/research/research-report/whats-in-the-chatterbox).

Pasquinelli, Matteo (2017): "Machines that Morph Logic: Neural Networks and the Distorted Automation of Intelligence as Statistical Inference." In: Glass Bead 1, (https://www.glass-bead.org/article/machines-that-morph-logic/).

Pasquinelli, Matteo (2023): "Labour, Energy, and Information as Historical Configurations: Notes for a Political Metrology of the Anthropocene." In: Journal for the Interdisciplinary History of Ideas 22/11 (https://doi.org/10.13135/2280-8574/7340).

Radovic, Alexander/Williams, Mike/Rousseau, David/Kagan, Michael/Bonacorsi, Daniele/Himmel, Alexander/Aurisano, Adam/et al. (2018): "Machine Learning at the Energy and Intensity Frontiers of Particle Physics." In: Nature 560/7716, pp. 41–48.

Rombach, Robin/Blattmann, Andreas/Lorenz, Dominik Lorenz/Esser, Patrick Esser/Ommer, Björn (2022): "High-Resolution Image Synthesis with Latent Diffusion Models." In: Proceedings of the IEEE/CVF Conference on Computer Vision and Pattern Recognition (CVPR 2022), New Orleans, LA, USA, pp. 10674–10685.

Schabacher, Gabriele (2022): Infrastruktur-Arbeit. Kulturtechniken und Zeitlichkeit der Erhaltung, Berlin: Kadmos.

Schmidhuber, Jürgen (2022): Annotated History of Modern AI and Deep Neural Networks, arXiv Preprint (https://arxiv.org/ftp/arxiv/papers/2212/2212.11279.pdf).

Schröter, Jens (2020): "Digitale Medien und Methoden. Jens Schröter zur Medienarchäologie der digitalen Medien.", July 10, 2020 (https://zfmedienwissenschaft.de/online/open-media-studies-blog/digitale-medien-und-methoden-6-schroeter).

Schüttpelz, Erhard/Gießmann, Sebastian (2015): "Medien der Kooperation. Überlegungen zum Forschungsstand." In: Navigationen – Zeitschrift für Medien- und Kulturwissenschaften 15/1, pp. 7–55.

Seising, Rudolf (2021): Es denkt nicht! Die vergessenen Geschichten der KI, Frankfurt a.M.: Büchergilde Gutenberg.

Simanowski, Roberto (2023): "Narrative der Weltbeglückung. Die neue Sprach-KI und die Mathematisierung der Ethik." In: Blätter für deutsche und internationale Politik 6, pp. 64–73.

Star, Susan L. (1999): "The Ethnography of Infrastructure." In: American Behavioral Scientist 43/3, pp. 377–391.

Suchman, Lucy (2006): Human-Machine Reconfigurations: Plans and Situated Actions, Cambridge: Cambridge University Press.

Sudmann, Andreas (2018a): "Zur Einführung. Medien, Infrastrukturen und Technologien des maschinellen Lernens." In: Christoph Engemann/Andreas Sudmann (eds.), Machine Learning. Medien, Infrastrukturen und Technologien der Künstlichen Intelligenz, Bielefeld: transcript, pp. 9–23.

Sudmann, Andreas (2018b): "Szenarien des Postdigitalen. Deep Learning als MedienRevolution." In: Christoph Engemann/Andreas Sudmann (eds.), Machine Learning. Medien, Infrastrukturen und Technologien der Künstlichen Intelligenz, Bielefeld: transcript, pp. 55–73.

Sudmann, Andreas (2023): "Zum Einsatz von KI als Methode und Werkzeug in der Medienwissenschaft." In: Sven Stollfuß/Laura Niebling/Felix Raczkowski (eds.), Handbuch Digitale Medien und Methoden, Wiesbaden: Springer (https://doi.org/10.1007/978-3-658-36629-2_16-2).

Varadi, Mihaly/Velankar, Sameer (2022:) "The Impact of AlphaFold Protein Structure Database on the Fields of Life Sciences." In: Proteomics, e2200128 (https://doi.org/10.1002/pmic.202200128).

Vaswani, Ashish/Shazeer, Noam/Parmar, Niki/Uszkoreit, Jakob/Jones, Llion/Gomez, Aidan N./Kaiser, Łukasz/Polosukhin, Illia (2017): "Attention Is All You Need." In: Advances in Neural Information Processing Systems 30, pp. 5998–6008.

Waibel, Alex (1987): "Phoneme Recognition Using Time-Delay Neural Networks." In: Proceedings of the Meeting of the Institute of Electrical, SP87-100, Information and Communication Engineers (IEICE), Tokyo, Japan.

Waibel, Alex/Sawai, Hidefumi/Shikano, Kiyohiro (1989): "Modularity and Scaling in Large Phonemic Networks." In: IEEE Transactions on Acoustics, Speech, and Sidnal Processing 37/12, pp. 1888–1898.

Research with Subsymbolic AI

Preliminary Theses

Andreas Sudmann, Anna Echterhölter, Markus Ramsauer, Fabian Retkowski, Jens Schröter

The current developments within information technology not only challenge scientific disciplines to study new phenomena; they also potentially alter and enhance research methods, practices, and outcomes across the natural sciences, social sciences, and humanities. Researchers have to negotiate interdisciplinary conceptual frameworks and access to new data infrastructures in order to participate in and benefit from the ongoing AI boom. Prima facie, data-intensive AI approaches, especially artificial neural networks (ANNs) but also other approaches of machine learning (ML), increasingly enable and support the production of knowledge across all disciplines.

To conceptualize these shifts as mere effects of technology, however, arguably oversimplifies the interrelationship between technology and society. Since its learning capabilities rely on datafication, AI-based research is always connected to society from the start: AI uses its data to classify, categorize, and cluster society. What is more, as modern societies come to increasingly rely on scientific knowledge, any change in scientific practices and research methods brought about by AI technologies is bound to affect society at large (DFG 2020; Zhang et al. 2021). Research on the influence of AI on science is therefore of the utmost importance in order to comprehensively understand the effects of AI on present and future society. The term "science" here refers to natural sciences, social sciences, and humanities, and "scientific research" to those research methodologies within the sciences which use empirical and quantifiable data.

One of the big challenges is to understand and conceptualize the relationship between new technologies, specifically AI, and the epistemologies they enable: Are AI-based methods basically more efficient tools which continue non-AI methods, merely extending them in terms of velocity and scope? Or do they allow research in new methodological ways, to ask and answer novel questions? Or both on different levels? Which socio-political implications does data-driven research with AI entail, compared to other long-standing data-based research practices? Is AI-enabled data science still confined to numerical and quantifiable problems, or does it also give access to qualitative problems (e.g., problems of fuzziness; Seising, 2009)? How do AI-based methods try to reduce the dependency on (big) data, e.g., by making use of pre-trained models, or specific approaches like transfer learning or one-shot learning (Duan et al. 2017; Weiss/Khoshgoftaar/Wang 2016)?

To answer such questions, it is insufficient to merely look at AI models in a narrower sense (i.e., "learning algorithms") and deduce their impact. Since scientific research always occurs within specific *epistemic cultures* (Knorr-Cetina 1999; see also Fleck 1980 [1935]; Latour/Woolgar 1979), the impact of AI-based methods on scientific research is only determinable by closely observing the interplay of technology and research practices. Their alleged opacity is an obstacle here: While AI-based approaches make it possible to process data in novel ways (i.e., identifying, classifying, categorizing), they partially or completely disable researchers' abilities to comprehend and track these AI-based processes (Adadi/Berrada 2018; Sudmann 2019). In light of these observations, we follow the hypothesis that data-driven and AI-based methods enable new epistemologies precisely by transforming one of the most long-standing scientific practices of all: *AI changes the way researchers interact with and relate to data.* Hence, we believe it is important to compare the specific impact of ANNs and ML procedures with existing findings on digital methods in the sciences, most notably simulation, big data, and statistical probability (Ash/Kitchin/Leszczynski 2019; Gramelsberger 2011; Leonelli 2016; Krüger/Daston/Heidelberger 1987).

Therefore, we currently investigate how AI-based methods are situated in concrete and specific research environments which draw together technologies and practices. We carefully develop our findings from firsthand knowledge of outstanding current AI research projects operating within the novel conditions of data infrastructures. We also take into account the history of the specific methods in question, including their affordances, and contextualize these techno-practical configurations within an in-depth history of data-based scientific practices. It is intended that our observations will eventually inform new research approaches, as our findings will be fed back into the development of an AI-based system that structures and comprehends scientific content from several modalities including text, speeches, and meetings. This includes a multitude of components such as automatic speech recognition, segmentation to automatically divide the content into coherent chapters, and text summarization. With this system, we hope to support the answering of meta-research questions of AI as part of our project.

To this end, and given the prevalence of data-driven research across disciplines, we have set-up a *transdisciplinary research project* combining the expertise of three disciplines: thinking through the complex entanglements of technologies, culture, and practices is one of the core assets of *media studies*; providing an in-depth history of data and modeling practices in various methodological traditions is the key contribution of the *history of science*; developing a profound mathematical understanding of AI models and using computational methods to engineer a cutting-edge tool for using AI to study AI is the current task of our project in computer science. We combine the expertise of these three disciplines to study the socio-technical uses of AI in three carefully selected external research projects. From an original sample of close to 150 research projects working with AI-based methods in Europe ("How is Artificial Intelligence Changing Science?" 2023), we have chosen three projects from three different disciplines as the center of our investigation (film studies, sociology and climatology/Earth sciences). To capture the current changes brought about by AI in general and ML and ANNs in particular, our working groups in *media studies* (MS), *history of science* (HS), and *computer science* (CS) will combine the specific strengths of the most advanced methods from their respective fields. The project will allow a unique documentation and investigation in this pivotal decade. Otherwise, many of the traces of this historical shift which are obtainable now, will be irretrievably lost.

In the following, we have compiled a selection of theses that address some of the central aspects and considerations of our research group while also illustrating the range of different disciplinary perspectives on the various dimensions of AI in science. Each thesis is preceded by a quote, pointing to a larger topic to be further investigated in the course of the research project.

Thesis I: AI revolution

> Major economies are on the 'cusp of an AI revolution' that could trigger job losses in skilled professions such as law, medicine and finance, according to an influential international organisation. (Milmo 2023)

One obvious rhetoric in the discourse of so-called "AI" is that we are dealing with a "revolution". This seems to imply that some fundamental change will occur with the advent of AI technologies. This is easier said than understood. Often the rhetoric of "revolution" is simply used for marketing – because a new product, be it an advanced toothbrush or a new type of AI software, sells better if it is claimed to be a brand-new breakthrough of some kind. In that case "revolution" is used as a synonym to the entrepreneurial buzzword "disruption". It is normally not meant that a fundamental societal upheaval is to be expected (and that is the idea connected with the word "revolution" in the twentieth century), but just that there is a new product that displaces other products on the market. Basically, this is also the meaning of the recent and somewhat disturbing announcements of AI-producing companies that their own products might put humanity in danger – and call for regulation. If they are in fact so dangerous, why don't they simply stop producing these programs? It is more likely that they want to direct attention to how powerful their brand-new products might be or that the established players want to impede competition.

Besides that and if you take the claim of "revolution" more seriously, it is often not very clear what exactly is meant by that. If we take the example above, mentioned in *The Guardian*, a "revolution" was indeed the case if the job losses triggered by AI would lead to the fundamental impossibility of our (capitalist) societies, based on wage labor, to reproduce itself (on the following, see Schröter 2019). This problem of "technological unemployment" is actually an effect that was predicted by certain strands of Marxian theory already long ago, and long before AI. For many authors, this means that capitalism has to be overcome, or at least that radical political solutions, like unconditional basic income, have to be sought for. But is this really meant by the headline of *The Guardian*? Even if *The Guardian* is left-leaning, it can be doubted that it really wants to say that the (often postponed) terminal crisis of capitalism is now really here – with AI. That would be a "revolution" indeed. It is more likely that the article wants to say that certain professions that seemed safe so far are now also under the threat of automation. Although this might be bad for the people involved, this is nothing new. Many technological transformations happened in the last 150 years, many people lost their jobs, but also many new professions appeared. In the current situation, one presumably new development is the destabilization of the position of 'knowledge workers' and creative workers, i.e., subjects whose tasks have up till now not been automatable to a satisfying degree. If this shift will evoke a fundamental change in the dynamics of the division of labor remains to be a point of investigation. However, drawing on a wider perspective it can be said that certain rhetoric and discursive figures always return with new technologies (see Kümmel/Scholz/Schumacher 2004). New technologies are very often accompanied by utopian and dystopian ideas regarding their possible effects. To name only one example related to AI, that has also been the case with the internet (Schröter 2004). In the end, neither the worst fears came true nor the utopian paradise started. No (technologically driven) revolution happened, but given social structures were extended, accelerated and thereby transformed (but not in an abrupt "revolutionary" way). The internet is a very good example for that: Instead of leading immediately (dystopian) to a totalitarian hive-mind or (utopian) to a "frictionless capitalism" (a term coined by Bill Gates, see Schröter 2012) or even to a post-capitalist society, it became integrated into real existing capitalism (quite full of frictions), extending (to "friendship", into every moment and place of life etc.) and accelerating it – step by step. There is no reason to expect that this will be fundamentally different with AI. Neither the dystopian (AI will take over the world and kill everyone, capitalism collapses and this leads to total social disaster, etc.), nor the utopian (AI will solve all problems, a wonderful post-capitalist society will be born, etc.) visions will come true – but as always some of the good and some of the bad prospects will be realized and a lot more things will happen which were not expected or predicted at all. But that they were not expected doesn't amount to a "revolution" – that's just what history is.

If we now turn to our project called "How is Artificial Intelligence Changing Science?" – can we say that there is a kind of "scientific revolution" caused by AI (that means today mainly machine learning)? Given the state of the research we have done, it is too early to give a clear-cut answer – but our preliminary research shows that it might make sense to be cautious here too. On the basis of our research on how machine learning (and computer simulation) is used in high-energy physics (Schröter 2021; Radovic et al. 2018), we were able to test a claim regarding an alleged "scientific revolution" in science – the case is formulated in Anderson's (2008) much discussed paper on the "end of theory". His argument does not address AI directly – but the role of detecting patterns in large amounts of data which is exactly the task of many AI-systems today. He argues that the classical procedure of the natural sciences is now obsolete: While it was, prior to "big data", necessary to formulate a theory which then has to be tested in experiments, now it is enough to observe patterns and correlations in data. Theory is not needed anymore – a "scientific revolution" indeed. But at least for the case studied, this argument turned out to be wrong. Theory, very complicated theory, is still needed in particle physics. It predicts effects. Based on the theory, simulation models are generated that show how the patterns of the predicted effects would "look" like in the particle accelerator that is used to conduct experiments. The machine learning systems are trained with these simulated patterns – and then they filter out possible fitting patterns from the gigantic data stream produced by the accelerator. In this way, the predicted Higgs boson was found in 2012. Although the basic epistemology seems unchanged, in detail there are differences:

> The traditional way to analyze, or generate simulated, data is to first develop algorithms based on domain knowledge, then implement them in software, and use the resulting programs to analyze or generate data. This process is labor intensive, and analyzing complex datasets with many input variables becomes increasingly difficult and sometimes intractable. Artificial intelligence (AI) and the subfield of machine learning (ML) attack these problems in a different way: instead of humans developing highly specialized algorithms, computers learn from data how to analyze complex data and produce the desired results. There is no need to explicitly program the computers. Instead, ML algorithms use (often large amounts of) data to build models with relatively small human intervention. These models can then be applied to predict the behavior of new, previously unseen data, to detect anomalies or to generate simulated data. (Bourilkov 2019: 1f.)

That means: The application of AI systems leads to continuities and discontinuities at the same time. Our thesis is also: "Revolution" is a too narrow concept to describe the coexistence of continuities and discontinuities in the process of the diffusion of AI. One needs more differentiated concepts from media historiography (Schröter/Schwering 2014) to describe the effects of AI, even when only focused on the use of AI in different fields of science.

It is, of course, not necessary to conclude that the application of machine learning in other scientific fields follows the same trajectory. While in physics it seems to change nothing on a fundamental epistemic level (except for making use of far larger datasets as before), this might turn out to be different in other disciplines. This is essentially what our project tries to find out.

Thesis II: AI embedded

> [I]n order to grasp [...] unwillingness to comply with mechanical innovation, we need to widen our perspective beyond machine technology. (von Oertzen 2017: 131)

For research on the implementation or the non-implementation of new technologies in various (scientific) fields, close attention should be paid to the infrastructures in which these technologies like machine learning are embedded. This allows for an analysis which refrains from conceptualizing abstract "technological enhancements" as the sole driving force of history. In this regard, non-implementations of AI technologies in scientific fields should not a priori be regarded as motivated by irrational conservatism or technophobia but must be researched within their specific political economy. Three examples from different scientific fields will serve to underpin this argument:

In the article quoted above, it is illustrated how the 19th-century Prussian census system by relying on "manual concepts, technologies, and practices of data power" (von Oertzen 2017: 129) managed to reach a similar level of effectiveness compared to other states which had implemented the Hollerith machine. Even after the eventual switch to machine-readable punch cards in the late 19th century which "enabled statisticians to accomplish tasks that were impossible to perform manually [...] they rested firmly on the concepts and paper tools developed for manual use." (ibid.: 132) This episode can serve as an incentive to investigate thoroughly the conditions of the possibility of implementing new technologies instead of focusing on tech companies' accelerating announcements of revolutions and breakthroughs in their technological products.

Another important aspect of institutions' rigidity or refusal to implement new technologies has to do with what STS scholars Sheila Jasanoff, Ulrike Felt and others have coined as "sociotechnical imaginaries", i.e., "collectively held, institutionally stabilized, and publicly performed visions of desirable futures, animated by shared understandings of forms of social life and social order attainable through, and supportive of, advances in science and technology." (Jasanoff 2015: 4). An episode from Western German computer history can serve as an example thereof: When Remington Rand delivered Europe's first large-scale computer system UNIVAC I to Frankfurt am Main in 1956, it caused considerable sensation about this "electronic brain". However, the plan to rent the machine to local companies for computational work ultimately failed and the UNIVAC I was shut down. As computer historian Corinna Schlombs argues, this failure was mainly due to Remington Rand's ignorance of "local customs and traditions" (Schlombs 2010: 98). This concerned the different organizations and sizes of German companies (139), European labor law (97), different infrastructural conditions, like electrical plugs (140) as well as the – proclaimed by a company report – "German users tend[ing] to be somewhat skeptical of the large scale systems" (ibid.: 139). Only after adapting to European conditions were Remington Rand's products able to gain a foothold in the European market. The story of the UNIVAC I points to the important question of why a technology does not fit into an environment, its types of organization, its infrastructures, and its sociotechnical imaginaries and which adaptive measures are taken to enable an implementation.

For the case of AI technology, such a focus on the infrastructural conditions of possibility (or impossibility) can further be helpful to grasp the different transformative speeds of human (scientific) practice. Instead of standing in awe of the daily releases and presentations and new models, close attention should be paid to a possible gap between modeling and implementation as to refrain from writing a mere history of ideas. As Urvi Sonawane and Matthieu Komorowski show in their contribution to this volume for the field of medical intensive care, "there is an increasing number of AI prototypes and early models being developed and trialled" (Sonawane/Komorowski 2023: 161). At the same time, "there seems to be a disproportionate disparity when it comes to translating these AI models from production to clinical evaluation." (ibid.: 161). Although the number of AI models released for the use in intensive care has risen significantly, implementation remains scarce. According to the authors, this is because, "the successful algorithms are less suited to be rolled out on a large-scale healthcare service or even across a country" (ibid.: 164) as well as the fact that "AI systems are notoriously difficult to integrate within and between systems" (ibid.: 164). Again, here it is the "problem" of different organizational systems and infrastructures being grown over a considerable amount of time, which yet complicates the AI models' widespread use in intensive care. This is of course not to foreclose the possibility that these systems can be fundamentally altered by the advent of new technologies. However, close attention should be paid to these different speeds of practical transformation by (AI) technology which can be addressed by an approach as outlined above.

Thesis III: Epistemological potentials

> Statistics is the study of uncertainty. (Lindley 2000: 301)

Machine learning is essentially a form of statistics, and AI applications clearly display a statistical anatomy (Alpaydin 2016: 27). Hence one might think that the main task of machine learning approaches of AI would be the study of uncertainty. Indeed, specifically, artificial neural networks (ANNs) allow a new technical level of dealing with problems of uncertainty, for example dealing with incomplete information or predicting future events. Nevertheless, it would be insufficient to simply describe the general capabilities of ANNs in the processing of uncertainty. As a predictive technology, ANNs are, of course, in some way always related to problems of uncertainty, yet this does not accurately describe their enormous epistemic-technical capability to deal with different forms of vagueness or fuzziness related to visual or acoustic challenges of pattern recognition.

Over the course of the 19th century, statistical methods and probabilistic approaches took a successful hold in sciences as diverse as psychology and paleontology, sociology and astronomy, evolutionary biology and economic reasoning about risk and crime, in insurance and gambling. This unanimous shift towards quantitative methods came at a price. At its core lay the acceptance of less precision – or the new form of evidence which was later deemed as probabilistic revolution (Krüger/Daston/Heidelberger 1987). Back then, questions of fuzziness and uncertainty were intensively discussed by scientists such as Gustav T. Fechner and Pierre-Simon Laplace, as a shift away from the ideal of determinism that still prevailed at the beginning of that century. Currently, the boom in statistical AI in the form of ANNs, among others, makes the discussion of questions of uncertainty and fuzziness seem particularly urgent.

Already Claude Shannon's mathematical theory of communication formulated as a theorem on what digital technology in the form of computers was soon to achieve, namely dealing with problems of uncertainty, be it in relation to communication and its encryption or decryption or in relation to the prediction of flying objects in the application of radar technology. What corresponding communications technology or, ultimately, a computer achieved in the one case as well as in the other, in information-theoretical terms, is to distinguish between information and noise, and to make this possible as an exact calculation. And it was precisely for this purpose that the principle of binary circuitry proved to be particularly effective.

However, as became apparent in the course of the second half of the 20th century, certain more sophisticated problems of fuzziness were very difficult to solve for decades, e.g., enabling a computer to visually perceive its environment and objects in it. No matter which AI methods were used to approach such tasks, whether with so-called symbolic, rule-based AI or with subsymbolic AI in the form of ANNs or with approaches of so-called fuzzy sets, in the end, all these methods, despite selective progress, remained quite far away from what current AI implementations are capable of, until the 2000s.

It was only about 15 years ago that the situation changed significantly, as the important fields of AI work, computer vision or natural language processing, exemplify. Only then were computers able to cope with technically more demanding problems of fuzziness of various types much better.

We would like to briefly highlight this epistemic potential once again: Machine learning methods in the form of ANNs are in any case not only able to recognize patterns in complex data that are difficult for us humans to recognize due to their size and complexity and therefore present themselves to us as fuzzy, but which are themselves fuzzy and/or incomplete as statistical patterns. ANNs can produce usable output despite incomplete data or on the basis of fuzzy patterns, and they can do this by calculating not exactly, but approximately, i.e., quasi-fuzzy themselves. Thus, already the iterative, optimisation-oriented training process of machine learning methods can ideally be understood as a process of successive reduction of the prediction error, thereby approximating the real data distribution. Moreover, in the context of ANNs, there is now a broad portfolio of methods for dealing with uncertainty problems including ensemble methods, data augmentation methods, dropout and transfer learning.

Thesis IV: Big tech and academia

> One important feature of AI's modern R&D trajectory is that private companies native to the digital economy such as Google and Facebook are playing an increasingly important role in basic research activities that used to be the domain of academia. (Jurowetzki et al. 2021: 3)

As in many other fields, the conditions of possibility for applying AI processes in the sciences are shaped more than ever by the big tech industry.

First, there is an industry-wide tendency to provide universal, i.e., non-domain-specific, infrastructural support and tools to users. This includes systems such as ChatGPT.

Second, the big tech industry as well as leading AI companies such as Open AI or Antrophic are selectively engaged in solving fundamental problems of science in different domains for which a) technical-epistemic approaches of AI are particularly suitable and which b) should also redeem the claim to be socially responsible AI. In this respect, it cannot be surprising that some of the most important AI developments in this regard have taken place in the field of medicine.

Third, and this seems to us to be a particularly revealing area, the big tech industry is also the addressee for requests for support or funding of scientific projects that are developed by universities or private research institutions.

Fourth, it can be assumed that the big tech industry will drive the development of domain-specific tools and infrastructure offerings even more strongly in the future. Meta's scientific language model Galactica (Taylor et al. 2022) and Google's language model for medicine dubbed Med-PaLM (Singhal et al. 2023) are indicative of this trend.

The points listed here are certainly not specific to the big tech industry. On the contrary, it can be assumed that corresponding activities are generally driven by the tech industry, including start-ups. However, the more successful the respective activities and developments are, the greater the likelihood that either corresponding start-ups will be acquired or the big tech industry will develop similar tools or offerings, even if this potentially leads to legal conflicts.

From our perspective, at any rate, there is a serious transnational dependence of the sciences on industry, the precise conditions of which, in turn, urgently need to be researched on an interdisciplinary basis.

There are already signs that universities in Europe want to strengthen their independence from the big tech industry with regard to their research activities, but also in terms of teaching, while at the same time and to a certain extent paradoxically there is a university policy interest in promoting alliances between science and industry in the development of AI projects, whereby local and regional funding aspects may be of importance here.

One of the problems concerning the relationship between industry and science also includes the fact that, on the one hand, the industry generally has an interest in ensuring that universities are able to train sufficiently qualified scientists, while at the same time, it also has a considerable share in the fact that particularly qualified scientists leave universities for industry and, in part, industry is increasingly moving to promote the internal training of IT specialists.

The aspects mentioned above thus concern fundamental questions about the conditions of digital science in the present. It seems important to us, however, that all the industrial and infrastructural problems and challenges mentioned above are causally linked to the specific potentials of subsymbolic AI.

The dependencies indicated here must be critically questioned, not only as a questionable contrast between a big tech industry worthy of criticism and university research supposedly independent from the outset, but in general with a view to the possibility of sustainably protecting technology from misuse, whether in the context of the private sector or with a view to public/state structures. Democratic states may lose their democratic or progressive status, much as the policies of large corporations may change drastically, and with it the question of what purposes AI is used for in the first place. Based on our observations so far, having to seriously consider the long-term consequences of AI is a relatively new phenomenon. Until recently, it seemed important to free AI from speculative discourse, and rather address problems of AI's present, but in light of recent developments, it does indeed seem necessary to extrapolate current developments and their speed to what problems will arise not just now, but in 5 or 10 years.

Thesis V: Expert crisis

> AI experts are in short supply. That's making the skills crisis worse. (Headline of an article on ZDnet, Hughes 2022)

One consequence of the fact that approaches of ANNs and other forms of machine learning could not really be used comprehensively or for advanced tasks, e.g., in many fields of science, is that a corresponding tradition of expertise was missing at the beginning of the current AI boom. Only a relatively small number of researchers in Europe and the US focused on or worked with such approaches in typical fields of AI research before 2016. Moreover, those who worked with ANNs in computer science in the 1990s and early 2000s, for example, had significant problems themselves at the time in being able to finance and publish their research at all. The establishment of the term 'deep learning' for ANNs from around 2006 onwards had a lot to do with the reputation of ANNs as ultimately being more or less a dead end in AI research, or as ultimately not being a target-oriented approach for many advanced problems in computer science. Accordingly, for a long time, it was an internationally very manageable community that continued unperturbed with ANN-based AI, in German-speaking countries, for example, research groups around Helge Ritter, Jürgen Schmidhuber and Alexander Waibel.

Against this background, the contrast to today's situation could hardly be more extreme. Especially from 2016 on, i.e., since the success of AlphaGo, a gold-rush mood has developed rapidly, which in turn not only affected the scientific field alone but more or less the society as a whole.[1]

From a disciplinary point of view, it is obvious that computer science in particular has benefited from the corresponding AI boom. In fact, it has not hurt the discipline in principle to have underestimated the epistemic potential of ANNs at the time.

Computer science is the big winner of the current AI boom in two respects: on the one hand, because of its historical core competence with respect to both the development and critical reflection of AI, and on the other hand, because of its now once again strengthened role as a collaborative or auxiliary discipline of other subjects. It may be that some disciplines, such as mathematics or physics, are not dependent on the external competencies of computer science to develop AI models for their purposes, but the humanities, cultural studies and social sciences are (even though in these areas knowledge on computer science has increased significantly in the recent past).

At the same time, the cultural sciences, social sciences, and humanities also benefit prima facie considerably from the sustained boom in AI. This applies, among other things, to philosophy, whose expertise has been called upon for some years now, especially for ethical issues in AI.

Finally, this concerns the interdisciplinary and transdisciplinary research field of digital humanities. Even if the corresponding orders of magnitude are difficult to estimate, one can certainly argue that AI, and especially forms of generative AI, have an important catalytic function in significantly expanding

1 A German platform currently lists 152 institutes and other institutions at German universities that conduct research with/on AI (Lernende Systeme 2022). Also see Huber/Huth/Alsabah (2020), a Bitkom survey according to which there are about 220 AI professorships in Germany at the time of the survey. Finally, one could point to the 100 AI Professorships Initiative, initiated in 2018 (BMBF 2022).

the possible uses of computers and Big Data, in the humanities and cultural studies, whether in research, or also in other areas such as teaching.

However, on the basis of our studies so far, we can state that certain subjects or research fields, such as climatology, are only gradually incorporating AI approaches to their research questions, and to a rather limited extent, and in some cases, there is also a rather great skepticism, perhaps even a certain conservatism, about using corresponding technologies.

Thesis VI: Sociological split seconds

> Computational social science is an interdisciplinary field that advances theories of human behavior by applying computational techniques to large datasets from social media sites, the Internet, or other digitized archives such as administrative records. Our definition forefronts sociological theory because we believe the future of the field within sociology depends not only on novel data sources and methods, but also on its capacity to produce new theories of human behavior or elaborate on existing explanations of the social world. (Edelmann et al. 2020)

A general phenomenon in the research landscape is the bifurcation of disciplines in a general and a computational branch. For instance, these twin disciplines have become a reality in many fields of the social sciences and are even traceable by citation analysis. Is computational sociology, which emerged towards the end of the 20th century, out of tune with the offline society or sociological theory? Will such disciplines eventually divide for good despite the integrative gestures the computational disciplines may provide? It is entirely conceivable that digital historians will seize to travel to the communities' main conferences (like the "Historikertag"). It is possible that computational Earth scientists find it easier to talk method with colleagues from the digital humanities than with colleagues returning from the field with samples and earth on their boots.

Technological advances provide new tools. Arguably, these provide new gravitational forces towards specific scientific methods and topics. The integration of new kinds of data practices across disciplines is all but new. Throughout the last centuries, the sciences, social sciences and humanities have benefited significantly from the availability of data. The empirical sciences of the 17th century, the social sciences of the 19th century and the digital humanities of the 21st century all profited from the influx of serialized and quantified types of information into their research methods. While it is probably the humanities which will be affected most by AI's conquest of the qualitative dimension, the social sciences present a puzzle.

For instance, the emergence of a separate field of computational sociology is a most interesting case, given the high affinity of sociology to data and empirical methods. This discipline can be said to have co-emerged with census taking and the intensified collection of data about the social from Adolphe Quetelet onwards. His "social physics" and his training with leading astronomers in Paris did not only lead him to stipulate neutral laws of the social, but let him develop moral statistics, a brand of criminology if not surveillance.

Given this high involvement of the discipline with social data, the division into separated disciplines of sociology may come as a surprise. This is not to say that sociologists do not embrace new data technologies. Attempts to map all articles in the Web of Science according to their level of AI-related methods show the social sciences almost as open to AI applications as the physical sciences or the life sciences (Gargiulo et al. 2022, fig. 1). Yet, this adaptability does not seem to appeal to all sociologists and at the moment several fields of knowledge split into computational versions of themselves. While this may prove to be a passing occurrence, it can also be an indication that AI-based methodologies are not perceived as empirical in the traditional sense.

Thesis VII: Data colonialism

> [...] I wonder whether data colonialism goes far enough to prompt a decolonial shift in thinking, assuming again that we are in the realm of Quijano and the modernity/coloniality school. Because the concept is more concerned with datafication as resource extraction, and seems less concerned with the key decolonial insight that Europe convinced itself and others that it has a privileged objective position from which it may make universal assertions and claims. (Mumford 2022: 1512)

In their widely discussed studies on data colonialism Nick Couldry and Ulisses Mejias (2019a; 2019b) claim that a new regime of data extraction has emerged. They see a logic of colonial dispossession at work because most data are collected for free. After all, this annexation is happening on a global scale, and results in huge profits for a very small group of people. Most AI applications still hinge on the availability of vast datasets and are part and parcel of the said colonialism. The scope of this new raw material is lamented and the problematic uses of a global surveillance capitalism are evoked. Some emphasize the correlation of systems of data extraction with systems of value (Thatcher/O'Sullivan/Mahmoudi 2016; Gray 2023). According to Denusa Mumford (2022), it seems questionable if the term colonization is already put to its best use in this discussion. The critiques mainly address capitalist strategies, especially the dynamics of primitive accumulation. While this dynamic is involved with the Global North as colonizer, the term points towards the fate of the Global South, but fails to show any specific engagement. Recent calls for epistemic decolonization remain unheard. A decolonial approach would entail efforts to decentralize one's position, to seek out "other worldings", to include specifics of the discussions from these regions, and to acknowledge the fundamental diversity of approaches. Thus, the diagnosis of data colonialism might not go far enough yet and would benefit from a deeper involvement with divergent perspectives from the outside, from the effects of data collection at the margins.

The history of data tries to make inroads and add to a fuller picture of the specific performative effects of data collection at various points of a data journey (Aronova/von Oertzen/Sepkoski 2017; Leonelli/Tempini 2020). The global data infrastructure is clearly built on an overexposure of marginalized and colonized bodies to various kinds of metrics (Hacking 1986; Lemov 2015; Radin 2017; TallBear 2013). Even these early colonial statistics and data collections already had radical effects. Indigenous communities could suddenly be shown to "go extinct" or dwindle under the curse of hunting parties of colonialists, hard physical labor and new contagious diseases (Rowse 2017; Malègue 2018; Renard 2021). The most widely discussed case of colonial statistics though is the performative effect and deep influences of the process of counting itself (Zimmerman 1999; Schlicht/Ledebur/Echterhölter 2021). The famous case in point is probably the Indian census, where the British used the categories of "caste" for the enumeration of all Hindus. Although castes existed in pre-colonial times, their statistical versions made them more rigid, scientifically defined, prominent, and publicly contested or lobbied for (Cohn 1987; Appadurai 1994; Dirks 2001). This shows that the stakes are high for any category, classificatory scheme, or label used on social data. The history of data classification is but one aspect of the decolonization of our rapidly growing data architectures.

During German colonialism, data collections and land surveys consistently relied on European legal notions and hence implemented foreign protocols.

To reverse this and similar processes, several initiatives are trying to arrive at more fitting frameworks for data collection in non-industrialized societies today (Abdilla 2021). To involve communities and to encourage participation is but one strategy to improve a technology that is perceived as "White" (Cave/Dihal 2020). Against this background, current initiatives for "indigenous data sovereignty" gain importance (Santos 2018; Kukutai/Taylor 2019; Lewis 2020), and fluid identities are being discussed as a blueprint to build better modeled and networked data infrastructures (Chun et al. 2019). Do indigenous perceptions of the non-human help to "breed" better algorithms? Who decides upon the categories used in clustering or classification in which region of the world? What would it mean, with Denusa Mumford, to arrive at decolonial data architectures?

Thesis VIII: The labor landscape shift

> ChatGPT and the like do improvise, promising to destabilize a lot of white-collar work, regardless of whether they eliminate jobs or not. (Lowrey 2023)

In stark contrast to our long-standing expectations and to great surprise, AI is not automating routine tasks or physical work as its first official act. Instead, generative AI impacts highly-skilled creative and knowledge workers by producing creative and knowledgeable output. In this respect, AI sets itself apart from previous technological developments.

This new realization is fueled by several contemporary developments in AI research. First and foremost, there has been remarkable progress in the area of language models leading to singular models capable of fulfilling diverse tasks such as creative content generation, summarization, translation, and code generation – to name a few of them. By defining instructions and prompts, many more tasks are conceivable that previously required meaningful investments and specialized systems. There are already several applications that give a first glimpse of the potential for and the impact on certain professions. Software engineers are fast to adopt new technical tools, such that it is no surprise that GitHub Copilot, a coding assistant, already amassed over a million users and is "behind an average of 46% of a developers' code" (Zhao 2023).

There are similar applications for other areas such as SciSpace Copilot, aiming to help with scientific literature research, or Casetext, which assists with legal research. For now, these systems are meant to be assistant to workers rather than to act in full autonomy. However, AI research is in rapid development[2]. The GitHub CEO, Thomas Dohmke, for example, claims that GitHub CoPilot will "sooner than later" (Scheffler 2023) write "80% of the code" (ibid.). It is an open question if this massive productivity gain will result in job losses but at the very least it reshapes the nature of some professions, adding oversight of and delegation to LLMs as a major bullet point to many job descriptions. Notably, Meta itself, as one of the leaders in AI research, announced to reduce hiring meaningfully going forward and focus instead on developer productivity, with AI toolings like code assistants and chatbots being a major part of that equation.

A relevant study in this context was conducted by Eloundou et al. (2023) in which the authors try to identify jobs that are exposed to LLM technology. While they neither provide a timeline nor make predictions on the impact on the labor market, it is found that higher-income occupations are more affected. Mathematicians, writers and authors, tax preparers, legal secretaries, or proofreaders are among a set of professions that are fully or close to fully exposed to LLMs; that means that LLMs take up to 100% of their occupational activities.[3]

At the same time, there is another area of AI research swiftly progressing and bringing completely different qualities to the debate. With the rise of recent diffusion models such as Adobe Firefly, Unity Muse, Midjourney, and Stable Diffusion, AI image synthesis is widely popularized, from the generation of digital art to photorealistic art. Point-E or Builder Bot are first approaches to 3D content generation while Imagen Video and Make-A-Video conceptualize video synthesis. Remarkably, the public outcry about this line of research has been much louder and more popular, leading to several copyright lawsuits and massive fears about potential job replacement. Presumably, that is, because visual output is more tangible and inaccuracies are not as crucial or obvious. Exemplary, Hollywood is one of several epicenters of this debate. While Disney just released the TV show *Secret Invasion* with an AI-generated opening, the actors guild SAG-AFTRA went on strike with one major contentious point being

---

2 Also see the following thesis, "AI's Self-Evolution".

3 The positive news for us in the research community is that professions related to science and critical thinking were found to have a low exposure to LLMs.

the digital replication of actors in films and shows. Again, the impacts on the labor market are hard to predict, however, in a *New York Times* (Roose 2022) article artists anecdotally describe the transformation of their work with the arrival of AI tools, while a report by *Rest of World* (Zhou 2023) already claims an effect of AI on the job market of game illustrators.

A recent McKinsey study (Chui et al. 2023) predicts a massive transformation of our economy and the labor market as a result of generative AI, encompassing both large language models and diffusion models. The report claims "that half of today's work activities could be automated" (ibid.). That is, however, on a largely uncertain timeline "between 2030 and 2060" (ibid.). In accordance with the study by Eloundou et al. and our own assumptions, the report eventually sees high-income workers as the most impacted group of this transformation. And while no one dares to make a definitive prediction on the impact of the labor market, "it's important to be honest that it's increasingly going to make some jobs not very relevant" (Altman 2022) as Sam Altman himself puts it in an article on his website. At the very least, AI is predestined to alter the labor landscape profoundly.

Thesis IX: AI's self-evolution

> [A] Large Language Model (LLM) is capable of improving its performance [...] by training on its own generated labels. (Huang et al. 2022)

The landscape of AI is undergoing a transformative shift as we witness the emergence of a cycle of self-evaluating and self-improving systems. Large language models are at the forefront of this development, demonstrating the ability to assess the quality of and improve upon their own generations. At the same time, generated output can be used to leverage and train the next generation of models or distill knowledge into more efficient ones allowing to deploy models on a large scale. Several recent developments are indicative of this development.

To begin with, current LLMs are cheaper and already produce comparable output compared to human labelers on crowdsourcing platforms such as Mechanical Turk in many tasks (Gilardi/Alizadeh/Kubli 2023). At the same time, studies like the one by Veselovsky, Horta and West (2023) found that crowd workers increasingly utilize LLMs themselves to complete their tasks. It is foreseeable that this will supercharge data collection in many cases, increasing the quantity of data while reducing the time needed.

Separately, there are a number of developments in AI research itself that suggest an increasingly fast cycle of self-evaluation and improvement. For example, the popular Alpaca model from Stanford (Taori et al. 2023) uses GPT-3 to produce training data to fine-tune and align Meta's LLaMA language model (Touvron et al. 2023). All this happened within two weeks of the release of LLaMA, further indicating the rapid speed at which improvements are achieved. Taori et al. use a recent trend in AI called "Self-Instruct" (Wang et al. 2023) to automatically generate training data where a number of seed tasks is defined and an LLM generates new instructions and corresponding instruction-answer pairs for them.

In addition, there are other ways to utilize current LLMs. Traditionally, LLMs but also other AI systems are evaluated by certain automated metrics. However, these metrics are generally not perfect and only offer correlation with human judgment to a certain degree. This prompted several researchers to develop performance metrics based on the judgment of LLMs, such as GPTScore (Fu et al. 2023) or GEMBA (Kocmi et al. 2023). It is also becoming more common to evaluate LLMs only relative to each other using rating mechanisms like ELO (see Chatbot Arena by Zheng et al. 2023). One pressing question that may arise now, is whether it is possible to use these capabilities in the training process of an LLM. Presently, reinforcement learning from human feedback (RLHF; Christiano et al. 2017) is used by state-of-the-art models like GPT-3. As part of this, a reward model is trained based on human feedback. This reward model can then be used as a proxy of human feedback while fine-tuning the LLM. It is conceivable though to use a fine-tuned LLM in place of humans to create synthetic ranking data for the training of reward models. This technique is further called reinforcement learning from AI feedback (RLAIF) and can potentially be applied in an iterative fashion (Bai et al. 2022).

It also becomes more prevalent to use these self-evaluation capabilities to improve the performance of current LLMs at inference time. Even without further training or fine-tuning, the model can reflect or critique itself by passing the output *again* into the model and refine its output (Gou et al. 2023; Shinn/Labash/Gopinath 2023; Xiao et al. 2023).

Now, with these trends reshaping AI, one could be tempted to speculate about a vicious self-improvement cycle leading to the "singularity". However, it is important to temper these advances with the recognition that current AI systems including LLMs remain narrow in nature and are far from achieving artificial general intelligence (AGI), thereby dispelling the notion of an imminent transformative "singularity" or emergence of "superintelligence" driven by existing AI paradigms. For this to happen, it might need a fundamentally different approach to AI as suggested by Yann LeCun (2022) and others.

List of references

Abdilla, Angie/Kelleher, Meghan/Shaw, Rick/Yunkaporta, Tyson (2021): Out of the Black Box: Indigenous Protocols for AI, Melbourne: Deakin University (https://hdl.handle.net/10536/DRO/DU:30159239).

Adadi, Amina/Berrada, Mohammed (2018): "Peeking Inside the Black-Box: A Survey on Explainable Artificial Intelligence (XAI)." In: IEEE Access 6, pp. 52138–52160.

Alpaydin, Ethem (2016): Machine Learning: The New AI, Cambridge, MA: The MIT Press.

Altman, Sam (2022): "DALL•E 2.", April 6, 2022 (https://blog.samaltman.com/dall-star-e-2).

Anderson, Chris (2008): "The End of Theory: The Data Deluge Makes the Scientific Method Obsolete.", June 23, 2008 (https://www.wired.com/2008/06/pb-theory/).

Appadurai, Arjun (1996): "Number in the Colonial Administration." In: Arjun Appadurai (ed.), Modernity at Large. Cultural Dimensions of Globalization, Minneapolis, MN: University of Minnesota Press, pp. 114–139.

Aronova, Elena/von Oertzen, Christine/Sepkoski, David (2017): "Introduction: Historicizing Big Data." In: Osiris 32/1, pp. 1–17.

Ash, James/Kitchin, Rob/Leszczynski, Agnieszka (eds.) (2019): Digital Geographies. Los Angeles, CA. et al.: SAGE.

Bai, Yuntao/Kadavath, Saurav/Kundu, Sandipan/Askell, Amanda/Kernion, Jackson/Jones, Andy/Chen, Anna/et al. (2022): Constitutional AI: Harmlessness from AI Feedback, arXiv Preprint (https://doi.org/10.48550/arXiv.2212.08073).

BMBF (Bundesministerium für Bildung und Forschung) (2022): "Stark-Watzinger: 100. zusätzliche KI-Professur wurde besetzt.", May 6, 2022 (https://www.bmbf.de/bmbf/shareddocs/pressemitteilungen/de/2022/05/030522-KI-Professoren.html).

Bourilkov, Dimitri (2019): "Machine and Deep Learning Applications in Particle Physics". In: International Journal of Modern Physics A 34/35, 1930019.

Cave, Stephen/Dihal, Kanta (2020): "The Whiteness of AI." In: Philosophy and Technology 33, pp. 685–703.

Christiano, Paul F./Leike, Jan/Brown,Tom/Martic, Miljan/Legg, Shane/ Amodei, Dario (2017): Deep Reinforcement Learning from Human Preferences, arXiv Preprint (https://doi.org/10.48550/arXiv.1706.03741).

Chui, Michael/Hazan Eric/Roberts Roger/Singla Alex/Smaje Kate/ Sukharevsky Alex/Yee Lareina/Zemmel Rodney (2023): "The economic potential of generative AI: The next productivity frontier.", June 2023 (https://www.mckinsey.com/capabilities/mckinsey-digital/our-insights/the-economic-potential-of-generative-AI-the-next-productivity-frontier).

Chun, Wendy/Apprich, Clemens/Steyerl, Hito/Cramer, Florian (2019): Pattern Discrimination, Minneapolis and London: University of Minnesota Press, and Lüneburg: meson press.

Cohn, Bernard S. (1987): "The Census, Social Structure, and Objectification in South Asia." In: Bernard S. Cohn (ed.), An Anthropologist among the Historians and Other Essays, New Delhi: Oxford University Press, pp. 224–254.

Couldry, Nick/Mejias, Ulises A. (2019a): The Costs of Connection: How Data Is Colonizing Human Life and Appropriating It for Capitalism, Stanford, CA: Stanford University Press.

Couldry, Nick/Mejias, Ulises A. (2019b): "Data Colonialism: Rethinking Big Data's Relation to the Contemporary Subject." In: Television & New Media 20/4, pp. 336–349.

Dastin, Jeffrey (2023): "Microsoft attracting users to its code-writing, generative AI software.", January 25, 2023 (https://www.reuters.com/technology/microsoft-attracting-users-its-code-writing-generative-ai-software-2023-01-25/).

Deutsche Forschungsgemeinschaft (2020): Digitaler Wandel in den Wissenschaften, Bonn: Deutsche Forschungsgemeinschaft (https://doi.org/10.5281/zenodo.4191345?).

Dirks, Nicholas B. (2001): Castes of Mind: Colonialism and the Making of Modern India, Princeton: Princeton University Press.

Duan, Yan/Andrychowicz, Marcin/Stadie, Bradly/Ho, Jonathan/Schneider, Jonas/Sutskever, Ilya/Abbeel, Pieter/Zaremba, Wojciech (2017): "One-Shot Imitation Learning." In: Advances in Neural Information Processing Systems (NIPS, 2017), Long Beach, CA, USA, pp. 1087–1098.

Edelmann, Achim/Wolff, Tom/Montagne, Danielle/Bail, Christopher A. (2020): "Computational Social Science and Sociology." In: Annual Review of Sociology 46/1, pp. 61–81.

Eloundou, Tyna/Manning, Sam/Mishkin, Pamela/Rock, Daniel (2023): GPTs are GPTs: An Early Look at the Labor Market Impact Potential of Large Language Models, arXiv Preprint (https://doi.org/10.48550/arXiv.2303.10130).

Fleck, Ludwik (1980 [1935]): Entstehung und Entwicklung einer wissenschaftlichen Tatsache, Frankfurt a. M.: Suhrkamp.

Fu, Jinlan/Ng, See-Kiong/Jiang,Zhengbao/Liu, Pengfei (2023): GPTScore: Evaluate as You Desire, arXiv Preprint (https://doi.org/10.48550/arXiv.2302.04166).

Gargiulo, Floriana/Fontaine, Sylvain/Dubois, Michel/Tubaro, Paola (2022): A meso-scale cartography of the AI ecosystem, arXiv Preprint (https://doi.org/10.48550/arXiv.2212.12263).

Gilardi, Fabrizio/Alizadeh, Meysam/Kubli, Maël (2023): "ChatGPT Outperforms Crowd-Workers for Text-Annotation Tasks." In: Proceedings of the National Academy of Sciences 120/30, e2305016120.

Gou, Zhibin/Shao, Zhihong/Gong, Yeyun/Shen, Yelong/Yang, Yujiu/Duan, Nan/Chen, Weizhu (2023): CRITIC: Large Language Models Can Self-Correct with Tool-Interactive Critiquing, arXiv Preprint (https://doi.org/10.48550/arXiv.2305.11738).

Gramelsberger, Gabriele (ed.) (2011): From Science to Computational Sciences Studies in the History of Computing and its Influence on Today's Sciences, Zürich: Diaphanes.

Gray, Catriona (2023): "More than Extraction: Rethinking Data's Colonial Political Economy." In: International Political Sociology 17/2, pp. 1–20.

Hacking, Ian (1986): "Making Up People." In: Thomas L. Heller/Morton Sosna/David E. Wellbery (eds.), Reconstructing Individualism. Autonomy, Individuality, and the Self in Western Thought, Stanford, CA: Stanford University Press, pp. 222–236.

Huang, Jiaxin/Shane Gu, Shixiang /Hou, Le/Wu, Yuexin/Wang, Xuezhi/Yu, Hongkun/Han, Jiawei/et al. (2022): Large Language Models Can Self-Improve, arXiv Preprint (https://doi.org/10.48550/arXiv.2210.11610).

Huber, Marco/Huth, Michael/Alsabah, Nabil (2020): KI-Forschung in Deutschland – Der schwere Weg zu 100 neuen KI-Professuren, Berlin: Bitkom.

Hughes, Owen (2022): "AI Experts Are in Short Supply. That's Making the Skills Crisis Worse." May 20, 2022 (https://www.zdnet.com/article/ai-experts-are-in-short-supply-thats-making-the-skills-crisis-worse/).

Jasanoff, Sheila (2015): "Future Imperfect. Science, Technology, and the Imaginations of Modernity." In: Sheila Jasanoff/Sang-Hyun Kim (eds.), Dreamscapes of Modernity. Sociotechnical Imaginaries and the Fabrication of Power, Chicago and London: The University of Chicago Press, pp. 1–33.

Jurowetzki, Roman/Hain, Daniel/Mateos-Garcia, Juan/Stathoulopoulos, Konstantinos (2021): The Privatization of AI Research (-ers): Causes and Potential Consequences–From university-industry interaction to public research brain-drain?, arXiv Preprint (https://doi.org/10.48550/arXiv.2102.01648).

Knorr-Cetina, Karin (1999): Epistemic Cultures: How the Sciences Make Knowledge, Cambridge, MA: Harvard University Press.

Kocmi, Tom/Federmann, Christian (2023): Large Language Models Are State-of-the-Art Evaluators of Translation Quality, arXiv Preprint (https://doi.org/10.48550/arXiv.2302.14520).

Krüger, Lorenz/Daston, Lorraine J./Heidelberger, Michael (eds.) (1987): The Probabilistic Revolution, Cambridge, MA: The MIT Press.

Kukutai, Tahu/Taylor, John (eds.) (2016): Indigenous Data Sovereignty: Toward an Agenda, Canberra: Australian National University Press.

Kümmel, Albert/Scholz, Leander/Schumacher, Eckhardt (2004): Einführung in die Geschichte der Medien, Paderborn: Fink.

Latour, Bruno/Woolgar, Steve (1979): Laboratory Life. The Construction of Scientific Facts, Princeton, NJ: SAGE.

LeCun, Yann (2022): A Path Towards Autonomous Machine Intelligence, OpenReview Preprint (https://openreview.net/forum?id=BZ5a1r-kVsf).

Lemov, Rebecca (2015): Database of Dreams: The Lost Quest to Catalog Humanity, New Haven, CT: Yale University Press.

Leonelli, Sabina (2016): Data-Centric Biology. A Philosophical Study, Chicago and London: University of Chicago Press.

Leonelli, Sabina/Tempini, Niccolò (eds.) (2020): Data Journeys in the Sciences, Cham: Springer Nature.

Lernende Systeme (2023): "KI-Landkarte.", continuously updated (https://www.plattform-lernende-systeme.de/ki-landkarte.html).

Lewis, Jason E. (ed.) (2020): Indigenous Protocol and Artificial Intelligence Position Paper, Honolulu, HI: The Initiative for Indigenous Futures and the Canadian Institute for Advanced Research (https://doi.org/10.11573/spectrum.library.concordia.ca.00986506).

Lindley, Dennis V. (2000): "The Philosophy of Statistics." In: Journal of the Royal Statistical Society: Series D (The Statistician) 49/3, pp. 293–337.

Lowrey, Annie (2023): "How ChatGPT Will Destabilize White-Collar Work.", January 20, 2023 (https://www.theatlantic.com/ideas/archive/2023/01/chatgpt-ai-economy-automation-jobs/672767/).

Malègue, Fanny (2018): "The Empire in a Census Table: Enumerating the Population and Governing the French Caribbean Colonies after the Seven Years' War." In: Histoire & Mesure 2, pp. 93–114.

Milmo, Dan (2023). "AI revolution puts skilled jobs at highest risk, OECD says.", July 11, 2023 (https://www.theguardian.com/technology/2023/jul/11/ai-revolution-puts-skilled-jobs-at-highest-risk-oecd-says).

Mumford, Denusa (2022): "Data Colonialism: Compelling and Useful, but Whither Epistemes?" In: Information, Communication & Society 25/10, pp. 1511–1516.

Radin, Joanna (2017): "'Digital Natives': How Medical and Indigenous Histories Matter for Big Data." In: Osiris 32, pp. 43–64.

Radovic, Alexander/Williams, Mike/Rousseau, David/Kagan, Michael/Bonacorsi, Daniele/Himmel, Alexander/Aurisano, Adam/Terao, Kazuhiro/Wongjirad, Taritree (2018): "Machine Learning at the Energy and Intensity Frontiers of Particle Physics." In: Nature 560, pp. 41–48.

Renard, Léa (2021): "Vergleichsverbot? Bevölkerungsstatistiken und die Frage der Vergleichbarkeit in den deutschen Kolonien (1885–1914)." In: Kölner Zeitschrift für Sozialpsychologie 73, pp. 169–194.

Roose, Kevin (2022): "A.I.-Generated Art Is Already Transforming Creative Work.", October 21, 2022 (https://www.nytimes.com/2022/10/21/technology/ai-generated-art-jobs-dall-e-2.html).

Rowse, Tim (2017): "The Statistical Table as Colonial Knowledge." In: Itinerario 41/1, pp. 51–73.

Santos, Boaventura de Sousa (2018): The End of the Cognitive Empire: The Coming of Age of Epistemologies of the South, Durham: Duke University Press.

Scheffler, Ian (2023): "GitHub CEO says Copilot will write 80% of code 'sooner than later'", June 17, 2023 (https://www.freethink.com/robots-ai/github-copilot).

Schlicht, Laurens/Ledebur, Sophie/Echterhölter, Anna (2021): "Data at the Doorstep. Intimacy and Formalization." In: Science in Context 34/4, pp. 1–18.

Schlombs, Corinna (2010): Productivity Machines: Transatlantic Transfers of Computing Technology and Culture in the Cold War, Ann Arbor, MI: UMI Dissertations Publishing.

Schröter, Jens (2004): Das Netz und die Virtuelle Realität. Zur Selbstprogrammierung der Gesellschaft durch die universelle Maschine, Bielefeld: transcript.

Schröter, Jens (2012): "The Internet and 'Frictionless Capitalism'." In: tripleC 10/2, pp. 302–312.

Schröter, Jens (2019): "Digitale Medientechnologien und das Verschwinden der Arbeit." In: Thomas Bächle/Caja Thimm (eds.), Mensch und Maschine: Freund oder Feind? Mensch und Technologie im digitalen Zeitalter, Wiesbaden: Springer, pp. 183–210.

Schröter, Jens (2021): "KI und die Wissenschaften. Das Beispiel der Teilchenphysik." In: Nicole Brandstetter/Ralph-Miklas Dobler/Daniel Jan Ittstein (eds.), Mensch und Künstliche Intelligenz. Herausforderungen für Kultur, Wirtschaft und Gesellschaft, München: UVK, pp. 51–58.

Schröter, Jens/Schwering, Gregor (2014): "Modelle des Medienwandels und der Mediengeschichtsschreibung." In: Jens Schröter (ed.), Handbuch Medienwissenschaft, Stuttgart and Weimar: Metzler, pp. 179–190.

Seising, Rudolf (ed.) (2009): Views on Fuzzy Sets and Systems from Different Perspectives: Philosophy and Logic, Criticisms and Applications, Berlin and Heidelberg: Springer (https://doi.org/10.1007/978-3-540-93802-6).

Shinn, Noah/Labash, Beck/Gopinath, Ashwin (2023): Reflexion: an Autonomous Agent with Dynamic Memory and Self-reflection, arXiv Preprint (https://doi.org/10.48550/arXiv.2303.11366).

Singhal, Karan/Azizi, Shekoofeh/Tu, Tao/Mahdavi, Sara/Wei, Jason/Won Chung, Hyung /Scales, Nathan/et al. (2023): "Large Language Models Encode Clinical Knowledge." In: Nature 620, pp. 172–180.

Sonawane, Urvi/Komorowski, Matthieu (2023): "Artificial Intelligence in Medicine. Potential Applications and Barriers to Deployment." In: Andreas Sudmann/Anna Echterhölter/Markus Ramsauer/Fabian Retkowski/Jens Schröter/Alexander Waibel (eds.), Beyond Quantity: Research with Subsymbolic AI, Bielefeld: transcript, pp. 155-178.

Sudmann, Andreas (2019): "On the Media-Political Dimension of Artificial Intelligence. Deep Learning as a Black Box and OpenAI." In: Andreas Sudmann (ed.), The Democratization of Artificial Intelligence. Net Politics in the Era of Learning Algorithms, Bielefeld: transcript. pp. 223–244.

Syed, Ali (2016): "A Brief Introduction to Decolonial Computing." In: Crossroads (Association for Computing Machinery) 22/4, pp. 16–21.

TallBear, Kimberly (2013): Native American DNA: Tribal Belonging and the False Promise of Genetic Science, Minneapolis, MN: University of Minnesota Press.

Taori, Rohan/Gulrajani, Ishaan/Zhang, Tianyi/Dubois, Yann/Guestrin, Carlos/Liang, Percy/Hashimoto, Tatsunori B. (2023): "Alpaca: A Strong, Replicable Instruction-Following Model.", March 13, 2023 (https://crfm.stanford.edu/2023/03/13/alpaca.html).

Taylor, Ross/Kardas, Marcin/Cucurull, Guillem/Scialom, Thomas/Hartshorn, Anthony/Saravia, Elvis/Poulton, Andrew/et al. (2022): Galactica: A Large Language Model for Science, arXiv Preprint (https://doi.org/10.48550/arXiv.2211.09085).

Thatcher, Jim/O'Sullivan, David/Mahmoudi, Dillon (2016): "Data Colonialism through Accumulation by Dispossession: New Metaphors for Daily Data." In: Environment and Planning D: Society and Space 34/6, pp. 990–1006.

Touvron, Hugo/Lavril, Thibaut/Izacard, Gautier/Martinet, Xavier/Lachaux, Marie-Anne/Lacroix, Timothée/Rozière, Baptiste/et al. (2023): LLaMA: Open and Efficient Foundation Language Models, arXiv Preprint (https://doi.org/10.48550/arXiv.2302.13971).

Veselovsky, Veniamin/Horta, Ribeiro Manoel/West, Robert (2023): Crowd Workers Widely Use Large Language Models for Text Production Tasks, arXiv Preprint (https://doi.org/10.48550/arXiv.2306.07899).

von Oertzen, Christine (2017): "Machineries of Data Power. Manual Versus Mechanical Census Compilation in Nineteenth-Century Europe." In: Osiris 32/1, pp. 129–150.

Wang, Yizhong/Kordi, Yeganeh/Mishra, Swaroop/Liu, Alisa/Smith, Noah A./Khashabi, Daniel/Hajishirzi, Hannaneh (2023). "Self-Instruct: Aligning Language Models with Self-Generated Instructions." In: Proceedings of the 61st Annual Meeting of the Association for Computational Linguistics (Volume 1: Long Papers), pp. 13484–13508.

Weiss, Karl/Khoshgoftaar, Taghi M./Wang, DingDing (2016): "A Survey of Transfer Learning." In: Journal of Big Data 3/1: 9.

Xiao, Wen/Xie, Yujia/Carenini, Giuseppe/He, Pengcheng (2023): ChatGPT-steered Editing Instructor for Customization of Abstractive Summarization, arXiv Preprint (https://doi.org/10.48550/arXiv.2305.02483).

Zhang, Daniel/Mishra, Saurabh/Brynjolfsson, Erik/Etchemendy, John/Ganguli, Deep/Sellitto, Michael/Shoham, Yoav/et al. (2021): The AI Index 2021 Annual Report, Stanford, CA: Human-Centered AI Institute, Stanford University (https://arxiv.org/pdf/2103.06312).

Zhao, Shuyin (2023): "GitHub Copilot Now Has a Better AI Model and New Capabilities.", February 14, 2023 (https://github.blog/2023-02-14-github-copilot-now-has-a-better-ai-model-and-new-capabilities/).

Zheng, Lianmin/Chiang, Wei-Lin/Sheng, Ying/Zhuang, Siyuan/Wu, Zhanghao/Zhuang, Yonghao/Lin, Zi/et al. (2023): Judging LLM-as-a-judge with MT-Bench and Chatbot Arena, arXiv Preprint (https://doi.org/10.48550/arXiv.2306.05685).

Zhou, Viola (2023): "AI is already taking video game illustrators' jobs in China.", April 11, 2023 (https://restofworld.org/2023/ai-image-china-video-game-layoffs/).

Zimmerman, Andrew (1999): "Anti-Semitism as Skill: Rudolf Virchow's "Schulstatistik" and the Racial Composition of Germany." In: Central European History 32/4, pp. 409–429.

When Achilles met the tortoise

Towards the problem of infinitesimals in machine learning[1]

Clemens Apprich

I would like to begin with a little story, a story that you probably already know. It's the story of Achilles and the tortoise.

One day, the hero of the Iliad met a tortoise whose mind was quicker than its legs. She challenged Achilles to a race, but asked him for a head start. Achilles willingly – and rather arrogantly – agreed to do so. The turtle crawled away. Achilles took his time, laced his sandals and finally started to run. In no time he covered the distance that had separated him from the turtle. In the meantime, however, the tortoise had also crawled on, and, while Achilles was catching up, she had again made a little progress. To cut the story short: no matter how fast Achilles ran, the tortoise always stayed a little way ahead – and so the famous hero could never catch up with the animal.

The story was told in this or a similar way – there is no exact record – by the Eleatic philosopher Zeno (around 490 to 430 BC) in order to present one of his paradoxes. The dichotomy paradox goes as follows: Because the world is one, movement is impossible. Every distance that a moving object has to cover can be broken down into an infinite number of partial distances (e.g. by continuous bisection: $\frac{1}{2}$, $\frac{1}{4}$, $\frac{1}{8}$, and so on), with one distance always remaining. As a consequence, no movement can ever be carried out completely, because there is always a distance remaining, no matter how small it may be.[2]

1 Parts of the argumentation in this article were developed in context of Simon Denny's collaborative exhibition *Proof of Stakes: Technological Claims* (Denny et al. 2022).

2 With this paradox, which exists in different variations (e.g., in the form of an infinite regression), Zeno wanted to (at least if you follow common introductions into philosophy) defend the teachings of his mentor Parmenides of Elea (born around 515 BC). According to Plato (1997), Parmenides was accompanied by Zeno, when he met Socrates around 445 BC in Athens and confronted him with the astonishing claim that beings (reality) are a holistic, unchangeable and unified entity (i.e. ontological/ontic monism).

Of course, such an idea completely contradicts our everyday experience, as it declares the immediate perception to be an illusion. Nevertheless – or precisely because of this – Zeno's paradox, passed down via Plato's dialogue "Parmenides" (1997), would not let go of Western philosophy for the next two and a half thousand years (from Archimedes to Giovanni Benedetti, to Isaac Newton, David Hume, Gottfried Wilhelm Leibniz, to Georg Cantor, Alfred North Whitehead and Gilles Deleuze – and most recently Gregory Chaitin with his algorithmic information theory). What it introduced and has since then haunted the history of science, in particular mathematics, is the problem of infinitesimals – with infinitesimals being distances in space or time that denote a smallest possible unit. It is assumed that an infinitesimal quantity is so close to zero that it has no numerical effect; it simply eludes any attempt to measure it, like sand trickling through your fingers.

Infinitesimals were crucial for the development of differential and integral equations – also known as calculus. As is well established, Gottfried Wilhelm Leibniz (1646 – 1716) and Isaac Newton (1642 – 1726) developed the mathematical branch of infinitesimal calculus independently of each other (or so the story goes) in the late 17th century.[3] Defining a systematic method for the calculation of surfaces and motion, it soon became a 'killer application' in modern mathematics as it geared to solve practical problems (e.g. ballistic calculations, motion of planets, the design of bridges). Calculus, eventually, turned out "to be the richest lode that the mathematicians have ever struck" (Kline 1977: 4). The development of calculus marked a new era in mathematics and its uses within the sciences have continued to the present day.

Not surprisingly, calculus is also at the heart of today's machine learning processes. Understood as optimization problems, machine learning-algorithms, in particular in the field of artificial neural networks, draw on calculus and, as a consequence, entail some of the paradoxes that come with it. Hence, by addressing the "quality issues" brought up in this volume, I want to argue that a machine learning-model, precisely because it is built on an exhaustive approximation as part of its optimization process, can never fully converge, and as a consequence does not yield any final result. This is of relevance because it shows that – contrary to widespread belief – machine learning is deeply entangled with mathematics and logics. What's more, such a paradoxical take on machine learning, which can also be seen as yet another iteration

3 In fact, the question of who invented calculus first became the subject of a huge controversy, now known as the calculus controversy (cf. Hall 1980).

of the "halting problem" (Turing 1936), resonates with recent debates around the incomputability of reality (Parisi 2013; Fazi 2018; Galloway 2021) as well as speculative attempts to overcome modern computation altogether (Amaro 2022). The goal of these interventions and by consequence the following article, is to highlight the necessity of moving beyond the limited imagination of (statistical) probability with regard to machine learning models in order to search for new "politics of possibility" (Amoore 2013).

1. Forever converging

In the beginning of Google's Machine Learning Crash Course,[4] Peter Norvig, Head of Google Research, makes the remarkable statement that – with machine learning – we are now moving from mathematics to natural science, from logics to statistics, and from coding to *growing* models:

> Machine learning changes the way you think about a problem. Software engineers are trained to think logically and mathematically [...]. With machine learning, the focus shifts from a mathematical science to a natural science: we're making observations about an uncertain world, running experiments, and using statistics, not logic, to analyze the results of the experiment. The ability to think like a scientist will expand your horizons and open up new areas that you couldn't explore without it. (Norvig 2020)

It is worthwhile to consider some of the deeper implications of Norvig's statement: What does it mean to move from mathematics to natural science? And, in the process, do we really leave logic behind? What the statement implies, is the fact that with machine learning, and respectively neural networks as the most recent implementation of machine learning systems, we are moving from deductive to inductive methods of data processing: the model learns a correlation pattern between input and output data in order to make predictions on unseen data. To do so, a loss function is calculated for each instance, which shows "how bad the model's prediction was on a single example" (Google Developers 2020).[5] Similar to the hot and cold play, the iterative strategy constitutes

4 Google's MLCC is one of the most popular machine learning (online) course with tens of thousands of users (Rosenberg 2018).

5 In the context of machine learning, unsupervised learning is often spoken of. However, when training a model (e.g. a recommendation system) most commonly a supervised

the essential thing of this learning approach, which, of course, corresponds to the aforementioned optimization process.[6]

Figure 1: Machine Learning Crash Course (screenshot from Google Developers 2020).

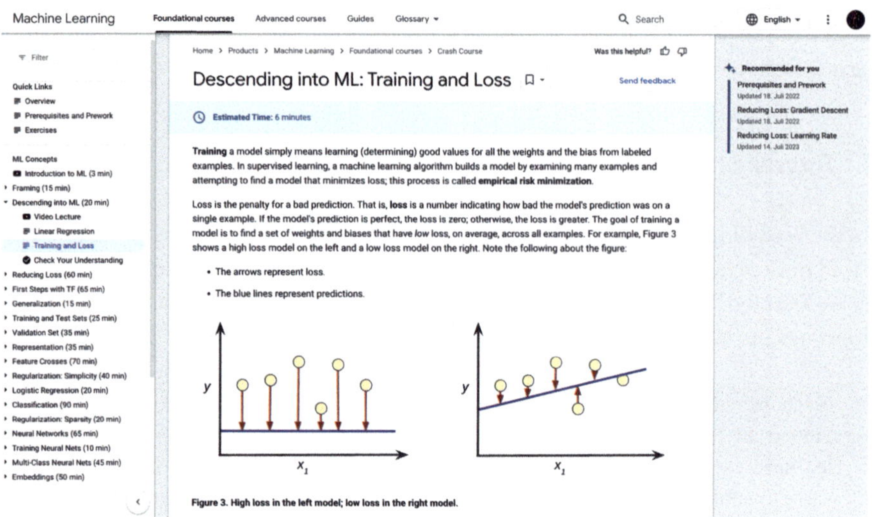

Suppose we had the time and the computing power to calculate the loss function for all possible learning parameters: the result would be a convex curve in which the rate of loss moves towards zero, that is the limit value to which the model converges. However, since the calculation of the loss function

---

procedure – or at least a mix of supervised and unsupervised learning – is used; this means that a data set is used containing both features (e.g. age, gender, search history of the user, temporal or geographical features in the data) and labels (what we want to predict).

6 The search for the optimal parameters constitutes machine learning. As Adrian Mackenzie writes in *Machine Learners* (2017): "[O]ptimization techniques are the operational underpinning of machine learning. Without their iterative process, there is no machine in machine learning" (95).

for every instance of a training set would take too long, a statistical method is used to solve the optimization problem: the gradient descent.[7]

Figure 2: Machine Learning Crash Course (screenshot from Google Developers 2020).

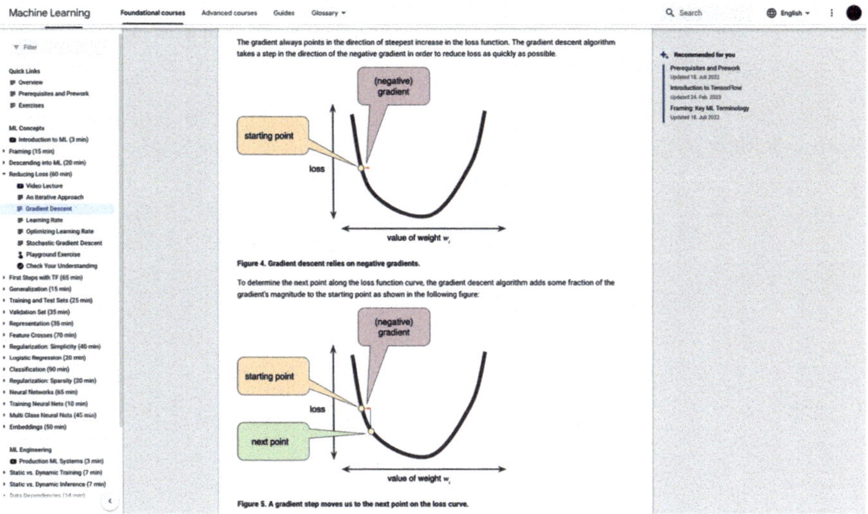

As is the case with mountaineering, the trick is to choose a descent direction and a step length in order to reach the valley of the curve as quickly as possible or, in other words, to reduce the loss as quickly as possible. The starting point is set arbitrarily, because it usually has no effect on the end result. To find the next point along the loss function curve, the learning algorithm then multiplies the gradient by a scalar quantity called the *learning rate* (sometimes also step length, although this can be misleading because the length of the step changes relative to the scalar). For example: if the amount of the (negative) gradient is 5 and the learning rate is 0.1, then the algorithm selects the next point

7 Gradient descent is one of many examples of optimization problems within machine learning systems. Others include coordinate descent, coordinate ascent or convex optimization.

0.5 units from the starting point and the next but one point at 0.05 units from this point.[8]

The correct setting of the learning rate makes up a good part of machine training. However, in practice, it is not necessary or possible to find the 'perfect' (or near-perfect) learning rate for successful training. The goal simply is to find a learning rate large enough for the model to converge in a timely manner, but not so large that it overshoots the target.

Figure 3: Machine Learning Crash Course (screenshot from Google Developers 2020).

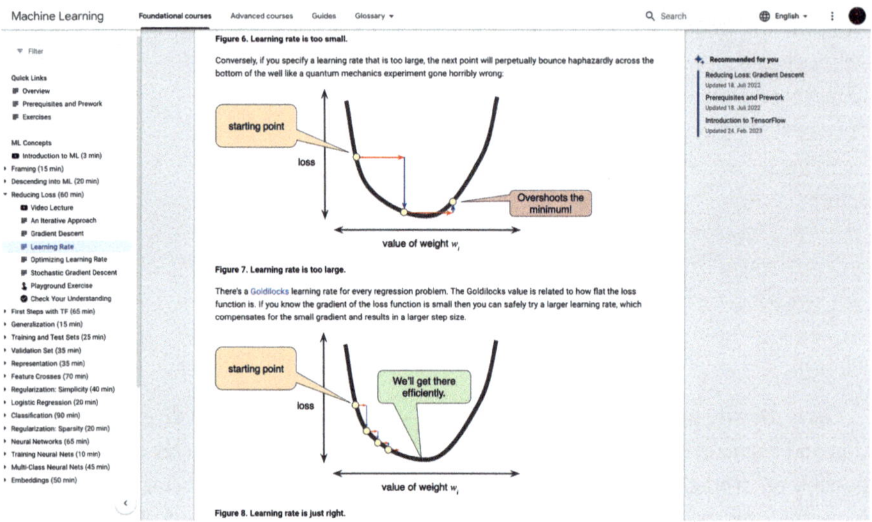

Hence, the idea behind gradient descent is to tweak the parameters iteratively until the algorithm converges to a minimum, that is to repeat the process "until the difference between the old value and the new value is very small" (Kansal 2020). Now, as you can already guess, behind the seemingly innocuous notion *very small* lurks the two and a half thousand-year-old paradox of Zeno. Because the learning steps gradually get smaller as the parameters approach

8 Another inspiration for this approach might be "fitness landscapes" (Wright 1932), a concept developed as part of evolutionary biology in the 1930s. I thank Claus Pias for this reference.

the minimum, each step can be divided into an infinite number of sections with the result that the model, at least in theory, never fully converges.

Peter Norvig argues that machine learning is no longer a logical problem, but an experimental one. That might be true if we follow the premise that with machine learning we are moving from a mathematical (i.e. deductive) to a natural (i.e. inductive) science. However, given the central role of mathematics in the natural sciences when converting observations into measurements (not to mention the creation of measurement itself), the statement seems to be at odds with its own premise. What it does though, is play right into the hands of similar attempts to *biologize* AI and machine learning (i.e. to *naturalize* and thus *normalize* the labor processes, the material infrastructures, but also the data politics behind it). The apologists of the new machine learning paradigm want to make us believe that the world of data is simply a natural phenomenon that does away with logical, that is theoretical, explanations (cf. Anderson 2008).

What is not mentioned in Norvig's statement, but is definitely an issue in computer or the data sciences, is the fact that a machine learning algorithm "must embody some knowledge or assumptions beyond the data it is given in order to generalize beyond it" (Domingos 2012: 81). A machine learning algorithm cannot see, hear, or perceive input examples (images, text, audio files, etc.) directly. Instead, a representation of the data has to be created in order to allow the model to *see* it. In other words, for the model to train, features have to be selected (often even created) which, in the eyes of the still very human trainer, best represent the data.[9]

Now this basic insight contradicts the common idea that we, and the models respectively, simply have to look at the data to get the desired outcome. What gets omitted, if not to say oppressed, in this rather naïve view, is the fact that the desired outcome (together with its logics) is always already inscribed in the process. With each iteration, the model gets more and more tweaked towards *good* property values (also called identity values) in order to filter out the *right* information from the data set.[10]

9 A common practice in machine learning is actually called 'feature engineering.'

10 This is in particular true for 'reinforcement learning from human feedback' (RLHF), a technique to train a reward model from human feedback that is central to current generative AI-systems such as GPT.

2. Never being

With the alleged shift from deductive to inductive reasoning in machine learning, a new kind of *identity politics* has entered the field. The problem is that the hidden assumptions about the data, which directly inform the machine learning models, correspond in so many ways to the rather old, historically grown social categories (e.g. race, class or gender). Reintroduced as *natural* representations, these categories bring about the much-discussed issues of data bias and algorithmic discrimination (cf. Apprich et al. 2019). In this process, normalizing standards such as Whiteness in algorithmic filtering and face recognition, become the default setting of machine learning models (cf. Katz 2020: 172f.).

Due to the fact that models learn from past data in order to be able to make predictions about the future, machine learning turns into a self-fulfilling prophecy. In her new book *Discriminating Data* (2021), Wendy Hui Kyong Chun makes that point clear when she explains how

> predictive algorithms [...] are verified as correct if they predict the past correctly, for they are usually cross-validated using past data that are hidden during the training period or out of sample data, similarly drawn from the past. (ibid.: 46)

By becoming the *ground truth* of (inductive) machine learning, limited and biased data from the past foreclose, rather than enhance, the future, with the effect that existing (racial, social and sexual) discrimination is perpetuated.

The usual answer to this problem is a call for better data or better models. However, as Ramon Amaro (2022) has shown, those well-intended attempts do not break away from the epistemic violence of current machine learning models. Instead, they merely *optimize* discriminatory practices. He writes, "What we experience today as algorithmic prejudice is the materialization of an overriding logic of correlation and hierarchy hidden under the illusion of objectivity" (ibid.: 61). Given the eugenic and biometric roots of correlation techniques, the past truly *overrides* our present and future by propagating a natural (i.e. eternal) truth through machine learning. Yet, to insist on the fact that machine learning models never fully converge, implies that they do not determine an ultimate truth or identity (cf. Cheney-Lippold 2011).

Beyond Quantity then also means that there always remains a surplus that cannot be calculated, because it does not fit into the (normalizing) norm of machine learning models. It means to – as Amaro and Khan (2020) propose – deploy a "calculus of variations", able to explore the liminal space between algorithmic calculations, the gaps and cracks that might open up to other, in particular non-white, versions of reality. Hence, exposing the internal limits of machine learning systems by confronting them with indeterminacy, incompatibility, as well as a "Black totality, always already in the process of transformation" (Amaro 2022: 62), might provide a way to work through those systems and put them to different ends.

The goal is to come up with machinic logics that break the shackles of merely inductive, but also deductive, reasoning. Rather than confirming what was already there, a generative (abductive) approach might allow for infinite possibilities. Machine learning, in this perspective, exposes the limits of computability in a productive way: To the same extent that learning algorithms are contingent on infinitesimals, the models themselves are not fixed by any preset identities or categories. On the contrary, the indeterminacy, in particular its inclusion in the calculation process, is what constitutes the ability to learn (cf. Parisi 2018). Hence, if the goal of machine learning is to *generalize* a model based on data, then generalization, when understood as an ongoing, open process, is at the core of machine learning; this concerns the central idea that concepts are not merely some abstract content that can be learned, but actually develop through learning as a discursive (i.e. social) practice.[11]

For machine learning to transform (and not merely repeat) the world, it is thus necessary

> to move from seeing an inert model as the machine learner to seeing the human researcher or developer – *along with*, and not separate from, his or her model and surrounding relations – as the machine learner. (Reigeluth/Castelle 2021: 104)

Because humans and machines are part of the same symbolic realm, they are, as learners, contingent on the same "*regular, discrete framework*" (Galloway 2021: 123). Acknowledging the social (not merely mechanical or cognitive) aspect of machine learning can help us better understand its ambiguity and contingency – moving back and forth between the formalization of real-word

11 The idea of 'concept-learning' as a social practice goes back to the Soviet psychologist Lev Vygotsky (1986).

problems and the actual implementation of such models to process those problems.

3. The incomputable

According do Hartmut Winkler, processing as the third media function of computing (besides storing and transmitting) implies to recognize the double character of regularity or repetition and innovation:

> Das Prozessieren – als Eingreifende Veränderung – scheint von vornherein auf die Seite des Neuen zu fallen, insofern es eben Eingriff und Veränderung, und mit Blick auf die Wiederholung die Verschiebung, betont. (Winkler 2015: 107)

The idea of an "interfering transformation" is crucial for machine learning as well. Precisely because it is characterized by variability and indeterminacy, it relies on repetitive steps. In this sense, machine learning, which is defined by the processing of data, also necessitates a formalization by means of programming (Python) and mathematics (Calculus).

In contradiction to Peter Norvig's statement, machine learning is deeply logical and heavily relying on mathematical science. To claim otherwise would be to promote a version of machine learning that is fetishized as a natural thing and, therefore, hides its inner workings (i.e. the processing steps) from its users. Consequently, Google TensorFlow as well as all the other machine learning-infrastructures, such as Amazon Web Services or Azure Machine Learning, depict themselves as mere *services*.[12] Similar to the Internet's client/server architecture (cf. Krajewski 2018), these hidden infrastructures are essential to how machine learning is presented to us and how these representations influence our understanding of it (cf. Luchs/Apprich/Broersma 2023).

Contrary to the common belief that machine learning algorithms simply process data until a final result is found, the actual process is rather messy. In fact, contingency, indetermination and uncertainty are at the center of modern mathematics and, therefore, computing. Luciana Parisi, by invoking

12 This becomes apparent in Google's MLCC itself, when the (Python) code to run the models is literally hidden in foldout boxes.

Gregory Chaitin's algorithmic information theory (Chaitin 2004),[13] explains that "[s]ince there are infinities that cannot be compressed into simpler postulates, theories, truths, it follows that there are realities that are logically irreducible" (Parisi 2021: 82). Accordingly, there are realities that cannot be computed, because they cannot be captured by today's algorithms. What we experience with machine learning is not simply a shift from deduction to induction, from mathematics and logics to natural sciences, but rather the introduction of the incomputable (i.e. negativity) at the heart of computation.

With Turing's "halting problem" (Turing 1936), which basically says that no algorithm (i.e. a finite step-by-step procedure) exists, which can determine in advance whether a machine will finish running a program, a fundamental shift within the logic of calculation has occurred. The inherent limit of the discrete-state machine opens it to dynamic forms of computation. Once more Parisi (2015): "the calculation of randomness or infinites has now turned what was defined as incomputables into a new form of probabilities, which are at once discrete and infinite." "In other words," she continues,

> whereas algorithmic automation has been understood as being fundamentally Turing's discrete universal machine, the increasing volume of incomputable data (or randomness) within online, distributive, and interactive computation is now revealing that infinite, patternless data are rather central to computational processing. (ibid.: 131)[14]

Applied to machine learning, this means that we are dealing with both, patternless data being processed *and* symbolic learning systems feeding on trial-and-error. Instead of a mere step-by-step procedure, those systems are adaptive, precisely because they have to deal with the contingency of messy data. Hence, the discrete framework of computation gets tainted by real-world applications with its infinite variations. Randomness, in this perspective, is not outside of computation or machine learning, but the very core of them.

---

13 With his Algorithmic information theory Chaitin wants to prove that there is no such thing as absolute certainty in mathematics. There are truths that cannot be proven, problems that are impossible to solve.

14 In a similar way, M. Beatrice Fazi (2018) argues for the incompleteness and, therefore, contingency of computation. Both see Kurt Gödel as a progenitor of the incompleteness problem and its productive application in mathematics and, consequently, computational thinking.

Rather than following claims about the end of logics and theoretical explanation, machine learning includes realities that cannot be proven, but are yet to be discovered. The immanent logic of those machines, therefore, offers a radical break with the inductive explanation of natural sciences, without falling back into a deductive predictability of classic form of computation. Allowing for "a *computational thought* that is contingent, and yet does not break away from structure" (Fazi 2018: 210) could yield a machinic logic that actually might take us by surprise. A new mode of thinking about the machinic based on its learning capacities and not as a one-sided solution for or against inductive or deductive reasoning.

This brings us back to the beginning of this article. What if Zeno did not simply use his paradoxes to confirm the ontological monism (i.e. the static identity of all things) taught by his teacher Parmenides,[15] but rather sought to defend the idea of motion by putting it to a test? In other words, what if he did not try to prove the one, but to problematize the many? That would bring him very close to the here discussed problematization of incomputability, in the sense that reality is less a question of true or false, but rather an affirmation of its (infinite) possibilities. A paradox, after all, always contains more than one perspective.

List of references

Amaro, Ramon (2022): The Black Technical Object. On Machine Learning and the Aspiration of Black Being, London: Sternberg Press.

Amaro, Ramon/Khan, Murad (2020): "Towards Black Individuation and a Calculus of Variations." In: e-flux 109 (https://www.e-flux.com/journal/109/330246/towards-black-individuation-and-a-calculus-of-variations).

Amoore, Louise (2013): The Politics of Possibility. Risk and Security Beyond Probability, Durham and London: Duke University Press.

Anderson, Chris (2008): "The End of Theory: The Data Deluge Makes Scientific Method Obsolete.", June 23, 2008 (https://www.wired.com/2008/06/pb-theory).

15 In fact, Plato (1997) writes about the encounter with Socrates that Zeno felt misunderstood by him – "you do not fully apprehend the true motive of the composition."

Apprich, Clemens/Chun, Wendy H. K./Cramer, Florian/Steyerl, Hito (2019): Pattern Discrimination, Minneapolis and London: The University of Minnesota Press, and Lüneburg: meson press.
Chaitin, Gregory J. (2004): Algorithmic Information Theory, Cambridge: Cambridge University Press.
Cheney-Lippold, John (2011): "A New Algorithmic Identity: Soft Biopolitics and the Modulation of Control." In: Theory, Culture & Society 28/6, pp. 164–181.
Chun, Wendy H. K. (2021): Discriminating Data. Correlation, Neighborhoods, and the New Politics of Recognition, Cambridge, MA: The MIT Press.
Denny, Simon/Beyes, Timon/Holt, Robin/Pias, Claus/Steinbrügge, Bettina (eds.) (2022): Proof of Stake: Technological Claims, Milano: Lenz Press.
Domingos, Pedro (2012): "A Few Useful Things to Know About Machine Learning." In: Communications of the ACM 55/10, pp. 78–87.
Fazi, M. Beatrice (2018): Contingent Computing. Abstraction, Experience, and Indeterminacy in Computational Aesthetics, Lanham: Rowman & Littlefield.
Galloway, Alexander (2021): Uncomputable. Play and Politics in the Long Digital Age, London: Verso.
Géron, Aurélien (2019): Hands-On Machine Learning with Scikit-Learn, Keras, and TensorFlow, London: O'Reilly Media.
Google Developers (2020): "Machine Learning Crash Course.", last modified September 26, 2022 (https://developers.google.com/machine-learning/crash-course/descending-into-ml/training-and-loss).
Hall, A. Rupert (1980): Philosophers at War. The Quarrel Between Newton and Leibniz, Cambridge: Cambridge University Press.
Kansal, Sahdev (2020): "Quick Guide to Gradient Descent and Its Variants.", August 18, 2020 (https://towardsdatascience.com/quick-guide-to-gradient-descent-and-its-variants-97a7afb33add).
Katz, Yarden (2020): Artificial Whiteness. Politics and Ideology in Artificial Intelligence, New York: Columbia University Press.
Kline, Morris (1977): Calculus. An Intuitive and Physical Approach, New York: John Wiley & Sons.
Krajewski, Markus (2018): The Server: A Media History from the Present to the Baroque, New Haven, CT: Yale University Press.
Luchs, Inga/Apprich, Clemens/Broersma, Marcel (2023): "Learning Machine Learning: On the Political Economy of Big Tech's AI Online Courses." In: Big Data & Society 10/1 (https://doi.org/10.1177/20539517231153806).

Norvig, Peter (2020): "Introduction to Machine Learning Crash Course.", last modified September 27, 2022 (https://developers.google.com/machine-learning/crash-course/ml-intro).

Parisi, Luciana (2013): Contagious Architecture. Computation, Aesthetics, and Space. Cambridge, MA: The MIT Press.

Parisi, Luciana (2015): "Instrumental Reason, Algorithmic Capitalism, and the Incomputable." In: Matteo Pasquinelli (ed.), Augmented Intelligence Traumas, Lüneburg: meson press, pp. 125–137.

Parisi, Luciana (2018): "Das Lernen lernen oder die algorithmische Entdeckung von Informationen." In: Christoph Engemann/Andreas Sudmann (eds.), Machine Learning. Medien, Infrastrukturen und Technologie der Künstlichen Intelligenz, Bielefeld: transcript, pp. 93–113.

Parisi, Luciana (2021): "Artificial Critique." In: Erich Hörl/Nelly Y. Pinkrah/Lotte Warnsholdt (eds.), Critique and the Digital, Zürich: diaphanes, pp. 75–108.

Plato (1997): The Dialogues of Plato, Volume 4: Plato's Parmenides, New Haven, CT: Yale University Press.

Reigeluth, Tyler/Castelle, Michael (2021): "What Kind of Learning Is Machine Learning?" In: Jonathan Roberge/Michael Castelle (eds.), The Cultural Life of Machine Learning. An Incursion Into Critical AI Studies, Cham: Palgrave Macmillan, pp. 79–115.

Rosenberg, Barry/Google Engineering Education Team (2018): "Machine Learning Crash Course", March 2, 2018 (https://developers.googleblog.com/2018/03/machine-learning-crash-course.html).

Turing, Alan. M (1936): "On Computable Numbers, with an Application to the Entscheidungsproblem." In: Proceedings of the London Mathematical Society, Volume 42/1, pp. 230–265.

Vygotsky, Lev (1986): Thought and Language, Cambridge, MA: The MIT Press.

Winkler, Hartmut (2015): Prozessieren. Die dritte, vernachlässigte Medienfunktion, Paderborn: Wilhelm Fink.

Wright, Sewall (1932): "The Roles of Mutation, Inbreeding, Crossbreeding and Selection in Evolution." In: Proceedings of the Sixth International Congress of Genetics 1, pp. 356–366.

From algorithmic thinking to thinking machines

Four theses on the position of artificial intelligence in the history of technoscience

Matteo Pasquinelli

1. AI and the historical epistemology of science and technology

When analysing the impact of AI on science it would additionally be important to clarify the position of AI in the history of science and technology. Rather than seeing it as a recent phenomenon, this paper aims in fact to contextualise AI as part of the large history of technoscience. It further intends to shed light on the relation of AI to the making of modern science and, in particular, to the paradigms of mechanical, statistical and algorithmic thinking. Right here, at the beginning, we should add an observation that is obvious to historians of science and philosophy, but not as widely supported by computer scientists, namely that the definition of intelligence is always historical: a universal definition of intelligence does not exist and this should be the perspective in which AI should be regarded. For this reason, the intention of writing the history of AI very quickly also turns into the project of a *historical epistemology of intelligence*, in which AI is not only a technical artifact, but also a project based on and affecting the definition and formalisation of human intelligence and knowledge.

In fact this paper would like to suggest to the field of AI studies, the incorporation of the method of *historical epistemology of science and technology*, which has been propagated, in different ways, by Boris Hessen, Henryk Grossmann, George Canguilhem and Gaston Bachelard and more recently by the work of the Max Planck Institute for the History of Science in Berlin and other institutions.[1] What is the approach of the historical epistemology of science and tech-

1 About the historical epistemology of AI, see Pasquinelli 2023; for a critique of social constructivism in technology studies, see Winner 1993; for an overview of historical and

nology? By way of introduction, we could say that while science and technology studies in general emphasize the influence of external factors on science and technology (unfolding different variants of *social constructivism*), historical epistemology on the other hand follows the dialectical interweaving of practice, knowledge and tools within a broader economic and historical dynamic. To paraphrase Boris Hessen's famous study of Newton's mechanics (Hessen 2009 [1931]), it could be said that historical epistemology is concerned with the investigation of the 'economic and social roots' of technoscience.

It should be noted that the method of the historical epistemology of science and technology has been pursued by a large number of historians without using this label. Feminist theorists such as Hilary Rose, Sandra Harding, Evelyn Fox Keller and Silvia Federici, for instance, have contributed to explaining the rise of modern rationality and mechanical thinking (to which AI also belongs) in relation to the transformation of women's bodies and the collective body in general into a productive and docile machine (see e.g., Rose/Rose 1976; Harding 1986; Keller 1985; Federici 2004). This paper attempts to illustrate the paradigm of *algorithmic thinking* at the core of AI in the same way (yet more modestly) in which the different schools of historical, critical, feminist and political epistemology have studied the rise of *mechanical thinking* in the modern age and, more in general, the social and economic genesis of the *abstractions of thought*, such as number, time, and space in the history of human civilisations.[2]

The following paper explores four theses:

I. *AI as the denial of epistemology.* In AI, the identification of machine output with human intelligence has to be questioned: algorithmic thinking has to be separated from material algorithms. In the history of science and technology, usually, the mental model of an artefact is distinguished from the material model, from the technical artefact that implements it.
II. *AI as symbolic representation vs. modelling.* The history of AI is not based on a single definition of intelligence being mechanised, but on competing models of intelligence and competing algorithms. AI algorithms are distinguished, for example, in those that aim at the direct implementation of

political epistemology see Omodeo 2019; Renn 2020; MPIWG 2012; Omodeo/Ienna/Badino 2021; Schmidgen 2011.

2 For mechanical thinking, see Damerow et al. 2004 [1991]; for the notion of number, see chapter 1 in this book and Damerow 2013 [1996]; for the notion of space, see Schemmel 2015.

logic (so-called GOFAI) and those that implement modelling techniques (i.e., artificial neural network, machine learning, etc.).

III. *AI as an experimental artefact.* AI algorithms did not emerge from the top-down application of mathematical ideas but through experimentation. Specifically, machine learning took shape at the confluence of two lineages of technoscience: electro-mechanical engineering and statistics.

IV. *AI as an epistemic scaffolding and meta-paradigm.* Rather than a project to automate intelligence in the abstract, AI should be considered a complex epistemic scaffolding and meta-paradigm in which social, technical, logical and ideological factors have to be constantly analysed in their historical imbrication and unfolding.

2. AI as the denial of epistemology

In the history of human civilization, tools have always emerged together with a system of explicit or less explicit technical knowledge associated with them, which is distinguished from the tools themselves. This aspect seems very confused in the artefacts of AI that are said to directly automate human intelligence. This epistemological dimension (or 'epistemic gap'), that is the obvious *distinction between technical knowledge and tools* exists, of course, also in the recent variant of AI, machine learning, as the distinction between the know-how to program an artificial neural network (e.g., in Python language) and their application (e.g., in image recognition). Yet this distinction seems to be continuously removed from the debate on AI that is fixated on an equation unique to the history of epistemology: machine output = intelligence. The faith in the direct implementation of human reasoning into a machine or an algorithm specifically belongs to the tradition of symbolic AI that has been canonically established in Alan Turing's essay 'Computing Machinery and Intelligence' and the Dartmouth workshop in 1956 in preparation of which McCarthy coined the term 'artificial intelligence' (Turing 1950; McCarthy et al. 2006 [1955]).

Traditionally, epistemology is a meta-reflection on the conditions of intelligent behaviour and knowledge making. It is based on the assumption that thinking is not immediate but *mediated* – by practices, tools, cultural techniques, language, physical properties of the brain, cognitive maps inside and outside the brain, etc. Epistemology is the self-awareness of the hiatus between reason and the medium of reason. When this canonical lesson is brought to the case of AI, an obstacle is perceived, as the main assumption is that AI is the straightforward implementation of intelligence. I would like to define provisionally as 'folk AI' (after the known expression 'folk psychology') the superficial identification of the output of a machine with intelligent behaviour and advance the hypothesis that such denial of the epistemological questions (and epistemology in general as a meta-discourse) has affected not only the scientific definition of AI but also its historiography since the 1950 and even earlier.

Ultimately, it should be noted that the propositional knowledge that symbolic AI aims at automating is not equivalent to scientific and experimental knowledge, that is a full process of knowledge making which is conventionally based on the progressive stages of observation, hypothesis, and testing. In short, back in the 1950s symbolic AI (as most of cybernetics) already represented a *reductionism of scientific mentality* and obliteration of the experimental method, whose consequences are yet to be studied.

Interestingly, it has not been the work of philosophers of mind but the industrial and commercial successes of deep learning in the automation of manual and mental labour which have forced scholars to look back at the history of computation, cybernetics and AI with a different perspective, prompting everyone to rediscover the fundamental difference between symbolic and connectionist AI. Even at this stage of widespread celebration of the powers of AI, the confusion remains: today we call 'artificial intelligence' what was actually the rival paradigm of artificial intelligence in the 1950s, namely artificial neural networks research, or connectionism. This terminological confusion and the current lack of a proper AI historiography is not related to the fact that AI is a novel field (it is at least half a century old), but to the cultural and philosophical hegemony of symbolic AI, which has obscured other readings and interpretations, especially regarding connectionism, statistics and modelling techniques.

3. AI as symbolic representation vs. modelling

Connectionism developed on the basis of different postulates than symbolic AI and it is actually even older. Connectionism was initiated by two historical papers by Warren McCulloch and Walter Pitts ('A Logical Calculus of the Ideas Immanent in Nervous Activity' from 1943 and 'How we know universals the perception of auditory and visual forms' from 1947). The term 'connectionist' itself was introduced by Donald Hebb to describe the organisation of neurons in his

1949 book *The Organization of Behavior*. This book was also crucial for introducing the so-called Hebbian rule of neuroplasticity 'Neurons that fire together, wire together', which would have a deep influence on the history of connectionism and cognitive science. Frank Rosenblatt adopted the term in 1958 to define his theory of artificial neural networks.

In which way is connectionism different from symbolic AI? According to symbolic AI, human thought can be formalised into mathematical or propositional logic, which can be then implemented into a deductive algorithm and successfully mechanised. Connectionism, on the other hand, is not concerned with human thinking per se rather the material processes of the brain that make thinking possible – in particular the functioning of neural networks, which were then seen and formalised as computing networks. According to connectionism, the brain thinks by building models of the world through the self-organisation of its neural networks and this process can be emulated by inductive algorithms and differential equations that describe the parameters of such models.

Folk AI and its specific form of epistemic reductionism should be understood in the background of the confrontation of these two paradigms of intelligence and computing. However, folk AI is not only based on the assumption (inherited from early symbolic AI) that a mechanism can fully implement and automate an act of reasoning, an inference, or rule, but also that a mechanism can implement the *interpretation* of the rule, as Wittgenstein already pointed out in his critique of Turing Machines (Wittgenstein 1958 [1953]: §§ 74, 77–81, 185, 193, 194, 199). According to Wittgenstein, there is a difference between 'mechanically following a rule' and 'following a mechanical rule', while according to symbolic AI, there is none (cf. Shanker 1998: 27–30). The fallacy derives also from the wrong expectation that the externalisation of a model of the mind can *exhaust* the act of modelling itself, while the principle of thinking implies the impossibility of the full identification of mind and world, of internal mental models and external technical models, such as tools, machines and algorithms.

The distinction between a direct logico-symbolic representation of the world and techniques of world modelling always existed in the AI debate, but has never properly come to the fore due to the cultural and academic hegemony of symbolic AI. A key essay from 1988 by Hubert and Stuart Dreyfus elucidated the development of AI according to the two paradigms of 'making the mind' (i.e., symbolic AI) vs. 'modelling the brain' (i.e., connectionism) (Dreyfus/Dreyfus 1988). As known, the project of symbolic AI (together with expert systems and knowledge databases) failed and machine learning gradually emerged from statistical techniques of data modelling pioneered by artificial neural networks. It should be noted that the power of machine learning derives precisely from its capacity to *automate statistical modelling* rather than logico-symbolic intelligence, as the early developers of AI argued.

The key moment in this history and confrontation of paradigms is the invention of the artificial neural network perceptron by Frank Rosenblatt in 1957, which attempted to perform pattern recognition through the automation of statistical tools of multivariable analysis rather than deductive logic (Rosenblatt 1957: 4; Rosenblatt 1958: 405; Rosenblatt 1961; cf. Pasquinelli 2023). The perceptron is considered, by convention, the first artificial neural network, prototype of deep learning and first algorithm of machine learning, yet an epistemological study of its foundation is still missing.[3] Although proceeding from quite different traditions and employing different techniques, both connectionism and statistics represent in fact paradigms and techniques of *modelling*. Avoiding to seek causal explanation, both statistical techniques and artificial neural networks compute models of world data based correlations and factor analysis. Machine learning gradually emerged as a spin-off of the tradition of statistics. Already in 2001, Leo Breiman distinguished the traditional technique of data modelling in statistics from *algorithmic modelling*, calling them the two cultures of statistics.

4. AI as an experimental artefact

The paradigm of connectionism, prototype of the current deep neural networks and large language models, did not emerge from the top-down application of mathematical ideas, but through experimentation, more precisely through building *experimental machines*. Connectionism took shape through the confluence of two lineages of technoscience: the tradition of *electro-mechanical engineering* and statistics. On the one hand, it belongs to the tradition that unfolded from modern mechanics into electro-mechanical engineering and

3 Rosenblatt, for example, was also influenced by the neoliberal economist Friedrich Hayek who published a tractate on connectionism, 'The Sensory Order', in 1952, which was already far more advanced than the definitions of AI that emerged from the 1956 Dartmouth workshop. Following the Austrian philosopher Ernst Mach and Gestalt theory, Hayek sketched the idea that the mind is made by material structures that model the world, rather than ideas that represent the world through propositional knowledge (Pasquinelli 2021).

digital computation (Babbage 1832; Turing 1936; Shannon 1938; von Neumann 1993 [1945]). On the other, it belongs to the controversial tradition of statistics that evolved from eugenics and the biometrics of intelligence (see the history of the IQ test) into the *analysis of multidimensional data* (as Stephen Jay Gould illustrated in his magisterial book *The Mismeasure of Man* from 1981). These two lineages merged together in a precise moment that the history of AI rarely acknowledges, which is Rosenblatt's invention of the artificial neural network perceptron.

The invention of the perceptron demonstrates (once again) the innovation proceeds by the continuous scaffolding of technical and logical paradigms on top of the previous ones, rather than by abrupt breaks and intuitions of solitary geniuses. Neither of these two lineages originated from the *top-down* application of pure mathematics, rather often *bottom-up* on the initiative of engineers, sociologists, psychologists, criminologists, cyberneticians responding to state and industrial drives for social control, information processing, and labour automation.

As just mentioned, multivariable analysis, for instance, originated from psychometric techniques that were part of eugenic and racist campaigns of class discrimination in Europe and North America. On the other hand, automated computation started with the Hollerith machine used to tabulate the punched card of the US census well before the Turing machine (which is perceived as the cornerstone of the information revolution) was conceptualised. Moreover, Thomas Haigh and Mark Priestley (2020) have clarified that the Turing machine did not help the actual design of the digital computer whose implementation von Neumann resolved in a different way.

The history of computation demonstrates once again that technological development drives scientific paradigms, rather than the other way around – also in the case of machine learning *invention predates theorisation*. This history also shows that the evolution of knowledge, techniques and technologies is a gradual implementation, stratification and scaffolding of *mental and technical models* on top of the previous ones. In this respect, AI can be truly illustrated as an *epistemic scaffolding* of social, technological, logical, and ideological forms. In such scaffolding, which is typical for the development of technoscience, (1) economic processes trigger (2) technological experiments and the invention of new machines that require (3) the formalisation of scientific paradigms, which all together influence also (4) mythologies and ideologies (see the cult of thinking automata). There is no deterministic development between levels, rather *each level models and is modelled back by the contiguous levels* in different ways.

5. AI as an epistemic scaffolding and meta-paradigm

The making of AI should be considered part of the general development of modern technoscience: this evolution shows no breaks or phenomena of 'singularity' as folk AI professes. Although it may appear highly 'abstract', 'artificial' and 'autonomous' to some, AI has gradually developed, just like other cultural techniques of humankind. The myth of machine autonomy shows an interesting parallel with intuitionism in mathematics and philosophy of mind and it would be interesting to discover how historians of science have already dealt with this problem. For instance, to contrast the illusion of *a priori* ideas in mathematics and to demonstrate their historical and material origins, the historian of mathematics Peter Damerow (2013 [1996]) proposed to frame the mind's activity as a continuous cycle of internalisation of actions with tools and externalisation of mental models, which is an intuition that this paper attempted to apply to the making of AI.

To explain the formation of the concept of number, then, Damerow suggested a scaffolding of technical and mental models that progressively unfold from *practices of counting* (e.g., reckoning with fingers) to *systems of numeration* (e.g., positional decimal system) to *techniques of computation* (e.g., algorithms) and eventually to *number theory* (e.g., arithmetic as a formal discipline). This process is not linear, but follows alternate movements of *representation* (the use of objects and signs a referent of other objects, signs and ideas) and *abstraction* (problem solving). This process of *reflective abstraction* (inspired by both Piaget's genetic epistemology and Hegel's dialectical logic) constitutes progressive stages of symbolic representation in which the passage from one order of representation to the following occurs via a new abstraction. In this reading, thought starts with labour that invents tools and technologies in order to solve problems mostly of economic and social nature and transform the world accordingly. Subsequently, these tools project new knowledge forms and scientific paradigms. In the Damerow scaffolding, technical and mental models evolved together and stimulated each other in a dialectical way. Tools, machines and algorithms are all forms of *material abstraction*.

The cycle of internalization and externalization of technical and mental models crosses the whole history of human civilisations and also includes advanced technology of automation, such as machine learning. As the historian of science Jürgen Renn has noted:

> After all, machine learning algorithms [..] are simply a new form of the externalization of human thinking, even if they are a particularly intelligent form. As did other external representations before them, such as calculating machines, for example, they partly take over – in a different modality – functions of the human brain. (Renn 2020: 398)

The Damerow scaffolding maintains together, in a consistent and historical way, material actions and mental models, praxis and abstraction and it can be useful to articulate the epistemic scaffolding of AI.

6. Conclusion

At the crossroads of different techniques and disciplines, AI has become one of the most crucial and complex paradigms of the present – a global meta-paradigm (such as the Anthropocene in other respects). Within the global economy, machine learning has become a key paradigm for data analytics, information processing, planning, forecasting and labour automation as much as management automation. Its production pipeline extends from the Global North to the South, involving multitudes of precarious gig workers and also 'ghost workers' (Gray/Suri 2019; Atanasoski/Kalindi 2019). A consistent analysis of contemporary AI requires the political understanding of its global scale and of the complex imbrication of social, technical, logical and ideological forms.

AI has been studied so far by a wide spectrum of AI Studies, which include Computer Science, Science and Technology Studies, Social History, Sociology of Labour and Automation, Semiotics, Philosophy of Mind and Language, Neuroscience, Media Theory, Visual Studies, etc. and in advancing a new methodology of research, we also have to consider the contributions and legacy of all these disciplines. It was in the search for a more comprehensive approach that the contribution of the historical, critical and political epistemology of science and technology has been advanced.

The approach of historical epistemology, however, can be received as a general methodology to syndicate the fields of AI studies and cover the numerous epistemic troubles that haunt AI. In conclusion, we could say that a basic historical epistemology of AI should be pursued according to three lines of inquiry: firstly, the investigation of the social and economic roots of AI (its relation to the current global economy and international division of labour); secondly, the comparison of AI to other knowledge models and forms of mental labour (learning, writing, design, scientific work, etc.) and thirdly, the positioning of AI in the long evolution of knowledge systems (extending the previous cultural techniques, 'information societies' and technologies of civilisation).

List of references

Atanasoski, Neda/Vora, Kalindi (2019): Surrogate Humanity. Race, Robots, and the Politics of Technological Futures, Durham: Duke University Press, 2019.

Babbage, Charles (1832): On the Economy of Machinery and Manufactures, London: Charles Knight.

Breiman, Leo (2001): "Statistical Modeling: The Two Cultures (with comments and a rejoinder by the author)." In: Statistical Science 16/3, pp. 199–231.

Damerow, Peter (2013 [1996]): Abstraction and Representation: Essays on the Cultural Evolution of Thinking, Dordrecht: Springer.

Damerow, Peter/Freudenthal, Gideon/McLaughlin, Peter/Renn, Jürgen (2004 [1991]): Exploring the Limits of Preclassical Mechanics. A Study of Conceptual Development in Early Modern Science: Free Fall and Compounded Motion in the Work of Descartes, Galileo and Beeckman, New York: Springer.

Dreyfus, Hubert/Dreyfus, Stuart (1988): "Making a Mind versus Modeling the Brain: Artificial Intelligence Back at a Branchpoint." In: Daedalus 117/1, pp. 15–43.

Federici, Silvia (2004): Caliban and the Witch: Women, the Body and Primitive Accumulation, New York: Autonomedia.

Gould, Stephen Jay (1981): The Mismeasure of Man, New York and London: Norton & Company.

Gray, Mary L./Suri, Siddharth (2019): Ghost Work: How to Stop Silicon Valley from Building a New Global Underclass, Boston and New York: Houghton Mifflin Harcourt.

Haigh, Thomas/Priestley, Mark (2020): "Von Neumann Thought Turing's Universal Machine Was 'Simple and Neat', But that Didn't Tell Him How to Design a Computer." In: Communications of the ACM 63/1, pp. 26–32.

Harding, Sandra (1986): The Science Question in Feminism, Ithaca and London: Cornell University Press.

Hebb, Donald (1949): The Organization of Behavior: A Neuropsychological Theory, New York: Wiley & Sons.

Hessen, Boris (2009 [1931]): "The Social and Economic Roots of Newton's *Principia*." In: Gideon Freudenthal/Peter McLaughlin (eds.), The Social and Economic Roots of the Scientific Revolution: Texts by Boris Hessen and Henryk Grossmann, Dordrecht: Springer Science & Business Media, pp. 41–101.

Keller, Evelyn Fox (1985): Reflections on Gender and Science, New Haven, CT: Yale University Press.

McCarthy, John/Minsky, Marvin L./Rochester, Nathaniel/Shannon, Claude E. (2006 [1955]): "A Proposal for the Dartmouth Summer Research Project on Artificial Intelligence." In: AI Magazine 27/4, pp. 12–14.

McCulloch, Warren S./Pitts, Walter (1943): "A Logical Calculus of the Ideas Ommanent in Nervous Activity." In: Bulletin of Mathematical Biophysics 5, pp. 115–133.

MPIWG (ed.) (2012): Epistemology and History: From Bachelard and Canguilhem to Today's History of Science, Max Planck Institute for the History of Science Preprint, no. 434 (https://www.mpiwg-berlin.mpg.de/preprint/conference-epistemology-and-history-bachelard-and-canguilhem-todays-history-science).

Neumann, John von (1993 [1945]): "First Draft of a Report on the EDVAC." In: IEEE Annals of the History of Computing 15/4, pp. 27–75.

Omodeo, Pietro Daniel (2019): Political Epistemology: The Problem of Ideology in Science Studies, Cham: Springer.

Omodeo, Pietro Daniel/Ienna, Gerardo/Badino, Massimiliano (2021): Lineamenti di Epistemologia Storica: Correnti e temi, MPG.PuRe Preprint (https://hdl.handle.net/21.11116/0000-0009-54BE-3).

Pasquinelli, Matteo (2021): "How to Make a Class: Hayek's Neoliberalism and the Origins of Connectionism." In: Qui Parle 30/1, pp. 159–184.

Pasquinelli, Matteo (2023): The Eye of the Master: A Social History of Artificial Intelligence, London: Verso.

Pitts, Walter/McCulloch, Warren S. (1947): "How We Know Universals: The Perception of Auditory and Visual Forms." In: Bulletin of Mathematical Biophysics 9, pp. 127–147.

Renn, Jürgen (2020): The Evolution of Knowledge: Rethinking Science for the Anthropocene, Princeton: Princeton University Press.

Rose, Hilary/Rose, Steven (eds.) (1976): The Radicalisation of Science, London and Basingstoke: MacMillan.

Rosenblatt, Frank (1957): "The Design of an Intelligent Automaton." In: Research Trends, Cornell Aeronautical Laboratory 6/2, pp. 1–7.

Rosenblatt, Frank (1958): "The Perceptron: A Probabilistic Model for Information Storage and Organization in the Brain." In: Psychological Review 65/3, pp. 386–408.

Rosenblatt, Frank (1961): Principles of Neurodynamics: Perceptrons and the Theory of Brain Mechanisms, Washington: Spartan Books.

Schemmel, Matthias (2015): Historical Epistemology of Space: From Primate Cognition to Spacetime Physics, Cham et al.: Springer.

Schmidgen, Henning (2011): "History of Science." In: Bruce Clarke/Manuela Rossini (eds.), The Routledge Companion to Literature and Science, London: Routledge, pp. 335–348.

Shanker, Stuart G. (1998): Wittgenstein's Remarks on the Foundations of AI, London: Routledge.

Shannon, Claude (1938): "A Symbolic Analysis of Relay and Switching Circuits." In: Electrical Engineering 57/12, pp. 713–723

Turing, Alan (1936): "On Computable Numbers, with an Application to the Entscheidungsproblem." In: Proceedings of the London Mathematical Society 42/1, pp. 230–265.

Turing, Alan (1950): "Computing Machinery and Intelligence." In: Mind 59/236, pp. 433–460.

Winner, Langdon (1993): "Upon Opening the Black Box and Finding It Empty: Social Constructivism and the Philosophy of Technology." In: Science, Technology & Human Values 18/3, pp. 362–378.

Wittgenstein, Ludwig (1958 [1953]): Philosophical Investigations, Englewood Cliffs, NJ: Prentice Hall.

A new canary in the coal mine?
On birds, AI and Early Warning Systems

Markus Ramsauer

In 1914, the *Coal Mining Institute of America* in Pittsburgh, Pennsylvania, discussed the susceptibility of living organisms to the toxins typically encountered under the surface of the Earth. The report "Experiments with Small Animals and Carbon Monoxide" suggests that "[o]f the common small animals, canaries are best adapted for exploration work" (Burrell/Seibert 1914: 244). In the case of a significant increase in carbon monoxide underground, canaries would express signs of distress, in the form of behavioral changes or collapse, much earlier than other species. Compared to mice or guinea pigs they show another advantageous capacity, namely to "recover quickly if exposed to fresh air" (ibid.: 243). The susceptibility of the birds would allow for coal workers to evacuate the mine before the toxic gas reaches a hazardous concentration. As for the origins of this practice, prior to its implementation in the United States, the authors point to the usage of canaries in England and "presumably in places on the continent also" (ibid.: 241) as well as to the late 19th century (self)experiments of John Scott Haldane. However, the canary in the coal mine is also discussed against the background of a much longer tradition of interpreting animalistic and especially avian behavior as signs for future developments (cf. Reif 2011; Keck/Lakoff 2013; Neo/Tan 2017; Keck 2020). If the whole system – including the mine inspectors and workers, the evacuation plans as well as the birds themselves – would be taken together as an ensemble, it could be addressed as a prototypical Early Warning System (EWS). These have been developed against various lethal threats from earthquakes to drought and in various scientific and infrastructural fields. Conservationist Ian Spellerberg refers to the canaries as a "biological early warning system" (2006: 157).

As canaries of the digital age, Early Warning Systems were prone to be augmented by the innovative powers of Artificial Intelligence. In this investigation, the genealogy of the data-heavy EWS is used as a starting point to observe – with reference to the editors' research project – how Artificial Intelligence is changing science (Echterhölter et al. 2021). The use of often large amounts of monitored data and the implementation of statistics can be seen as cornerstones of these technologies for crisis detection and prediction, therefore the application of Machine Learning Technology, deployed as prediction machines, comes as little surprise and is underway in several international and national agencies.[1] Key for the implementation of these systems is how scientists and institutions conceptualize the impending crisis by relating the future to the threatened self in a specific way. To suggest the crucial elements at play in EWS and to assess the role of AI in this field of disaster research, we use a broad notion of EWS, introduce and compare various kinds of analogue, digital and AI-based systems in various fields and highlight their respective epistemological potential.

Initially, the argument is made that Early Warning Systems contribute to the perception of a constant state of crisis, with signs detectable to those capable of interpreting them. The use of sensors or sentinels, such as birds or AI, is seen as a means of mitigating the impact of potential hazards. Following this logic, the development of digital Early Warning Systems since the 1970s can be described as technologies of *preparedness* (Lakoff 2008; Lakoff 2017). To guarantee preparedness, EWS models with necessity hinge on one crucial aspect: signals have to be detected in large amounts of data about natural states or social behavior, and for this, thresholds have to be set. This presupposes a conceptualization of what constitutes a signal point to processes of 'normalization', in the sense of what is seen as a catastrophic development and what is not worth issuing a warning for. The promise of the whole procedure is to detect patterns of threat in the environment and to intervene long before the environment becomes lethal.

As a second step, three examples of early warning models, which build on the trope of bird behavior as signals for an impending systematic crisis, will be introduced. These should serve as illustrations of how institutions make use of

---

1 Cf. Lamsal/Kumar (2020); for disaster mitigation see the UNDRR collection on "Artificial Intelligence for Disaster Risk Reduction" (https://www.preventionweb.net/collections/artificial-intelligence-disaster-risk-reduction); for a current EWS project with explicit use of AI methodology in Germany see "Daten- und KI-gestütztes Frühwarnsystem zur Stabilisierung der deutschen Wirtschaft" by Fraunhofer Heinrich Hertz Institut (http://www.daki-fws.de).

detection potential found in birds but also in statistical machines, in order to acquire more timely future knowledge and enable better preparation for crises. In a sense, the studies presented show how AI takes the place of the bird in timely warning concerns.

The examples of canaries as animalistic intelligences of birds or machinic intelligences like AI can furthermore serve as an incentive for a reflection on the discourse revolving around the intelligence and the 'knowledge' of AI. It is argued that instead of concentrating on the question whether a machine is able to 'pass as human', the limitations of human abilities in sensation and cognition, as revealed by animals or AI, can provide guidance for analyzing the discursive construction of 'the human'.

1. An epistemology of Early Warning Systems

Early Warning Systems appeared most prominently in the 1960s and 1970s. An attempted genealogy of these technologies can take on two (mutually informing) directions. One of them leads to the military context of WWII, where information EWS were implemented in order to predict attacks via the use of intelligence data (Austin 2004: 4). This 'birthplace' might also serve as an explanation for the functional similarities of EWS and radar technologies – these byproducts in the search for a laser beam gun (Pircher 2010: 52–54). In the literature on EWS, other traces of direct interference from the military context to other scientific fields are easily found, as for example the "Weak Signals" approach by Igor Ansoff (1975) – a US mathematician and former member of the RAND Corporation which served as a blueprint for EWS in business administration (Hammer 1998: 216–225).

A second genealogical thread for EWS is taken up by Irasema Alcántara-Ayala and Anthony Oliver-Smith in their article "Early Warning Systems: Lost in Translation or Late by Definition?" (2019). They trace the origins of EWS back to the devastating famines in Ethiopia and Sudan in the 1980s. As a consequence of the death of more than one million people caused by starvation, the 'Famine Early Warning System' (FEWS) was established by USAID. It operated via the constant monitoring of data of different kinds, enabling a mapping of impending famines which should lead to a timely response (ibid.: 321–323). The authors consider the FEWS a prototype for EWS in other areas like disaster risk reduction for earthquakes, floods, storms and more.[2] In epidemiology, another field where EWS have gained prominence, significant efforts were made during the early 2000s with the establishment of WHO's Global Outbreak Alert and Response Network (GOARN) or the Program for Monitoring Emerging Diseases (ProMED) (Hall 2020).

Even though it is important to stress that EWS in different fields do not necessarily consist of the same constituents, certain dynamics, such as the importance of monitoring changes in data or behavior, are shared by most EWS. The United Nations Office for Disaster Risk Reduction (UNISDR) defines EWS as an "integrated system of hazard monitoring, forecasting and prediction, disaster risk assessment, communication and preparedness activities systems" (2016: 2). The *Berghof Handbook for Conflict Transformations* utilizes the term "Early Warning System" to refer to "any initiative that focuses on systematic data collection, analysis and/or formulation of recommendations, including risk assessment and information sharing" (Austin 2004: 129). By relying on this logic, EWS share many characteristics and constituents with other forms of predictive and anticipating technologies like forecasting, sentinels, barometers, risk assessments or scenarios.[3] Given these shared epistemological features and the timing of EWS technologies' emergence, it is possible to consider them as integral components of a shift in the operational mode of governance, as articulated by anthropologist Andrew Lakoff (2008). Based on Foucault's analysis of different modes of *Gouvernementalité*, Lakoff holds that in the mid-20th century there has been a shift in state rationale when confronted with threats of different kinds. While 17th-century monarchies, in their fight against adversaries, relied on a logic of interdiction that was followed by the 19th-century reliance on prevention (especially with the emergence of the hygienic movement and its use of statistics), the mid-20th century saw a shift to *preparedness* for the emergence of threats. For this latter paradigm, Lakoff identifies the use of scenarios as decisive technologies against threats by "unpredictable, potentially catastrophic events" (Lakoff 2008: 403). However,

2 According to the authors, the development of EWS in these fields went hand in hand with a departure from long long-term perspective in favor of technicistic solutions for "shorter-term occurrences of events" (Alcántara-Ayala/Oliver-Smith 2019: 322). The Indian Famine Codes of 1880 are sometimes considered historical forerunners of the FEWS (Enten 2008: 13–15).

3 The genealogies of EWS could of course in principle be prolonged into analogue times, when disaster warning had other names, for instance with the history of human observers acting as seismographs. Cf. Coen 2012; Pietruska 2017; Edwards 2013.

and this is important to note, these different governing rationales should not be viewed as mutually exclusive (ibid.: 421). The emergence of the EWS concept with its reliance on the use of data analysis and statistics does fall into the period of the shift to preparedness, which is also acknowledged by Lakoff himself, saying that important building blocks of the preparedness apparatus were found in “more exercises, more vulnerability assessments [and] improved early warning systems” (Nucho 2022; cf. Lakoff 2017). EWS can thus be located within the preparedness paradigm although they should not be regarded as tantamount to scenario technologies. Whereas the latter “function […] to authorize knowledge claims in the absence of actual events” (Lakoff 2008: 419), the rationale of EWS is to deprive a potential threat from its ‘event character’ as an irruptive catastrophe and instead conceptualize it as a trend-like deterministic development. The threat can be detected ‘early’, i.e., ‘early enough’, or ‘earlier than last time’ (Hall 2020) with the use of the right instruments.

As one commenter on the FEWS noted in *Science*: “The signs are there if they can be recognized. As stress occurs, behavior changes.” (Walsh 1986: 1146)[4] Catastrophe in this rationale is always latently present and can be detected by using the right instruments. The implementing institution must know ‘what to look for’, i.e., which parameters to monitor, and where to set the threshold for triggering an alarm. Sometimes the ability of parameter and threshold setting depends on experience: what kind of behavior, or what change in behavior, is interpreted as a signal of an impending crisis? This ability to detect the right information is exemplified by J.S. Haldane’s experimental work as discussed in Burrel and Seibert (1914):

> The authors of this paper do not hesitate to say that, because of his greater experience in experimenting with small animals, Dr. Haldane might detect outward symptoms in a mouse that would escape the authors’ attention. (ibid.: 242f.)

Despite the morally questionable approach of exposing living creatures (including the scientists themselves) to potentially lethal concentrations of poisonous gasses, the usage of their sensory abilities went hand in hand with an

4 For the FEWS, behavioral changes which are considered to be signals (or signifiers) of an impending crisis are e.g., an increase in the sale of jewelry or a rise in the consumption of roots, grasses and berries (Walsh 1986).

intimate relationship with the animal and knowledge about what constitutes a symptom. What is needed for an EWS to be effective is likewise double or multi monitoring. At a first stage, the bird or the machine monitors changes in the environment, which leads to a change in their behavior. At a second stage, the EWS consists of anomaly detection, i.e. monitoring the bird's or the machine's behavioral changes and interpreting them accordingly. Thereby, EWS contribute to the determination to which changes can reasonably be said to constitute a crisis and to which developments can still be considered as 'non-critical' or 'normal'. This dynamic is especially prevalent for EWS in the field of the social sciences.

As part of the preparedness paradigm, these technologies "bring the future prospect of catastrophes into the present as an object of knowledge and intervention" (Lakoff 2008: 23). They thereby contribute not only to the question of 'what is a crisis' but epistemologically shift the onset time of crises towards an earlier point in time.

The following presentation of three (partly) AI-based EWS further illustrates some important constituents of EWS and highlights the functional role of AI technology. Before that, however, it is necessary to recapitulate some of EWS' characteristics as being a) often implemented in the aftermath of crises, b) part of a preparedness logic, c) reliant on data/environment monitoring, signal detection and threshold setting, d) contributors to the question of what counts as a crisis, respectively as normal e) conceived as triggering a precise and effective warning.

2. Quasi-avian Early Warning Systems

In computer science, the trope of the canary as an early warning mechanism was introduced in the 1990s by Cowan et al. (1998; 1999). Here, the canary is a mere name for a function of programming, yet recognizably the function is the one of signaling danger. The security system *Stackguard* protects against buffer overflow attacks in a way which "seeks not to prevent stack smashing attacks from occurring at all, but rather to prevent the victim program from executing the attacker's injected code" (Cowan et al. 1999: 3). The programme thereby follows a logic of preparedness by mitigation. Concerning the functioning of this technology, what is essential to grasp for the purpose of this article is that by storing more data in a buffer (a region of memory used to hold data temporarily) than it can handle, hackers can cause that buffer to 'overflow' with extra data. This potentially enables them to overwrite the return address of a program. Normally, after executing a function like a calculation task, e.g., the processor should go back to the return address. In the case of a stack buffer overflow, "[w]hen the function returns, instead of jumping back to where it was called from, it jumps to the attack code" (Cowan et al. 1998: 64). This can lead to the attackers gaining administrative authority over a computer system.

As a solution to this threat, the authors present the security mechanism of the 'stack canary' which they jestingly introduce as: "[a] direct descendent of the Welsh miner's canary" (ibid.: 3). The canary is a 'value' (a number or a word) which is placed next to the respective return address. In the case of an attempted overwriting of the return address, the canary word is overwritten and thereby changed "before jumping to the address pointed to by the return address word" (Cowan et al. 1999: 3). This change constitutes a warning signal which should cause the program to display an error or to terminate before the attack can cause significant harm to the computer system. The signal thereby relies on a shift in 'code behavior'. What is absent in this digital application is the aspect of data collection and threshold setting, since the overwriting of the code is not gradual but follows an either-or logic.

As an inducement for their efforts to enhance security when using stack canaries, the authors point to the Morris Worm of 1988. This is considered to be one of the first major malware attacks, infiltrating approximately 10 percent of all internet systems, thereby revealing their vulnerability (Furnell/Spafford 2019: 31). The emergence of the stack canary after the launch of the Morris worm illustrates the 'productive force' of catastrophes: EWS and other infrastructures of preparedness tend to be modelled and built primarily in the aftermath of system failures. Vulnerabilities are revealed and consequently followed by attempts to mitigate the damage in case of a future occurrence.

A further application of the 'canary-logic' in the area of computer science is a technique called 'canary release'[5]: When introducing a new version of a software, instead of presenting the new version as a whole to a general audience, only some users are chosen to test the innovation. With this technique, the software company can track and collect data on how the new version affects the production environment (Sato 2014). For this example, one could say that the users become the birds whose behavior is to be monitored. It therein bears a similarity to the second example of an Early Warning System study titled

5 Also 'phases rollout' or 'incremental rollout'.

"Earthquake Shakes Twitter Users: Real-time Event Detection by Social Sensors" (Sakaki/Okazaki/Yutaka 2010).[6]

This study represents a comparatively early model of using social media behavior as data for detecting and predicting catastrophes. Similar approaches have gained considerable public recognition, especially in the field of predicting epidemic events like in the case of Google's Flu Trends (Cukier/Mayer-Schönberger 2013: 1–32).[7] In the Japanese earthquake study "each Twitter user is regarded as a sensor and each tweet as sensory information" (Sakaki/Okazaki/Yutaka 2010: 852). Like for the literal canary in the coal mine, here, a change in tweeting behavior is interpreted as a signal for an impending catastrophe. While for the former, this catastrophe is a hazardous rise in CO concentration, Sakaki, Okazaki and Yutaka propose a model to mitigate the effects of earthquakes via the issuing of early warnings. They do so by analyzing event-relevant tweets and trying to localize them with the use of an algorithm; thereby trying to determine the epicenter of an earthquake. Of course, this system can only detect earthquakes that are felt by a considerable number of people with access to the internet. The event-relevant tweet words are rather obvious ones like 'shaking' or simply 'Earthquake!' (ibid.: 852). The earthquake warning can be rolled out only after a large number of Twitter users have already experienced the ground shaking, wherefore it cannot be regarded as a technology of latency. The authors argue that the model still has the quality of an *early* (or earlier) warning system due to its inbuilt earthquake reporting system. They argue for sending out personal messages (e-mails in this case) as warnings to people in the region, instead of using TV broadcasting. By applying this method, the warning time could allegedly be reduced significantly (ibid.: 857f.). Overall, the study suggests that Twitter can be a valuable tool for earthquake detection and response and highlights the potential of social media as a source of real-time information in emergency situations.

As a third recently published study, the "Spark Streaming-Based Early Warning Model for Gas Concentration Prediction" by Huang et al. (2023) shall be introduced. It illustrates the practice of threshold setting through the

---

6 For the timely detection of earthquakes there exists a long tradition discussing the potential use of animal behavior monitoring. Cf. Tributsch 1978; Pschera 2016: 63–65; Liu/Dhakal 2020; critical of this idea: Hough 2016.

7 For a critical account on the usefulness of Google's tool, respectively its methods, cf. Lazer et al. 2014.

optimization of parameters by using data for training and testing purposes. It can furthermore be seen as an instance of supersession of animal-supported EWS-labor by algorithm-supported EWS-labor. The model is intended for usage in Chinese coal mines. Having the biggest mining industry worldwide, the need to reduce systemic malfunction caused by gas exposure in China is evident. Building upon neural network-based gas concentration prediction models, the "Spark Streaming framework (SSF)" should "provide [...] a new way of thinking for intelligent gas prediction and early warning" (Huang et al. 2023: 2). It operates by using data sets of gas concentration collected from the mine's 'face' (ibid.: 6f.).[8] Throughout the training process, an optimization of the prediction parameters – number of neurons in hidden layers; number of hidden layers, batch size, time steps – is established (ibid.: 6–9). The resulting prediction model together with the gas sensors at the face is used to determine the gas thresholds whose transgression should trigger a warning. Gas concentration below the set threshold is labelled 'normal'; transgressions are classified as level 1 and level 2 warnings (ibid.: 9–11). Hence, the EWS determines the conditions of the normal and the abnormal state. The quality of the gas concentration prediction model is measured by comparing it with real-world data of gas diffusion, resulting in an accuracy level above 90 percent (ibid.: 14). The authors assess this value to be sufficiently high as to guarantee "accurate predictions and graded warnings of gas concentrations [...] for the safe production of coal mines" (ibid.: 15).

This study suggests a supersession of the bird's gas-detecting body by electronic sensors and the neural network's architecture. The use of canaries (besides mice and ponies) in coal mining, however, was already brought to a halt in the 1980s. "Modern technology is being favored over the long-serving yellow feathered friend of the miner in detecting harmful gasses", the BBC reported in 1986. "Miners are said to be saddened by the latest set of redundancies in their industry but do not intend to dispute the decision" (ibid.). The birds' designated successors were electronic monitoring and detection devices referred to as 'electronic noses', analyzing gas concentration data and displaying it on a digital screen. All three of them, the canaries, the gas nose and the proposed technology by Huang et al., should contribute to bringing a (for humans) latent danger to the surface. They can be interpreted as created systems with readable symptoms as warnings. One of the main differences between the use of the

8 This refers to the surface where mining operations are currently progressing.

animals compared to the later auxiliaries is that the latter operate with quantified data on gas concentration. In doing so, they contribute to the delivery of Gabriel Tarde's prediction, taken up and complemented by Bruno Latour: "[Thanks to statistics] public broadsheets will be to the social world what the sensory organs are to the organic world." (Latour 2010: 115; comment in original) In this logic, statistical tools could for example be seen as a help for detecting social upheaval before the breakout of political crises. This suggested use of statistics as auxiliaries for making quantifiable data 'the sensory organs' of the social world should be seen as an epigraph for the following argument, which builds up on the epistemological 'closeness/similarity' of animals, birds in this case, with statistical data analysis (not only) in the field of EWS.

3. EWS, AI, and Kinds of Intelligence[9]

The asserted epistemological 'closeness' of animals and statistical machines may appear paradoxical, since, of course, in many ways these are not alike; it becomes clearer when considering their proclaimed ability to predict danger. Both animals and statistics can offer knowledge about the (otherwise unknown) future for the human, if the latter is able to use them; thereby extending his sensory functions as well as his future-knowledge. "The signs are there, if they can be recognized. As stress occurs, behavior changes." (Walsh 1986: 1146) Considering the examples of birds as early detectors of hazards, as in the case of gas concentration, often goes hand in hand with the metaphysical notion of (these) animals having a 'sixth sense', which allows for them to be used as EWS. The same can be said about snakes or elephants which change their behavior, e.g., fleeing the area or producing sounds prior to an earthquake before it can be recognized by seismologic sensors or humans (Tributsch 1978). Their abilities point to a limitation of the human which calls for their utilization by the latter in order to be better prepared for environmental risks.

Concerning the case of statistics as important tools in the *Taming of Chance* (Hacking 2010), the metaphysical aspect of the knowledge obtained by it is less apparent. After all, the quantification of human behavior served the purpose of introducing a law-like structure – "the law of large numbers" (ibid.: 95–104) – into social affairs. However, the subject of prediction or anticipation, even if it

9 Compare the project "Kinds of Intelligence" by the Leverhulme Center for the Future of Intelligence (http://lcfi.ac.uk/projects/kinds-of-intelligence/).

is based on the usage of statistical correlation and probability, in many cases carries a metaphysical, uncanny or magical baggage with it. For example, it could be noteworthy to mention the conception of statistical knowledge attributed to Florence Nightingale, herself a founding figure of statistics: "[T]o understand God's thoughts, [...] we must study statistics, for these are the measure of His purpose." (Pearson 1924: 415) Or, to invoke a more recent example from the stream of Big Data correlation: Schönberger and Cukier (2013) discuss the uncanny anecdote of a retail company analyzing a woman's shopping behavior which indicates a high probability of her being pregnant. This allows the company to 'know' about the pregnancy before the woman's parents do (Schönberger/Cukier 2013: 57f.).

However, common ground between different cases of animals detecting hazardous gases, based on physiognomy and sensory functions, in relation to a company's detection of the pregnancy, based on the use of algorithms and large amounts of data, might be that both are used to bring to the surface potentially significant environmental changes. They deal with something which lies beyond the scope of human cognition. This constitutes a knowledge that is unlike human intelligence, unless the human learns to make use of it. Its utilization leads to an extension of the 'human senses' for detecting latent but yet impending danger, which can only be accessed by collaboration with e.g. animals like the canary or information machines like statistics; or (more recently) by relying on the application of AI with its "statistical anatomy" (Alpaydin 2016: 27). In this logic, the threat is already there, only the right senses to detect it have not yet been found.

The notion of an expansion of the human senses, and thereby future-knowledge about danger, can serve if not as a lens then at least as an inducement for an argument about the knowledge and the 'intelligence' of AI. The two probably most prominent tropes called upon when discussing the question of whether or not computers and machines can *reasonably* be called 'intelligent', are the proposal for the Dartmouth Conference of 1956 with its proclaimed conviction "that every aspect of learning or any other feature of intelligence can in principle be so precisely described that a machine can be made to simulate it" (McCarthy et al. 2006 [1955]: 12) as well as the famous 'Imitation Game' proposed by Alan Turing six years earlier. This thought experiment, which later came to be known as 'Turing Test', relies on a computer's ability to imitate human-like behavior in a way that makes it impossible for the human dialogue partner to distinguish between human and machine. If this imitation is successful, the machine can be deemed as intelligent (Turing 1950). The critique on this proposed conception of intelligence is well known and need not to be rolled out again. The trope of a 'sixth sense' etc. for catastrophe prediction in animals and statistic-based EWS invites us to shift the focus away from the question, if applications like Chat GPT can pass a Turing Test, which would justify them being labelled as 'intelligent'. Instead of concentrating on the mimicking of human thinking by artificial neural networks, we can 'reverse' the question and highlight the way the concept of intelligence is evolving in the course of its contestation vis à vis other forms of knowledge; namely those forms of knowledge which are always already discursively excluded from speaking truth and thereby excluded from knowing. This approach is in line with Benjamin Bratton's critique of the 'intelligence' in the Turing test, when he writes

> The threshold by which any particular composition of matter can be said to be 'intelligent' has less to do with reflecting human-ness back at us than with testing our abilities to conceive of the variety of what 'intelligence' might be. (Bratton 2015: 75)

The analysis of (catastrophic) future prediction points to two knowledge-related discourses for grasping the concept of intelligence – artificial or not.[10] The first one obviously revolves around the question what kind of knowledge statistics have to offer, respectively what kind of world-knowledge is 'revealed' by the use of quantification and statistical analysis. Historical research on *The Rise of Statistical Thinking* (Porter 2020) shows us that it is not only since the coining of the term 'AI' that these technologies were "associated with an impressive extension of the domain of knowledge and not with its limitations" (ibid.: 163). It can thereby shed light on the discourse about the (statistics and data-based) artificial intelligence.

Apart from this, the preoccupation with EWS, based on animalistic as well as non-animalistic signal detection, opens up a second realm of possibly fruitful analyses concerning the question of what kind of knowledge AI 'has', or better 'offers'. Instead of concentrating on the question whether AI can pass as having acquired human-like intelligence, we can turn our attention to the ways the knowledge of those has been discussed (and created), which most certainly don't pass as 'intelligent', since they constitute the necessary 'Other' of 'human intelligence'. This concerns, to various extents, the thinking of children, non-

10 Whatever non-artificial intelligence might be.

European indigenous groups, people who are differently abled mentally as well as non-human animals. The psychological attempts to grasp and possibly utilize these other forms of sensing and knowledge can shed light on the construction of intelligence. Not least because of the ways artificial intelligence is repeatedly brought into connection with children, non-human animals etc., by comparing their problem-solving abilities with each other. Turing himself proposed: "Instead of trying to produce a programme to simulate the adult mind, why not rather try to produce one which simulates the child's?" (1950: 456) But also, in media reports dealing with scientific developments in AI, we regularly come across headlines in the manner of "AI had IQ of four-year-old child" (BBC 2015). For the case of animals, a good example would be the recently published study by Wasserman, Kain and O'Donoghue (2023), which deals with the learning mechanisms of pigeons that are said to bear significant similarities with the type of learning of machine learning algorithms, particularly reinforcement learning. The authors point to BF Skinner's planned usage of pigeons as 'brains' for his experimental guidance system for directing ballistic missiles to possible WWII military targets. Skinner himself justified this choice as follows: "We have used pigeons, not because the pigeon is an intelligent bird, but because it is a practical one and can be made into a machine, from all practical points of view." (Capshew 1993: 851). Although the usage of birds in this example cannot be interpreted as a defensive EWS but rather served as a measure of attacking the enemy, it illustrates the deployment of non-human cognition and sensing by humans and at the same time makes a comparison to machines. The human makes use of these abilities of the other and thereby expands, to invoke Tarde again, their 'sensory organs'. This rationale also applies to the implementation of Early Warning Systems of various sorts. Concentrating on the reliance of catastrophe prediction abilities, be it via the monitoring of small animal behavior in coal mines or deviations in 'tweeting behavior' via the use of AI, cannot only contribute to an investigation into the gears of the preparedness-apparatus (Lakoff), it can furthermore, as it was argued above, help shed light on the question of 'knowing the human'.

To conclude this investigation into Early Warning Systems and their potential transformation via the use of machine learning, it will be useful to again invoke the report on "Experiments with Small Animals and Carbon Monoxide". Considering the differences between men (not humans) and small animals in feeling distress when exposed to dangerous concentrations of carbon monoxide, Burrell and Seibert assert that "a man is in an excellent position to determine effects upon himself [whereas] small animals may feel distress but not show it." (1914: 243). The reasoning here implies that there are traces to be found in the animal's 'feelings' beneath the behavioral surface. The human, via interacting with the animals and monitoring their behavior, can utilize these feelings by 'making the animal speak', i.e., detecting symptoms even before the animal becomes 'aware' of them. For EWS models like in Sakaki/Okazaki/Yutaka (2010), where the users become birds, whose tweeting behavior is monitored, it is the algorithm's job to identify behavioral patterns as indicators for catastrophes; ideally, even before the users explicitly show their distress. By gathering ever more data about environment-monitoring sensors, be they avian, human, or other, and analyzing them ever more effectively, they will potentially become utilizable for hazard detection even easier and, most importantly, earlier. What will remain unaltered by this extension of the 'sensory organs' via implementing machine learning technology in EWS, however early the signs for danger might be detected (or created), is the determination of what is even perceived as a danger to be prepared for and further: a danger for whom? We can remain skeptical whether it will be the birds having the final say in this matter.

List of references

Alcántara-Ayala, Irasema/Oliver-Smith, Anthony (2019): "Early Warning Systems: Lost in Translation or Late by Definition? A FORIN Approach." In: International Journal of Disaster Risk Science 10/3, pp. 317–331.

Alpaydin, Ethem (2016): Machine Learning: The New AI, Cambridge, MA: The MIT Press.

Ansoff, Harry Igor (1975): "Managing Strategic Surprise by Response to Weak Signals." In: California Management Review 18/2, pp. 21–33.

Austin, Alex (2004): "Early Warning and the Field: A Cargo Cult Science?" In: Alex Austin/Martina Fischer/Norbert Ropers (eds.), Transforming Ethnopolitical Conflict: The Berghof Handbook, Wiesbaden: Springer VS, pp. 129–150.

Bratton, Benjamin (2015): "Outing Artificial Intelligence: Reckoning with Turing Tests." In: Matteo Pasquinelli (ed.), Alleys of your Mind: Augmented Intelligence and its Traumas, Lüneburg: Meson Press, pp. 69–80.

British Broadcasting Corporation (BBC) (1986): "Coal Mine Canaries Made Redundant.", December 30, 1986 (http://news.bbc.co.uk/onthisday/hi/dates/stories/december/30/newsid_2547000/2547587.stm).

British Broadcasting Corporation (BBC) (2015): "Intelligent Machines: AI Had IQ of Four-Year-Old Child.", October 7, 2015 (https://www.bbc.com/news/technology-34464879).

Burrell, George A./Seibert, Frank M. (1914): "Experiments with Small Animals and Carbon Monoxide." In: Journal of Industrial & Engineering Chemistry 6/3, pp. 241–244.

Capshew, James H. (1993): "Engineering Behavior: Project Pigeon, World War II, and the Conditioning of B.F. Skinner." In: Technology and Culture 34/4, pp. 835–857.

Cowan, Crispin/Beattie, Steve/Day, Ryan Finnin/Pu, Calton/Wagle, Perry/Walthinsen, Erik (1999): Protecting Systems from Stack Smashing Attacks with StackGuard, LinuxExpo Report (https://www.cs.jhu.edu/~rubin/courses/sp03/papers/stackguard.pdf).

Cowan, Crispin/Pu, Calton/Maier, Dave/Walpole, Jonathan/Bakke, Peat/Beattie, Steve/Grier, Aaron/et al. (1998): "StackGuard: Automatic Adaptive Detection and Prevention of Buffer-Overflow Attacks." In: Proceedings of the 7th USENIX Security Symposium, San Antonio, TX, USA, pp. 63–78.

Cukier, Kenneth/Mayer-Schönberger, Viktor (2013): Big Data: A Revolution That Will Transform How We Live, Work, and Think, Boston: Houghton Mifflin Harcourt.

Echterhölter, Anna/Schröter, Jens/Sudmann, Andreas (2021): How Is Artificial Intelligence Changing Science? Research in the Era of Learning Algorithms, MediArXiv Preprint (https://doi.org/10.33767/osf.io/28pnx).

Enten, Francois (2008): "Food Aid and The Politics of Numbers in Ethiopia (2002–2004).", May 15, 2008 (https://msf-crash.org/en/publications/humanitarian-actors-and-practice/food-aid-and-politics-numbers-ethiopia).

Fraunhofer-Institut für Nachrichtentechnik (2021ff.): "DAKI-FWS: Daten- und KI-gestütztes Frühwarnsystem zur Stabilisierung der deutschen Wirtschaft.", December 1, 2021ff. (https://daki-fws.de/).

Furnell, Steven/Spaffor, Eugene H. (2019): "The Morris Worm at 30." In: ITNOW 61/1, pp. 32–33.

Hacking, Ian (2010): The Taming of Chance, Cambridge: Cambridge University Press.

Hall, Kevin (2020): "Über die Möglichkeiten und Grenzen von Frühwarnsystemen.", September 3, 2020 (https://www.soziopolis.de/ueber-die-moeglichkeiten-und-grenzen-von-fruehwarnsystemen.html).

Hammer, Richard (1998): Strategische Planung und Frühaufklärung, Berlin et al.: de Gruyter.

Hough, Susan Elizabeth (2016): Predicting the Unpredictable: The Tumultuous Science of Earthquake Prediction, Princeton: Princeton University Press.

Huang, Yuxin/Li, Shugang/Fan, Jingdao/Yan, Zhenguo/Li, Chuan (2023): "A Spark Streaming-Based Early Warning Model for Gas Concentration Prediction." In: Processes 11/220, pp. 1–17.

Keck, Frédéric (2020): Avian Reservoirs: Virus Hunters and Birdwatchers in Chinese Sentinel Posts, Durham: Duke University Press.

Keck, Frédéric/Lakoff, Andrew (eds.) (2013): Limn 3 (Sentinel Devices) (https://limn.it/issues/sentinel-devices/).

Lakoff, Andrew (2008): "The Generic Biothreat, or, How We Became Unprepared." In: Cultural Anthropology 23/3, pp. 399–428.

Lakoff, Andrew (2017): Unprepared: Global Health in a Time of Emergency, Oakland, CA: University of California Press.

Lamsal, Rabindra/Kumar, T. V. Vijay (2020): "Artificial Intelligence and Early Warning Systems." In: T. V. Vijay Kumar/Keshav Sud (eds.), AI and Robotics in Disaster Studies, Singapore: Palgrave Macmillan, pp. 13–32.

Latour, Bruno (2010): "Tarde's Idea of Quantification." In: Mattei Candea (ed.), The Social after Gabriel Tarde: Debates and Assessments, London: Routledge, pp. 145–162.

Lazer, David (2014): "The Parable of Google Flu: Traps in Big Data Analysis." In: Science 343/6176, pp. 1203–1205.

Leverhulme Centre for the Future of Intelligence (n.d.): "Kinds of Intelligence.", (http://lcfi.ac.uk/projects/kinds-of-intelligence/kinds-intelligence/).

McCarthy, John/Minsky, Marvin L./Rochester, Nathaniel/Shannon, Claude E. (2006 [1955]): "A Proposal for the Dartmouth Summer Research Project on Artificial Intelligence, August 31, 1955." In: AI Magazine 27/4, p. 12.

Neo, Jacqueline Pei Shan/Tan, Boon Huan (2017): "The Use of Animals as a Surveillance Tool for Monitoring Environmental Health Hazards, Human Health Hazards and Bioterrorism." In: Veterinary Microbiology 203, pp. 40–48.

Nucho, Joanne R. (2022): "Succeeding through Failure: Andrew Lakoff on Preparing for Emergencies.", June 8, 2022 (https://www.publicbooks.org/andrew-lakoff-on-preparing-for-emergencies/).

Pearson, Karl (1924): The Life, Letters and Labours of Francis Galton, Cambridge et al.: Cambridge University Press.

Pircher, Wolfgang (2010): "Regelkünste: Über Kybernetik und Operational Research." In: Günther Freisinger (ed.), Geist in der Maschine: Medien, Prozesse und Räume in der Kybernetik, Wien: Turia & Kant, pp. 51–71.

Porter, Theodore M. (2020): The Rise of Statistical Thinking, 1820–1900, Princeton: Princeton University Press.

Pschera, Alexander (2016): Animal Internet: Nature and the Digital Revolution, New York: New Vessel Press.

Reif, John S. (2011): "Animal Sentinels for Environmental and Public Health." In: Public Health Reports 126/suppl. 1, pp. 50–57.

Sakaki, Takeshi/Okazaki, Makoto/Yutaka, Matsuo (2010): "Earthquake Shakes Twitter Users." In: Proceedings of the 19th International Conference on World Wide Web (WWW'10), Raleigh, NC, USA, pp. 851–860.

Sato, Danilo (2014): "Bliki: CanaryRelease." June 25, 2014 (https://martinfowler.com/bliki/CanaryRelease.html).

Spellerberg, Ian (2006): Monitoring Ecological Change, Cambridge: Cambridge University Press.

Tributsch, Helmut (1978): Wenn die Schlangen erwachen: Mysteriöse Erdbebenvorzeichen – Endlich ernstgenommene Naturphänomene führen zu einem Frühwarnsystem, Stuttgart: Deutsche Verlags-Anstalt.

Turing, Alan M. (1950): "Computing Machinery and Intelligence." In: Mind 59/236, pp. 433–460.

United Nations General Assembly (2016): Report of the Open Ended Intergovernmental Expert Working Group on Indicators and Terminology Relating to Disaster Risk Reduction, A/71/644 (https://digitallibrary.un.org/record/852089).

United Nations Office for Disaster Risk Reduction (n.d.): "Artificial Intelligence for Disaster Risk Reduction." (https://www.preventionweb.net/collections/artificial-intelligence-disaster-risk-reduction).

Walsh, John (1986): "Famine Early Warning Closer to Reality: Chronic Lag in Response to Food Emergencies in Africa Spurs Use of Remote Sensing to Improve Reaction Time." In: Science 233/4769, pp. 1145–1147.

Wasserman, Edward A./Kain, Andrew G./O'Donoghue, Ellen M. (2023): "Resolving the Associative Learning Paradox by Category Learning in Pigeons." In: Current Biology 33/6, pp. 1112–1116.

Yanbing, Liu/Dhakal, Sanjev (2020): "Animal Behavior Abnormality and Earthquake Disaster Early Warning." In: Revista Cientifica 30/3, pp. 1371–1380.

Cross-interactions between AI and epistemology

Jean-Gabriel Ganascia

1. Introduction

The aim of this paper is to show that the mutual epistemological stakes of artificial intelligence (AI) and sciences, both 'hard sciences' and Human and Social Sciences (HSS), are multiple. It specifically addresses two of them. The first is reflexive: it concerns the epistemology of AI itself, which, as a scientific discipline, deserves a philosophical and historical look at its foundations. It is a question of specifying the nature of this discipline, which cannot just be reduced to a technology and which, as a science, is neither a theoretical science, even if it has originally been founded by mathematicians, nor a 'science of nature' strictly speaking, nor really a 'science of culture' that is a discipline of the humanities. To clarify these different issues, we shall first recall the genesis of AI, its history and its definitions, before trying to approach its epistemological status.

The second issue is related to the uses of AI, machine learning and data processing in different scientific disciplines and the major changes that these uses induce in these disciplines by automating tedious tasks. In doing so, our aim here is to show that AI techniques do not only allow to automate certain tasks, but that they also contribute to designing new interpretation operators, new proof procedures and, more generally, new scientific approaches such as *in silico sciences* (cf. Ganascia 2008). In other words, the contribution of AI is not only practical; it introduces into these scientific disciplines what Gaston Bachelard calls an epistemological rupture (cf. Bachelard 1938), that is, a dissociation between the primary evidence of observation and the scientific facts resulting from experimentation.

To address these different epistemological issues, the paper is divided into two main parts framed by this introduction and its conclusion. The first part is dedicated to the genesis, the history and the epistemology of AI, while the second concerns the impact on the theoretical sciences, on the sciences of nature and on the sciences of culture, i.e., on the humanities.

2. AI groundings

2.1 Prehistory of AI

Attempts to formalize the laws of thought and to automate reasoning are ancient. Born in antiquity, logic aimed to give the laws of right thinking; to do so, it characterized, by means of formal-mechanical rules, valid reasoning as being sequences of inferences — inferences being formal manipulations of symbolic expressions — that correspond to elementary figures listed as being valid themselves.

In classical formal logic, i.e., in Aristotelian or in Stoic logic, the set of elementary inferences was determined from 'regulatory' syllogisms, i.e., figures that lead from two propositions to a third. Later, in the second half of the 17th century, Leibniz tried to mathematize logic, i.e., the laws of correct thinking, in order to prove the validity of an argument by a calculation, without having to memorize all the valid elementary syllogisms, as in traditional Aristotelian logic. This was immediately followed by the desire to automate this calculation on a machine. Therefore, we can say that Leibniz is a forerunner of artificial intelligence since he tried — unsuccessfully — to draw the plans of a machine capable of reasoning by itself. This project was taken up in the 19th century by George Boole who created binary algebra to account for the laws of logic and then by William Stanley Jevons, who actually built a machine, the 'logical piano', that could mechanically deduce the consequences of logical premises based on the work of George Boole.

Several other attempts to automate reasoning were made in the early 20th century. Let's think for instance of the mechanical chess player machines of Torres y Quevedos that were built in 1912 and 1920. We must also mention the cybernetic movement with Warren McCulloch and Walter Pitts (cf. McCulloch & Pitts 1943), Claude Shannon and Norbert Wiener, among others, because it was also at the origin of several attempts to reproduce thought on electronic computers. Finally, we must not forget Alan Turing, who wondered, in his famous article *Computing Machinery and Intelligence* published in 1950 (cf. Turing 1950), what it means for a machine to think and how to build such a thinking machine.

2.2 Birth and epistemic assumptions of AI

However, although the idea that it's possible to build a machine that reproduces thought had been around for a long time, the term 'artificial intelligence' did not appear for the first time until 1955 in a summer school proposal submitted by four researchers, John McCarthy, Marvin Minsky, Nathanael Rochester, and Claude Shannon to the Rockefeller Foundation for a grant to organize a summer school at Dartmouth College, New Hampshire, in 1956. For the promoters of this summer school, artificial intelligence was a scientific discipline that aimed to study intelligence with computers. More precisely, to quote them, "The study is to proceed on the basis of the conjecture that every aspect of learning or any other feature of intelligence can in principle be so precisely described that a machine can be made to simulate it" (McCarthy et al. 1955: 1).

This means that all cognitive faculties, in particular reasoning, calculation, perception, memorization and even scientific discovery or artistic creativity, could be described with such precision that it should be possible to reproduce them using a computer. Let us insist on the epistemological importance of this conjecture: it draws a horizon of tasks to be accomplished, just like Galileo's postulate according to which the Book of Nature is "written in mathematical language". Since then, despite the considerable progress made and the changes in the technologies used, from those based on symbolic logic to numerical and emergent connectionist approaches, and despite the various debates about the parallels between the nature of intelligence itself and the way it is simulated by machines, the study of artificial intelligence has always been based on the same conjecture, which nothing has yet been able to disprove or prove irrefutably. To clarify, it's important to specify that what the philosopher, John Haugeland, has mistakenly called GOFAI ("good old fashioned artificial intelligence", Haugeland 1985) is, as some AI researchers like Drew McDermott have mentioned, a myth, for many reasons, in particular because the seminal text on AI, the Dartmouth College Summer School proposal (cf. McCarthy et al. 1955), explicitly mentioned neural networks as methods that had to be developed by AI. This doesn't mean that the way the machine simulates "every aspect of learning or any other feature of intelligence" is similar to the way they are implemented in nature. For example, just as Frederick Jelinek beautifully put it in his famous "airplanes don't flap their wings", so the learning styles of children and machines are not at all the same at all and the way the computers work has nothing to do with the way our brains work, even if they make use of neural networks.

Nevertheless, the idea that it is possible to reproduce all the cognitive functions of any intelligent being still constitutes the epistemological horizon of AI.

The same scientists, who were trying to reproduce various cognitive abilities such as reasoning, theorem proving, image or speech recognition, knowledge representation in memory, etc. on computers, were in parallel tempted to take a practical advantage of these simulations and to incorporate them into many technological devices. Very soon, for example, Herbert Simon and Alan Newell wrote papers both on the performance of general-purpose problem-solving computer programs (cf. Newell/Simon 1956) and on human problem-solving using AI tools as cognitive models to study human reasoning (cf. Newell/Shaw/Simon 1958).

The incorporation of AI simulations in technologies has been very popular in recent years, giving AI the privilege of being one of the most active fields of applied research in many areas such as medicine, agriculture, geology, etc. Today, when people talk about AI, they almost always mention the various technological applications of AI. This most often corresponds to the current meaning of the term AI.

Finally, it should be noted that, among the general public, the success of the term 'artificial intelligence' is often due to a damaging misunderstanding according to which AI would produce artificial entities endowed with intelligence and which, as a result, would compete with human beings. This idea, which refers to ancient myths and legends such as that of the Golem, has recently been revived by contemporary personalities such as Stephen Hawking or Elon Musk, by engineers such as Ray Kurzweil, or by the proponents of what is now called 'strong artificial intelligence' or 'general artificial intelligence'. We will not discuss this meaning here, because it only attests to an abundant imagination inspired more by science fiction than by a tangible scientific reality confirmed by experiments and empirical observations.

2.3 Very brief history of AI

Since its birth, even if the seminal definition of the Dartmouth College Summer School has always remained valid, AI has undergone many evolutions that we can summarize in six stages.

2.3.1 The time of the prophets

A few achievements, in particular the Logic Theory Machine (cf. Newell/Simon 1956), which automatically proved logic theorems, the seminal work of Arthur

Samuel on reinforcement learning applied to the game of checkers (cf. Samuel 1959) and the first efficient neural network learning process, the so-called Perceptron (cf. Rosenblatt 1958), aroused enthusiasm. In the euphoria that followed, the researchers let themselves go to some rather unconsidered declarations that they have been much reproached for afterwards. For example, on November 14th, 1957, Herbert Simon delivered a speech at the banquet of the Twelfth National Meeting of the Operations Research Society of America in which he said:

> I am willing to make the following predictions, to be realized within the next ten years:
> 1. That within ten years a digital computer will be the world's chess champion, unless the rules bar it from competition.
> 2. That within ten years a digital computer will discover and prove an important new mathematical theorem.
> 3. That within ten years a digital computer will write music that will be accepted by critics as possessing considerable aesthetic value.
> 4. That within ten years most theories in psychology will take the form of computer programs, or of qualitative statements about the characteristics of computer programs.

The lecture was then transcribed and the paper was co-signed with Alan Newell and published in the journal *Operation Research* (Simon/Newell 1958: 7).

2.3.2 The dark years

In the mid-1960s progress was not as fast as expected. In particular, a chess-playing machine was defeated by a ten-year-old boy in 1966, which made the first point of Herbert Simon's statement suspect and by contaminating the others, so AI received some bad press, which resulted in some dark years for AI. This corresponds to what is now called the AI winter, a period during which AI research became less popular, although contrary to popular belief, work never stopped altogether. For example, the first chatbot named Eliza was created by Joseph Weizenbaum at MIT between 1964 and 1966 (cf. Weizenbaum 1966), and later, Terry Winograd (cf. Winograd 1971), still at MIT, developed a famous program called SHRDLU for natural language understanding, i.e., for translating simple sentences into logical formulas. Note finally that during that AI winter, Marvin Minsky and Seymour Papert (cf. Minsky/Papert 1969) showed the intrinsic limitations of the Rosenblatt's Perceptron learning algorithm, because it was restricted to two-layers neural networks while Warren McCulloch and Walter Pitts, in their seminal paper (McCulloch/Pitts 1943) show that only three-layers neural networks were universal, i.e., able to implement any Boolean logic function.

2.3.3 Semantic artificial intelligence

Nevertheless, as previously said, during that AI Winter, work never stopped. Researchers were then focused on new directions and inspired by works in psychology and linguistics, which gave birth to the first cognitive science approaches. Note that interest in human cognition is far older and that cybernetics had already attempted to model social and cognitive processes with information processing mechanisms. However, new interdisciplinary approaches combining artificial intelligence, psychology and linguistics began in the mid-1970s. This corresponds to what has been called the 'semantic turn'. It led to an increased interest in modeling memory, in the mechanisms of comprehension, which was tried to be simulated on a computer as well as in the role knowledge plays in reasoning. This is what gave rise to knowledge representation techniques (cf. Bobrow/Winograd 1976) with semantic networks (cf. Collins/Quillian 1969) and frames (cf. Minsky 1974), to object-oriented programming and to so-called expert systems, because they used the knowledge of human experts to reproduce their reasoning. The latter raised enormous hopes in the early 1980s.

2.3.4 Neo-connectionism and machine learning

In parallel with the rise of artificial intelligence in the early 1980s, the techniques derived from cybernetics and connectionism were perfected, freed from their initial limitations and made the object of multiple mathematical formalizations. More specifically, as mentioned above and as Marvin Minsky and Seymour Papert had shown (cf. Minsky/Papert 1969), Rosenblatt's Perceptron learning algorithm was restricted to elementary logic functions. In the mid-1980s, this algorithm was generalized to multi-layer neural networks (cf. Rumelhart/Hinton/Williams 1986), giving rise to the backpropagation learning algorithm, which wasn't subject to such limitations. This led to distributed parallel processing, which enabled the use of neural networks in many supervised machine learning tasks.

2.3.5 From artificial intelligence to 'animistic informatics'...

Since the late 1990s, artificial intelligence has often been coupled with robotics and human-machine interfaces to produce intelligent agents that suggest the presence of another, whether it be human, or just an abstract entity. This trend of artificial intelligence can be sketchily characterized as a form of computer animism insofar as it seeks to elicit the projection of a breath of life onto the everyday objects of our environment. The current successes of Chatbots and, more recently, of ChatGPT, testify to the vitality, popularity and fashion of this trend.

2.3.6 The renaissance of artificial intelligence

With the massive development of the Web it became necessary to deal with large amounts of data. More specifically, since the rise of Web 2.0 at the turn of the century, the economics of the Web giants were based on targeted advertising, which made profiling critical. It follows that, based on information about individual behavior such as search queries, websites visited, etc., profiling had to scale to the size of the Web, which required dealing with massive amounts of data. This became known as 'Big Data'. The computational power of machines gradually made it possible to use large corpora of data with machine learning techniques, such as SVM, Kernel Machines, or Random Forests, which made AI very popular. Then, since the 2010s, the extension of Neural Network architectures to Convolutional Neural Networks (CNN) corresponding to the techniques currently called Deep Learning (cf. LeCun/Bengio/Hinton 2015), has produced impressive results that have tremendously accelerated the efficiency and the use of AI techniques.

Later, the Generative Adversarial Nets (GAN) enabled significant advances in image generation techniques (cf. Goodfellow et al. 2014), and the notion of transformers (cf. Vaswani et al. 2017) enabled the construction of Large Language Models (LLM) with hundreds of billions of parameters and impressive text generation techniques of which chatGPT is a popular example.

2.4 Epistemology of AI

The different steps of the evolution of AI corresponded to different epistemological views of this discipline that can be characterized as follows.

2.4.1 Logical-mathematical approach

The first works of artificial intelligence in the fifties and sixties were based on mathematical modeling, in particular on statistics and logic. This has been the case of automatic theorem provers (cf. Newell/Simon 1956), problem solving (cf. Newell/Shaw/Simon 1958) and the first attempts at machine learning, in particular reinforcement learning (cf. Samuel 1959). This gave rise to a science of models, to a 'science of the artificial', to use the title of a book by a pioneer of artificial intelligence, Herbert Simon (cf. Simon 1969), which is distinct from both the natural and the cultural sciences.

2.4.2 Semantic approaches

At the same time, there was a scientific current that used behavioral psychology to evaluate the plausibility of cognitive models of thinking or learning. From the end of the 1960's onwards, a new trend was inspired by other approaches from psychology (cf. Collins/Quillian 1969), in particular Charles Bartlett's schema theory and the theory of prototypes, and from linguistics, with the transformational grammars stemming from Chomsky's theories on the one hand and Fillmore's case grammars or Montague's semantic grammar on the other hand, in order to better understand human cognitive abilities before modeling them. Knowledge representation techniques (cf. Bobrow/Winograd 1976), in particular semantic networks (cf. Collins/Quillian 1969), frames (cf. Minsky 1974) and knowledge-based systems or expert systems are directly derived from these works. This led to a tension between two views of artificial intelligence, one focusing more on the logical-mathematical properties required to simulate cognitive processes on machines to be possible, the other on the study of the psychological processes to be modeled (cf. Newell 1982). This tension was resolved in the early 1980s with the logical formalization of knowledge representation techniques, in particular with description logics, that now form the basis of so-called formal ontologies (cf. Brachman/Fikes/Levesque 1983).

2.4.3 Learning theories and deep learning

From the eighties and the implementation of many learning models (Top-Down Induction of Decision Trees, Genetic Algorithms, Reinforcement Learning, Neural Networks, in particular Back-prop algorithms that generalized the perceptron (cf. Rumelhart/Hinton/Williams 1986), Inductive Logic Programming, etc.), there were attempts to theorize machine learning with, in particular, Leslie Valiant's work on the theory of learnability (cf. Valiant

1984) and Vladimir Vapnik's on statistical learning (cf. Vapnik 1999). These approaches were at the origin of new approaches, in particular ensemble methods ('bagging' and 'boosting') and support vector machines (SVM), which appeared to be prominent in AI since the mid-1990s.

2.4.4 Big Data

Since the beginning of this century, the Web Giants have been using aforementioned machine learning techniques such as SVM to process very large masses of data that are counted in gigabytes (10^9 bytes), terabytes (10^{12} bytes), and even petabytes (10^{15} bytes). Some claim that huge amounts of data solve all problems, without the need for theory or knowledge representation (cf. Anderson 2008), although this is highly debatable from an epistemological point of view. However, since 2010, pragmatic approaches using formal neural networks organized in multiple layers, the so-called Deep Learning techniques (cf. LeCun/Bengio/Hinton 2015), have produced statistical results far superior to previous models, without having any mathematical theory to explain them. This seems to be of great interest from an epistemological point of view, which is ours in this paper. However, nothing says that such a theory will not be available in the future.

3. Impacts of AI on sciences

AI does not only aim at better understanding intelligence by breaking it down into cognitive functions, simulating each of them and exploiting these simulations for technological purposes. It also transforms the scientific activity itself. This is the question we will address in the second part of this article.

3.1 Impact on the natural sciences: *In silico* experimentations

Today, almost all facts can be reduced to huge data sets. It follows that it is possible to induce and test theories directly from data using AI and Machine Learning (ML) techniques, without having to conduct experiments in the outside world. These data sets come from collecting information issued from sensors, or from automated analysis such as the sequencing of macromolecules like proteins or DNA. In addition, computer models make it possible to simulate parts of the physical world and conduct experiments on the results of these models. Undoubtedly, this kind of experimentation is changing scientific activity, at least in part. This is obviously the case in the natural sciences, since many real-world experiments no longer need to be performed, which seems highly desirable for both economic and ecological reasons...

At the end of the 1980s, biologists who wanted to give a name to this type of experiment performed with computers, or more precisely, with the silicon microchips that make up the core of computers, invented a new Latin idiom: *in silico* (see http://en.wikipedia.org/wiki/In_silico). The term was constructed by analogy with — and in contrast to — *in vivo* experiments, i.e., experiments on living organisms, and *in vitro* experiments, which relate biological mechanisms to chemical processes reproduced in glass test tubes. Of course, this term reflects the growing role of computers in the sciences in general. But a careful study shows that computers are not just new tools here, but represent an epistemological turn in the empirical sciences in general, because they change the status of the experiment.

To be more precise, let us recall that in ancient times, science was first and foremost a question of observation and for Plato the most important sense was that of sight. Later on, in modern times, touch took over from sight: people wishing to understand the natural world spent more and more time provoking the subjects they were studying. Thus, in the 16th century, Andreas Vesalius (1514–1564) renewed human anatomy by dissecting the corpses of people condemned to death. Scientific experimentation in its modern meaning corresponds to this reversal: it is not enough just to observe; a scientist will intervene in the world in order to first understand it and then to transform it. This active intervention in the real world continued relentlessly: soon, autopsies no longer satisfied naturalists, who chose to provoke natural phenomena on the living body in order to understand the life springs. They then went further and started performing what are known as *in vivo* experiments because they are carried out on living beings. And so it went on: investigation was not only a question of touching and provoking nature, but also of reconstructing it. This led to the idea of reproducing *in vitro*, i.e., in glass test-tubes, the chemical reactions that are at the origin of the elementary physiological functions.

Today, this trend continues, not only with glass test tubes, but also with computers: we now think we can imitate all natural mechanisms, especially those of the living, reducing them not uniquely to chemical processes, but also to information processing. This gives rise to *in silico* experiments, which are experiments of a singular form in the sense that they no longer call upon the external senses, whether sight or touch, but only upon the temporal unfolding of logical and/or mathematical operations.

Insofar as the *in silico* experiments take place virtually, without touching their object of study, but by operating only on transformations of its representations, they are similar to 'thought experiments' (cf. Mach 1976; Sorensen 1992), even if they clearly can't be assimilated to them, since they provide objective results. And the detailed examination of *in silico* experiments seems to confirm this intuition. Indeed, their role in contemporary scientific activity is twofold.

The first role is to validate hypotheses on large amounts of pre-recorded data such as those obtained from the sequencing of genomes or proteins or from simulation of physical phenomena. Any experiment is, of course, the confrontation of a hypothesis with reality, but, in the case of *in silico* experiments, the observations are collected before the hypothesis is put forward, whereas in classical experiments, the scientific hypothesis led to the construction of an experimental apparatus through which data was collected to validate or invalidate the initial theory. The *in silico* experiments are thus presented as imaginary experiments in which hypotheses are tested on facts that are stored in memory. Note that, in addition to hypothesis validation, AI techniques can automatically generate many plausible hypotheses from data sets which can then be tested for facts. This led to the partial automation of scientific discovery. More precisely, being given an ontology, the machine becomes able to generate hypotheses and to test them on data (cf. Kings et al. 2004).

The second role of *in silico* experiments concerns the simulation of natural processes: just as, in any mental experiment, we reproduce real phenomena in our imagination, so, in many *in silico* experiments, the computer mimics material processes by transforming representations. The *in silico* experiment corresponds then to a virtual intervention on a fictitious world.

What's new today is the central role that *in silico* experiments play in contemporary scientific activity. Whereas in the past, many philosophers — including Karl Popper, one of the most famous — have criticized the role of 'thought experiments' in science (cf. Popper 1959), because they did not provide a strong scientific justification, today, *in silico* experiments, which are the computational equivalent of 'thought experiments', are now scientifically defendable, because they provide some tangible results and they are refutable. In other words, and in conclusion, the extensive use of *in silico* experiments in the natural sciences represents an epistemological turn that deserves attention.

3.2 Impacts on the humanities

This revolution in the natural sciences is accompanied by a major transformation in the humanities, i.e., in the disciplines that study human works. In this case, it is no longer a question of extracting general laws from data by induction, but of interpreting individual cases, for example literary works, on the basis of a large variety of data. In literature, we can try to identify markers of influence in the writings of great authors. It will then be possible to validate certain hypotheses, thereby renewing the traditional disciplines of scholarships.

In order to understand this specificity of the epistemological changes of the humanities, by distinguishing them from the transformations that have taken place in the so-called 'hard' sciences, we will draw on the opposition introduced by neo-Kantian philosophers, in this case Heinrich Rickert (cf. Rickert 1921) and Ernst Cassirer (cf. Cassirer 1923;1942), at the beginning of the 20th century, between the 'sciences of the nature', which deal with the world as it appears to us, and the 'sciences of the culture', which study human works. They — and particularly Ernst Cassirer (cf. Cassirer 1942) — show that both the natural sciences and the cultural sciences are empirical sciences, i.e., based on observable facts, but that the logic of each is different. The sciences of nature aim mainly to construct general laws by induction from observations and forgetting individual cases, while the sciences of the culture focus principally on the individual cases to give them meaning by explaining them. In this case, however, it is no longer a matter of extracting general laws by induction from data, but of interpreting individual cases, for example literary works or historical episodes, by using a great variety of data in order to understand them, or, more precisely, to give them meaning. To do this, an approach based on what logicians call abduction must be adopted, that is to say, on the search for explanations in the light of general theories. Thus, in the case of literature, we can try to identify markers of influence in the writings of great authors. From then on, it becomes possible to validate certain hypotheses empirically, which renews the traditional disciplines of scholarship.

Note that, in practice, the distinction between the 'sciences of the nature' and the 'sciences of the culture' is not so abrupt, since there are many cases where 'sciences of the nature' are also, at least in part, 'sciences of the culture' and vice versa. For example, medicine and health sciences are obviously 'sciences of the nature', while the nomenclature reflects medical traditions that depend on culture. Similarly, geography, which is clearly a 'science of the culture' is also, and in part, a 'science of nature', since it is based on many hard sciences.

Moreover, many epistemologists note that the logic of most of the 'sciences of nature' is not strictly inductive, since the process of discovery has sometimes been seen as mainly abductive, and that the deduction obviously has a place in any scientific reasoning. Symmetrically, the logic of the 'sciences of the culture' is not strictly abductive; deduction plays a role and it may happen that induction be used in some disciplines. This may be the case in literary studies when characterizing the style of an author (cf. Jockers 2013), or the figure used in a particular genre (cf. Boukhaled/Ganascia 2015) or again the expression of a character in a theater play. Nevertheless, it is clear that abduction plays a major role in 'sciences of culture', while induction is prominent in many 'sciences of nature'. Our goal, here, is to show that AI can be useful both for the 'sciences of nature' by mainly providing tools for automatic induction, and for the 'sciences of culture' by helping to interpret individual cases.

To aid in this search for interpretation, a certain number of tools have been developed and deployed that perform multiple operations, such as comparing textual states (cf. Ganascia 2011) or searching for reuses (cf. Ganascia/Glaudes/Del Lungo 2014), or, in archaeology, reconstructing pottery or buildings in three dimensions. These tools do not simply automate existing tasks. They propose new interpretive operators that completely transform the disciplines of scholarship. To illustrate, in the literary domain, Franco Moretti (cf. Moretti 2005) introduces the notion of distant reading, where he identifies general characteristics on large corpora, such as sentence length or punctuation. Similarly, we can characterize quotations or borrowings on large corpora, still in the literary domain. Note that, in both cases, whatever the size of the corpora may be, the inferences are clearly not inductive, but abductive, since they don't generate knowledge by themselves, but help interpretation.

These new interpretation operators have a double contribution. Some have a purely heuristic function by suggesting new avenues of research that need to be explored. They then help bring to light hitherto hidden phenomena, allowing human works to be seen under new conditions. These lines of research then require more rigorous investigation, with proven methodologies.

Others bring empirical elements of validation or invalidation of working hypotheses, for example, in the literary field, by highlighting certain influences, or on the contrary by showing the absence of explicit and/or implicit references and citations. In the latter case, the very scientific basis of certain disciplines is strongly modified, since, as in the natural sciences, the very notion of proof evolves with the introduction of AI in the cultural sciences.

4. Conclusion

Finally, let us recall that one of the pioneers of AI, Herbert Simon, wrote a book entitled *The Sciences of the Artificial* (cf. Simon 1969), in which he discusses scientific approaches to modeling and the function of models in science. This could lead to the question of what characterizes AI as a science: is it exclusively a theoretical science, based on mathematics, or is it an empirical science? And in the latter case, is it more akin to the natural sciences or, to use the terminology of the neo-Kantian philosophers mentioned earlier, such as H. Rickert or E. Cassirer, to the 'sciences of the nature' or to the HSS, i.e., to the 'sciences of the culture'? What makes us lean toward the former possibility is that Machine Learning is inherently inductive, aiming to generate general rules from particulars. What makes us lean toward the latter possibility is that AI is largely concerned with the modeling of deliberate individual practices that are the result of conscious activities and thus can be seen as human works. As a study of human works, it is therefore a science of culture, in the sense that the term has been defined above. Undoubtedly, the methods it uses are essentially based on mathematical and statistical approaches. At the same time, from a logical point of view, a large part of the activity of AI consists in calculating for and simulating tasks that are the fruit of some human practices, such as those mentioned here, and that, as such, belong to culture. Thus, the study of the relations between AI and HSS leads not only to showing the historical interest of AI for HSS, to highlight the use of AI by HSS and the modifications of the latter, with AI, or what we call the 'computational turn' of the latter, but also to show, in this respect, the proximity between AI and HSS.

List of references

Anderson, Chris (2008): "The End of Theory: The Data Deluge Makes the Scientific Method Obsolete." June 23, 2008 (https://pdodds.w3.uvm.edu/files/papers/others/2008/anderson2008a.pdf).

Bachelard, Gaston (1938): La Formation de l'esprit scientifique, Paris: Presses Universitaires de France.

Bobrow, Daniel G./Winograd, Terry (1976): An Overview of KRL, A Knowledge Representation Language, Stanford Artificial Intelligence Laboratory Memo AIM-293, Stanford: Stanford University.

Boukhaled, Mohamed Amine/Ganascia, Jean-Gabriel (2015): "Computational Study of Stylistics: a Clustering-based Interestingness Measure for Extracting Relevant Syntactic Patterns." In: Proceedings of the 16th International Conference on Intelligent Text Processing and Computational Linguistics, Cairo, Egypt, April 14–20.

Brachman, Ronald/Fikes, Richard/Levesque, Hector (1983): "KRYPTON: A Functional Approach to Knowledge Representation." In: IEEE Computer 16/10, pp. 67–73.

Cassirer, Ernst (1923): Substance and Function, Chicago: Open Court.

Cassirer, Ernst (1942): "Zur Logik der Kulturwissenschaften." In: Göteborgs Högskolas Årsskrift 47 [Translated into English under the title The Logic of the Humanities (1961)], New Haven, CT: Yale University Press.

Collins, Allan M./Quillian, Ross (1969): "Retrieval Time From Semantic Memory." In: Journal of Verbal Learning and Verbal Behavior 8/2, pp. 240–247.

Ganascia, Jean-Gabriel (2008): "'In silico' Experiments: Towards a Computerized Epistemology." In: Newsletter on Philosophy and Computers 7/2, pp. 11–15.

Ganascia, Jean-Gabriel (2011): "MEDITE – A Unilingual Text Aligner for Humanities. Application to Textual Genetics and to the Edition of Text Variants." Supporting Digital Humanities (SDH 2011), Copenhagen, Denmark (http://www-poleia.lip6.fr/~ganascia/Medite_Project?action=AttachFile&do=get&target=SDH2011.pdf).

Ganascia, Jean-Gabriel (2015): "The Logic of the Big Data Turn in Digital Literary Studies." In: Frontiers in Digital Humanities 2/7 (https://doi.org/10.3389/fdigh.2015.00007).

Ganascia, Jean-Gabriel/Glaudes, Pierre/Del Lungo, Andrea (2014): "Automatic Detection of Reuses and Citations in Literary Texts." In: Literary and Linguistic Computing 29/3, pp. 412–421.

Goodfellow, Ian/Pouget-Abadie, Jean/Mirza, Mehdi/Xu, Bing/Warde-Farley, David/Ozair, Sherjil/Courville, Aaron/Bengio, Yoshua (2014): "Generative Adversarial Nets." In: Proceedings of the International Conference on Neural Information Processing Systems (NIPS 2014), pp. 2672–2680.

Haugeland, John (1985): Artificial Intelligence: The Very Idea, Cambridge: The MIT Press.

Jockers, Matthew (2013): Macroanalysis: Digital Methods and Literary History, Champaign: University of Illinois Press.

King, Ross D./Whelan, Kenneth E./Jones, Ffion M./Reiser, Philip G./Bryant, Christopher H./Muggleton, Stephen H./Kell, Douglas B./Oliver, Stephen G. (2004): "Functional Genomic Hypothesis Generation and Experimentation by a Robot Scientist." In: Nature 427/6971, pp. 247–252.

LeCun, Yann/Bengio, Yoshua/Hinton, Geoffrey (2015): "Deep Learning." In: Nature 521, pp. 436–444.

Mach, Ernst (1976): "On Thought Experiments." In: Knowledge and Error. Sketches on the Psychology of Enquiry, Dordrecht: D. Reidel Publishing Company.

McCarthy, John/Minsky, Marvin L./Rochester, Nathaniel/Shannon, Claude E. (1955): "A Proposal for the Dartmouth Summer Research Project on Artificial Intelligence.", August 31, 1955 (http://raysolomonoff.com/dartmouth/boxa/dart564props.pdf).

McCulloch, Warren/Pitts, Walter (1943): "A Logical Calculus of the Ideas Immanent in Neuron Activity." In: Bulletin of Mathematical Biophysics 5/13, pp. 115–133.

Minsky, Marvin (1974): A Framework for Representing Knowledge, MIT-AI Laboratory Memo 306, Cambridge: MIT.

Minsky, Marvin/Papert, Seymour (1972): Perceptrons: An Introduction to Computational Geometry, Cambridge: The MIT Press.

Moretti, Franco (2005): Graphs, Maps, Trees: Abstract Models for a Literary History, London: Verso.

Newell, Alan (1982): "The Knowledge Level." In: Artificial Intelligence 18, pp. 87–127.

Newell, Allen/Shaw, J.C./Simon, Herbert (1958): "Elements of a Theory of Human Problem Solving." In: Psychological Review 65, pp. 151–166.

Newell, Alan/Simon, Herbert (1956): The Logic Theory Machine. A Complex Information Processing System, P-868, Santa Monica: The Rand Corporation.

Rickert, Heinrich (1921): Kulturwissenschaft und Naturwissenschaft, Tübingen: J.C.B. Mohr.

Rosenblatt, Frank (1957): The Perceptron, A Perceiving and Recognizing Automaton, 85–460-1, Buffalo: Cornell Aeronautical Laboratory.

Rosenblatt, Frank (1958): "The Perceptron: A Probabilistic Model for Information Storage and Organization in the Brain." In: Psychological Review 65, pp. 386–408.

Rumelhart, David/Hinton, Geoffrey/Williams, Ronald (1986): "Learning Representations by Back-Propagating Errors." In: Nature 323/6088, pp. 533–536.

Samuel, Arthur L. (1959): "Some Studies in Machine Learning Using the Game of Checkers." In: IBM Journal of Research and Development 44, pp. 206–226.

Simon, Herbert (1969): The Sciences of the Artificial, Cambridge: The MIT Press.

Simon, Herbert/Newell, Alan (1958): "Heuristic Problem Solving: The Next Advance in Operations Research." In: Operations Research 6, pp. 1–10.

Sorensen, Roy (1992): Thought Experiments, New York: Oxford University Press.

Turing, Alan (1950): "Computing Machinery and Intelligence." In: Mind 59/236, pp. 433–460.

Valiant, Leslie (1984): "A Theory of the Learnable." In: Communications of the ACM 27/11, pp. 1134–1142.

Vapnik, Vladimir (1999): The Nature of Statistical Learning Theory, New York: Springer.

Vaswani, Ashish/Shazeer, Noam/Parmar, Niki/Uszkoreit, Jakob/Jones, Llion/Gomez, Aidan N./Kaiser, Lukasz/Polosukhin, Illia (2017): "Attention Is All You Need." (https://doi.org/10.48550/arXiv.1706.03762).

Weizenbaum, Joseph (1966): "ELIZA – A Computer Program for the Study of Natural Language Communication Between Man and Machine." In: Communications of the ACM 9, pp. 36–35.

Winograd, Terry (1970): Procedures as a Representation for Data in a Computer Program for Understanding Natural Language, MIT AI Technical Report 235, Cambridge: MIT.

AI and the work of patterns

Recognition technologies, classification, and security

Gabriele Schabacher

The connection between AI and patterns is so self-evident that addressing it might seem downright redundant. Nevertheless, I hope to make this connection a little less self-evident and to identify some aspects of what I will call the work of and on patterns in AI. While Kaufmann, Egbert and Leese (2019) limit the "politics of patterns" solely to questions of policing applications, I will understand the political dimension of patterns in a broader cultural-historical sense, asking for the politics specifically associated with the work of patterns. This means examining the power and agency of patterns and including contexts and discourses that at first glance seem far removed from current AI issues. In doing so, I will contribute to the question that interests this volume, how changes 'beyond quantity' occur in the context of artificial neural networks, that is, how ways of knowing are affected by AI technologies and vice versa.

I begin with the assumption that AI epistemologically finds itself in a middle, and an ambiguous, position in at least three ways. Firstly, from a disciplinary perspective, it is situated between the sciences of nature and the sciences of culture (Ganascia 2010: 71), rendering AI an intermediary realm between the two (ibid.: 68, with reference to Rickert 1926: 101). Secondly, its theoretical-methodological status oscillates between science and tool (cf. Russell/Norvig 2021), which makes it both an object of academic research and an agent in economic application contexts (product, service), thus generating a kind of 'scientific economic complex' that is accompanied by specific affordances. For while the supposed 'AI winter(s)' were related to the impossibility of adequately representing intelligence in machines in a rule-based way (symbolic AI), the current success owes much to the displacement of this question in favor of the broad applicability of AI technologies operating on the basis of machine learning and increased computational power (subsymbolic AI) (on the genealogy of

AI cf. Crevier 1993; Sudmann 2019). Thirdly, from the cultural and media studies approach of this paper, AI technologies are to be understood as media (in the broader sense).[1] This means to take them seriously in their role as mediators and to ask which inherent logics they go hand in hand with, which forms of knowledge and power they express, which genealogies they entail and how they transform social and societal relations and institutions (intimacy, education, health, security etc.).

The argument will proceed in four steps. The first two will focus on pattern formation and on pattern detection, respectively. I will here take a closer look at the role of patterns in general and explore their agency and effects: What exactly is the power of patterns in contexts of cognition or application, what exactly do patterns 'do' in this process, how does resorting to the notion of pattern inform processes of understanding? In doing so, I will (culturally and historically) distinguish between two forms of patterns, or more precisely two ways of conceptualizing them, namely *template* in the sense of 'stencil' (German: Schablone) on the one hand, and *correlation* (respectively *emergence*) on the other. In the further course, it will become apparent how these two forms are peculiarly intertwined in the horizon of AI technologies. Thus, the paper does not discuss a historical development or translation from template to correlation, but the specific layering of these two understandings in today's AI systems. In a third step, using the application domain of security research, I will look at what the concrete experimental settings and setups of activity recognition reveal about the status of patterns and show how the blending of template and correlation works out here. The focus will be on German pilot projects in Berlin and Mannheim that test the use of intelligent video analysis. And finally, I will comment on the statistical creativity of AI image generators such as DALL-E, highlight four overarching aspects associated with the work of patterns of AI technologies, and describe their effects on scientific understanding, but also on culture and society in general: These concern the connection between promised simplification and actual complication by AI technologies, the

---

1 Such a perspective assumes that not only communication media (mass media, social media), but also scientific instruments, technical apparatuses, means of transportation, infrastructural networks, and bodies can be understood as media insofar as they are instances of mediation and transmission. Evidence can already be found in the history of the term, according to which *medium* in classical Latin equally meant "middle", "intermediary", and "means" (OED 2023: medium; Seitter 2002: 19–32).

politics of rationalization and familiarization going along with them, their legitimization by scientific application contexts, and the invisibilization of their normative aspects.

1. Pattern formation

According to the German sociologist Armin Nassehi (2019), the success of digitization – and for him this implies the use of AI systems – is that it makes the regularities of societies visible again. Thus, in Nassehi's eyes, digitization does not produce anything radically new, but rather it represents a fundamental irritation for the self-understanding of 20th century modernity in terms of freedom and plurality: For it makes us aware of the extent to which types, regularities and categorizations are in operation (ibid.: 50–51), even if, as Andreas Reckwitz puts it, "the society of singularities" (2020) does not want to admit this. Although this article will not follow a systems theory approach, Nassehi's suggestion to understand what digital (and AI) technologies are doing as a kind of 'rediscovery' of patterns of order seems worth considering.

Figure 1: Haeckel's art forms of nature. Taken from Haeckel (1904: plate 84).

Patterns are central structures in the cultural history of mankind, because they are essential for the fact that something like cognition can take place at all. Patterns are regular structures, which are characterized by repetition (be it spatially or temporally) and (self-)similarity (cf. Stewart 2001: 28–37). Human perception as well as information theory mainly operate with patterns of medium entropy, that is with such structures that are neither mere noise nor completely identical. For patterns understood in this way, one could think of the simple organisms in the field of fauna and flora described by Ernst Haeckel as "art forms in nature"; it is the geometry of their basic shapes ("Grundformen", 1904: 9) that Haeckel emphasizes as aesthetic and as accessible to a morphological observation (also through techniques of microscopic magnification). The plates for the organisms in question are therefore always displayed twice: The diatoms, for example, are shown (fig. 1) once in a realistic fashion and once only as schematic outlines, which makes the patterned nature of the forms (symmetries, repetitions) even more obvious.

In a broader sense, familiar phenomena like waves, dunes or clouds also exhibit pattern formation. Here, one could also think of the fractals described by Benoit Mandelbrot (1982), which imply self-similarity in a recursive logic and consist of reduced copies of themselves. Thereby, he illustrates the differences between the different types of self-similarity (fig. 2).

Figure 2: Self-similarity, standard and fractal. Taken from Mandelbrot (1982: 44).

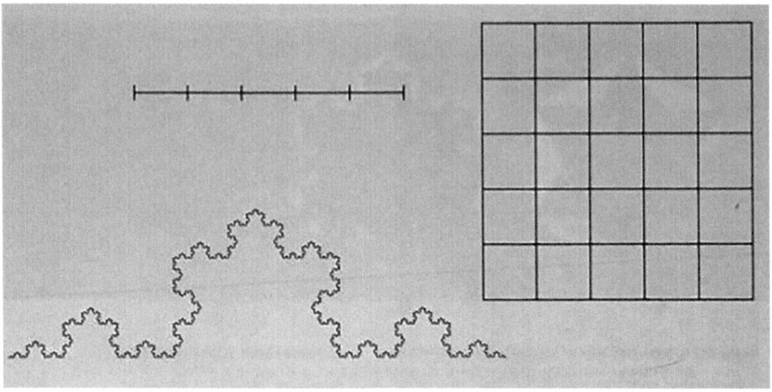

Figure 3: Tilings from Portugal, 15th century. Taken from Grünbaum/Shephard (2016 [1987]: 7).

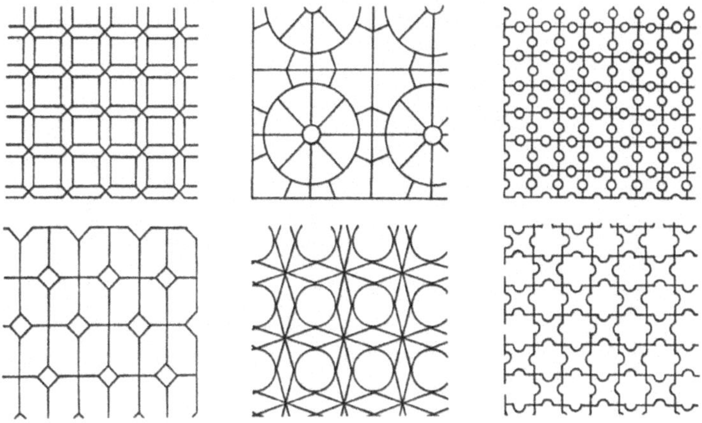

Patterns, however, do not only arise naturally, they can also be actively manufactured. One may think of tropes and figures in rhetoric, forms of parquetry and tiling (fig. 3), wallpaper, fabric and knitting patterns (fig. 4), but also of architectural ornaments (cf. Gombrich 1984) or patterns in music (in the sense of recurring rhythmic or harmonic structures).

Figure 4: Instructions for a baby cap from a Victorian knitting book, each row indicating the stitches to be knitted. Taken from Riego de la Branchardière (1848: 44f.).

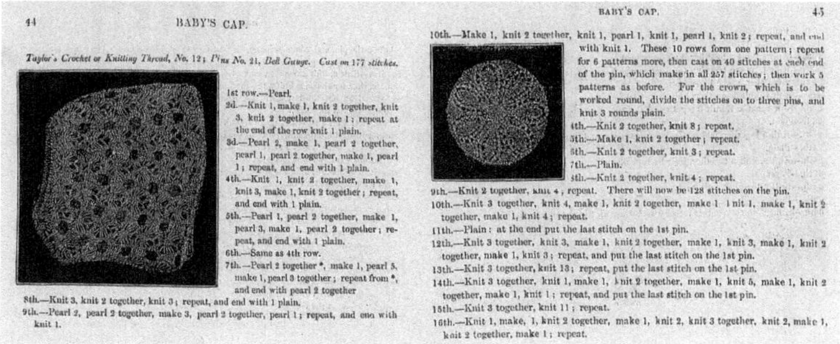

44 BABY'S CAP.

Taylor's Crochet or Knitting Thread, No. 12; Pins No. 21, Bell Gauge. Cast on 177 stitches.

1st row.—Pearl.
2d.—Knit 1, make 1, knit 2 together, knit 3, knit 2 together, make 1; repeat at the end of the row knit 1 plain.
3d.—Pearl 2, make 1, pearl 2 together, pearl 1, pearl 2 together, make 1, pearl 1; repeat, and end with 1 plain.
4th.—Knit 1, knit 2 together, make 1, knit 3, make 1, knit 2 together; repeat, and end with 1 plain.
5th.—Pearl 1, pearl 2 together, make 1, pearl 3, make 1, pearl 2 together; repeat, and end with 1 plain.
6th.—Same as 4th row.
7th.—Pearl 2 together *, make 1, pearl 5, make 1, pearl 3 together; repeat from *, and end with pearl 2 together
8th.—Knit 3, knit 2 together, knit 3; repeat, and end with 1 plain.
9th.—Pearl 2, pearl 2 together, make 3, pearl 2 together, pearl 1; repeat, and end with knit 1.

BABY'S CAP. 45

10th.—Make 1, knit 2 together, knit 1, pearl 1, knit 1, pearl 1, knit 2; repeat, and end with knit 1. These 10 rows form one pattern; repeat for 6 patterns more, then cast on 40 stitches at each end of the pin, which make in all 257 stitches; then work 5 patterns as before. For the crown, which is to be worked round, divide the stitches on to three pins, and knit 3 rounds plain.
4th.—Knit 2 together, knit 8; repeat.
5th.—Make 1, knit 2 together; repeat.
6th.—Knit 2 together, knit 3; repeat.
7th.—Plain.
8th.—Knit 2 together, knit 4; repeat.
9th.—Knit 2 together, knit 4; repeat. There will now be 128 stitches on the pin.
10th.—Knit 3 together, knit 4, make 1, knit 2 together, make 1 knit 1, make 1, knit 2 together, make 1, knit 4; repeat.
11th.—Plain: at the end put the last stitch on the 1st pin.
12th.—Knit 3 together, knit 3, make 1, knit 2 together, make 1, knit 3, make 1, knit 2 together, make 1, knit 3; repeat, and put the last stitch on the 1st pin.
13th.—Knit 3 together, knit 13; repeat, put the last stitch on the 1st pin.
14th.—Knit 3 together, knit 1, make 1, knit 2 together, make 1, knit 5, make 1, knit 2 together, make 1, knit 1; repeat, and put the last stitch on the 1st pin.
15th.—Knit 3 together, knit 11; repeat.
16th.—Knit 1, make, 1, knit 2 together, make 1, knit 2, knit 3 together, knit 2, make 1, knit 2 together, make 1; repeat.

Figure 5: Simple geometric forms. Taken from Day (1887: plate 3).

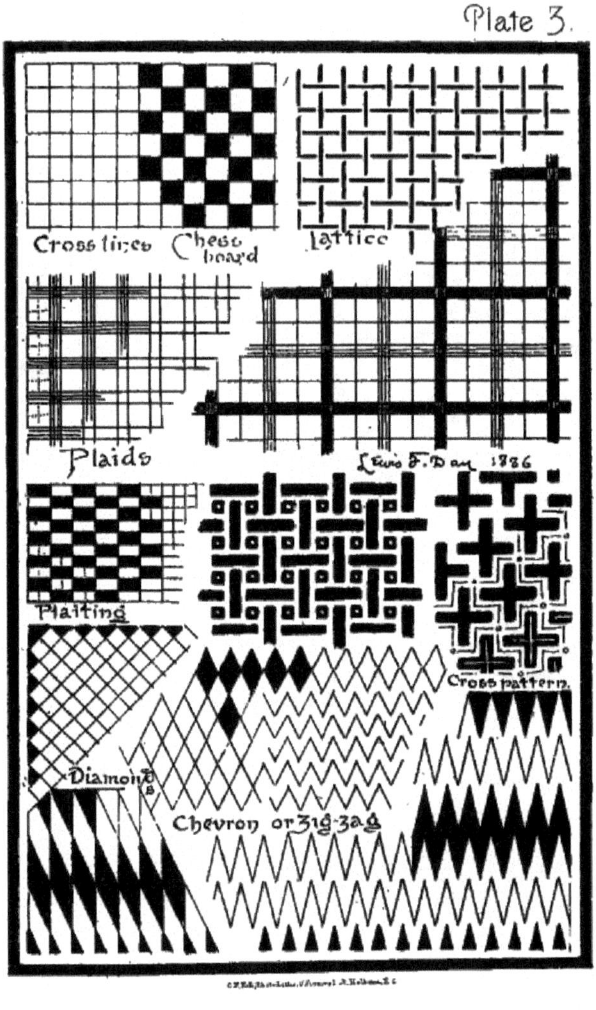

Figure 6: Jacquard loom, Musée d'art et d'industrie de Saint-Étienne, France. Photograph © Hélène Rival (2012).

Although patterns might at first be understood as primarily visual and spatial, reference to sounds, music, and speech reveals their equally acoustic and temporal dimensions. In all these cases, patterns can be generated because they are calculable due to their properties of iterability and regularity, and thus usable for diverse kinds of compositions. They can be written down and made available as instructions and plans. Variation is possible on the basis of simple, geometric forms and operations (fig. 5).[2]

2 If space fillings (tessalation) are designed in such a way that basic shapes are repeated symmetrically, one speaks of periodic patterns: they are created by mirroring, shifting

Errors, irritations and other forms of disturbance can be identified as deviations from the specified pattern. These man-made patterns are thus based on a form of programming, i.e., an operation in individual steps that takes the form of algorithmic processing. It is therefore no coincidence that the history of computing refers to mechanical weaving as a predecessor, and especially the programming of looms with punch cards since the late 18th century (cf. Schneider 2007) (fig. 6).[3]

The important role of regularity of patterns is seen early on. Lewis F. Day, a British artist of the Arts and Crafts movement, records in his little instruction booklet on ornamental design, *The Anatomy of Pattern* (1887):

> The very repetition of parts, then, produces pattern; so much so, that one may say wherever there is ordered repetition there is pattern. Take any form you please, and repeat it at regular intervals, and you have, whether you want it or not, a pattern, as surely as the recurrence of sounds will produce rhythm or cadence. (ibid.: 2)

For Day, the process of creating patterns is fundamentally accompanied by operations of differentiation, grouping, and classification, which refer to the fundamental regularity of the world:

> [A]nd just as the physiologist divides the animal world, according to anatomy, into families and classes, so the ornamentist is able to classify all pattern-work according to its structure. Like the scientist, he is able even to show the affinity between groups to all appearance dissimilar; and, indeed, to point out how few are the varieties of skeleton upon which all this variety of effect is framed. (ibid.: 3f.)

Patterns, thus, relate not only to regularity, but are tied to basic operations of classification.

---

and rotating geometric figures (rhombuses, triangles, quadrilaterals), which regularly fill surfaces as basic elements. If such symmetries are not provided, one speaks of aperiodic tiling patterns such as Penrose tiling (cf. Grünbaum/Shepard 2016 [1987]).

3 Lorraine Daston, in her book on rules, describes the transition from "rule-as-model" to "rule-as-algorithm" (2022: 21) as the consequence of the division of labor in the 19th century, which decomposed processes into calculable single steps and in this way made possible the transition from uncertain to controllable, fixed contexts (ibid.: 120f.).

2. Pattern detection

Corresponding to their regular formation, the recognition of patterns requires the ability to identify such similarities, regularities, repetitions or rules in given corpora of data. This insight is equally applied in cognitive science and perceptual psychology (Eysenck/Keane 2015; with reference to Gestalt psychology Koffka 1935: 106–177; Ehrenfels 1890), but also in computer science where the focus is on automating such recognition processes. The recognition of patterns here is considered to be the detection of feature complexes, which are (after a training phase) automatically assigned to certain categories. Pattern recognition is thus always accompanied by tasks of classification. However, it does not only concern the assignment of objects to already existing classes, but also the assignment of feature complexes to different classes, which are thus created in the first place. Even if today's computer scientists consider the term "input-output mapping" to be more accurate and prefer it to that of pattern recognition in order to avoid the comparison to biological systems and visual perception, Matteo Pasquinelli states: "Nonetheless, the construction of a relation between an input *x* and output *y* is still fundamentally the search for a pattern" (2019: 8; original emphasis).

This means that before a pattern can be recognized by information technology, it must be produced: In machine learning, an AI system has to first learn on the basis of training data what it is supposed to recognize at all. Since – analogous to natural neural networks – it is a matter of experiential learning, it is important which 'experiences' the AI makes.[4] The objects of these recognition processes (be they images, objects, activities) are nothing more than "statistical distributions of a pattern" (Pasquinelli 2019: 8). In principle, all forms of machine learning work with the three operations *training*, *classification*, and *prediction*, which are fundamentally related to patterns: The training phase concerns "*pattern abstraction*", the algorithm learns to associate an input with a certain output (for example, a label); classification can be understood as "*pattern recognition*" in the literal sense: new input data are compared with the learned statistical distribution in order to see if they fall within its range and have to be

4 Different types of machine learning can be distinguished (Russell/Norvig 2022: 670f.): The AI can learn by defining input/output pairs (*supervised learning*), by defining only the input and letting the neural networks come to results themselves (*unsupervised learning*), or by implementing a kind of self-optimization that works with reinforcing feedback (*reinforcement learning*).

assigned the corresponding output label; finally, prediction can be understood, as Pasquinelli puts it counterintuitively, as "*pattern generation*" (ibid.: 8f.; original emphasis). Here, new input data is used to "predict their output value *y*", that is, the statistical model "is run *backwards* to generate new patterns rather than recording them" (ibid.; original emphasis).

However, training data for AI systems is a "scarce resource" (Mühlhoff 2020: 1873), since their production is labor, time and computationally intensive and therefore causes high costs. For this reason, the same large benchmark data sets are used again and again, rendering them "the alphabet on which a *lingua franca* is based" that is used and expanded in the competition between the different companies for the best performance (Crawford 2021: 97; original emphasis) and which generates "[g]enealogies of data collections [...], each building on the last—and often importing the same peculiarities, issues, or omissions wholesale" (ibid.: 102). For example, the image data set ImageNet, published in 2009 with 14 million images and 20,000 categories, relies on taxonomies derived from the WordNet lexical database, which has been under development since the mid-1980s and dates back to the 1961 Brown Corpus (ibid.: 136). But although it would be crucial for classification systems and the political-social institutions that relate to and depend on them, to this day there are "no standardized practices to note where all this data came from or how it was acquired" (ibid.: 103). The history of science, however, recently turned to such questions of data re-use, asking, from both a theoretical as well as methodological perspective, what effects the mutability and mobility of data – their "data journeys" – have on the respective disciplines, the knowledge produced, and the politics associated with the data (Leonelli 2020). This is all the more relevant because economic factors play an important role. Part of the story of the production of ImageNet, for example, was that for the first time data labeling was outsourced to poorly paid crowdworkers on Amazon Mechanical Turk, from which significant errors in the data resulted, not least because of the immense time pressure (50 frames had to be labeled per minute). Training data is thus accompanied by various forms of bias, which is discussed as discriminatory AI or "discriminating data" (Chun 2021). Distortions can be found on at least three levels: Firstly, the implementation of already existing stereotypes in the AI systems (*world bias*), then the way the training data is produced (capturing, formatting, labeling) and, for example, whether it includes older (more conservative) taxonomies to save costs (*data bias*), and finally computational errors and "information compression" that make already existing inequalities even more unequal (*algorithmic bias*) (Pasquinelli 2019: 9f.). As we have seen with respect to the genealogies of data sets for training, it is especially data bias that is important here. In the field of biometric recognition, errors can be found with respect to feature extraction (cleansing, reduction, and incompleteness of data), but also with regard to the inaccuracy of annotations or the non-representative weight of groups (gender, race, class, age, origin, etc.) leading to multiple forms of discrimination (gender bias, racial bias, age bias etc.), when supposedly detected features are assigned to certain classes (cf. Boulamwini/Gebru 2018; Benjamin 2019; Nobel 2019).

AI systems trained on a sufficiently large data set with correctly labeled data should subsequently be able to correctly classify new data according to the learned pattern. They thus proceed inductively and generalize, starting from their training data. In supervised learning, two types of generalization errors can occur (*bias-variance tradeoff*) (Samutt/Webb 2011: 100): in one case, the system does not learn correctly, i.e., does not establish the correct relationships between input and output (*underfitting*); in the other, it is too sensitive to variations in the training data (*overfitting*). Thus, while in the first case it learns the 'wrong' patterns, in the second it cannot sufficiently distinguish between pattern and background (Pasquinelli 2019: 11). We all know such forms of irritation from the field of human perception, when faces are seen in things where there are none, in the well-known case of pareidolia. Pictures of the Hungarian-French artist Brassaï can illustrate the point in question: On his wanderings through nocturnal Paris around 1930, Brassaï photographed walls with indentations and holes (fig. 7) which all seem face-like, as they invoke the culturally and historically anchored schematicity of the face (dot, dot, comma, stroke) (Weigel 2017: 126; on faciality cf. Deleuze/Guattari 1987: 167–191). Overfitting to training data generates similar effects, which is why such phenomena are also described as "data paranoia" (Apprich 2018) or apophenia (Steyerl 2018).

But if all these irritations, errors, and deviations occur in the context of machine learning, why do the fundamental problems of classification and taxonomy in and by AI systems receive such little attention? According to Kate Crawford, "the issue of bias in artificial intelligence has drawn us away from assessing the core practices of classification in AI, along with their attendant politics" (Crawford 2021: 128). The companies concerned see forms of bias as a purely technical problem – "a bug to be fixed" (ibid.: 130) – rather than a call to debate "*why* these forms of bias and discrimination frequently recur and whether more fundamental problems are at work than simply an inadequate underlying dataset or a poorly designed algorithm" (ibid.: 129; original emphasis). However, this leads to a self-reinforcing logic that confirms the supposed neutrality of the technical (cf. ibid.: 131), thus normalizing the underlying worldviews and classifications: "[T]raining datasets pass as purely technical, whereas in fact they contain political interventions within their taxonomies: they naturalize a particular ordering of the world which produces effects that are seen to justify their original ordering" (ibid.: 139).

Figure 7: Brassaï, Series "La Naissance du Visage" and "Masques et visages" (around 1930). © Estate Brassaï 2023.

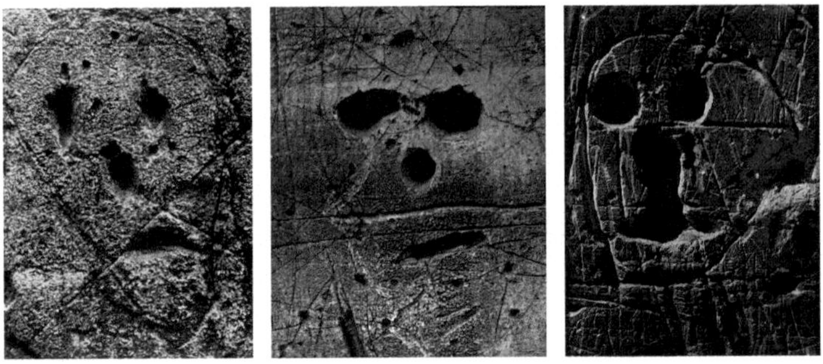

What, then, do patterns from the fields of cultural and natural history have to do with those of computer science? What do we gain by relating them to each other? I argue that at least three aspects can be highlighted concerning the use of patterns in the context of AI technologies: the crossing of two different forms of patterns (template and correlation), the visibility, and respectively invisibility, of patterns, and the temporal dimension of patterns.

Regarding the first aspect, there is a crossing of two models of patterns, which I heuristically call the model of the *stencil* (template) (German: Schablone) on the one hand and the model of *correlation* or emergence on the other. In his *Oekonomische Enzyklopädie* (1805), Johann Georg Krünitz had already distinguished between different kinds of patterns (German: Muster): the model or prototype (German: Vorbild), in the physical as well as moral sense, then the sample of goods, the sample piece, and finally the pattern in the sense of a figure (for instance in gardening), referring to the dimension of showing and making see (*monstrare*) (ibid.: 219f.; cf. OED online 2023b). I argue that in the dimension of the prototype (in the broadest) sense as well as in the sample piece, a normative dimension of the pattern in the sense of the stencil is revealed, while the showing of a figure emphasizes the emergent dimension of the pattern. Now, in AI systems, I further argue, a blending of these two understandings of pattern occurs. For AI systems work with predefined patterns (such as annotated features and categories in training data), which they are not only supposed to recognize in use cases, but to enrich with further data to generate new patterns when it comes to prediction or, for example, image generation. On the basis of first-order patterns (stencils in the sense of templates), second-order patterns (correlations) are generated.

Secondly, such higher-order patterns – I refer once again to Armin Nassehi's argument – make something visible that modern society does not want to know about itself, namely how typifiable, classifiable and regular it is. In contrast, then, to the culturally and historically familiar visible surface divisions in the realm of mosaics, knitted fabrics, or wallpaper, we are dealing here with phenomenally invisible patterns, which, analogously to statistical surveys, only become visible with the use of mass data. Thus, patterns are centrally concerned with the question of their visibility and invisibility. Already in the case of the single-celled organisms analyzed by Haeckel, visibility was not immediately given, but had to be established first, for example by microscopic magnification. And in the case of parquetry, it has always been a matter of calculability. In the case of AI systems, the paradoxical situation arises that the comprehension of the calculation is not made available, so that the patterns *appear* as pure emergences.[5]

Thirdly, the patterns produced by AI technologies are accompanied by a shift in the temporal vector of cognition. The goal is not *re*-cognition alone (as it was in cultural and natural history), but rather *pre*-cognition. On the basis of a principal calculability of all conceivable correlations between myriads of categories, AI systems model expectations of consumption, behavior, but also of security. In the following I will briefly discuss the security domain, because AI systems are on the agenda here less to contain current problems than to preemptively deal with future ones. On the basis of stenciled training data, AI

5 In fact, there is no such thing as 'pure' emergence. As Boris Groys has shown for the field of art, the new is always based on re-combinations of what already exists, revaluations of contexts, and new comparisons (1992, 2000). In contrast to previous forms of the 'new', AI technologies can recombine any number of elements in any number of subtle ways without making any mistakes or forgetting anything; they therefore no longer have the possibility of negation, intervention, or deviation.

systems do not only anticipate the future, but also shape it quasi-automatically through the policies that accompany them.

3. Security in crowded settings

In the aftermath of 9/11, security issues have become a preferred domain for the application of AI systems, coupling forms of visual surveillance with the control of data flows (dataveillance) and giving rise to systems of intelligent video surveillance (Stanley 2019; in broader perspective Andrejevic 2020). The focus is on biometric recognition systems (face, iris, gait, etc.) as well as on object and activity recognition, which are believed to have decisive gatekeeping functions for regulating traffic flows and correlating security regimes. In the following, I would like to refer to image recognition methods that are used in video analytics, in which images of surveillance video feeds are automatically analyzed. In particular, I concentrate on two German pilot projects in Berlin and Mannheim that are testing such AI systems in public spaces, that is at a train station and in further urban areas. At first glance, train stations seem to be much less security-critical settings than, for example, border regimes of states. Nevertheless, such constellations are a good illustration of the "'becoming environmental' of surveillance" by today's AI systems (Andrejevic 2020: 84). The focus on Germany is interesting against this background, since the stricter data protection laws allow the use of AI technologies in public spaces only in test constellations (cf. Schabacher/Spallinger forthcoming), which on the one hand makes the conditions and problems of their use comparatively explicit and on the other hand represents a strategy of familiarization with these systems. However, even though facial recognition is highly controversial, also in Germany the Covid pandemic has driven general datafication and normalized facial recognition technologies as systems of automated identification "at a distance" (Andrejevic 2021: 150). The respective AI companies see this as a gateway to generalized data networking, operating simultaneously at individual and biopolitical levels in the sense of "granular biopower" (ibid.: 153), thus exhibiting the basic promise of AI systems: "to modulate the milieu at the level of the individual" (ibid.: 152f.).[6]

---

6 Louise Amoore also emphasizes this biopolitical dimension when she analyzes the increasing datafication of border regimes – "biometric borders" – that make the pris-

With regard to patterns and AI technologies, I would like to discuss two questions in particular: What can be understood as security patterns in the first place and what problems are encountered in their conceptualization and implementation? Following test phases on facial recognition, in 2019, a pilot test on situation and behavior recognition systems was carried out at the train station Berlin Südkreuz for several months (for further details on both tests, cf. Schabacher 2021; forthcoming). The tests took place as a collaboration of the Federal Ministry of the Interior, the Federal Police, the Federal Criminal Police Office, and Deutsche Bahn AG, which labelled Berlin Südkreuz a "security station" ("Sicherheitsbahnhof") (Federal Ministry of the Interior and Community 2017) and announced the testing of intelligent surveillance systems. While the first test in 2017 and 2018 was concerned with the identification of individuals, which was accompanied by a great deal of public interest and triggered many critical debates, the public perceived the second test as supposedly less critical. This was because it did not use facial recognition, but aimed at detecting dangerous situations under anonymity conditions. In order to generate training material for the AI systems, corresponding scenarios were performed by actors and recorded (including demarcation scenarios) on several days of the week on site at the station. The trained scenarios referred to four predefined patterns ("lying (helpless) person", "entering defined areas", "flows or gatherings of people", "abandoned object"), the possibility of "counting people" as well as two additional functions, namely the tracking of persons or objects as well as the "retrograde evaluation" of video material (Federal Police 2019).

Already the selection and naming of the dangerous situations show the operation of stenciling in the sense of first-order patterns. Certain types of movements are, in a sense, cut out and set apart from the background and are thus made relevant in terms of security compared to supposedly normal situations at the station. They concern the posture of people – compared to standing and walking, lying down represents a deviation, which can refer both to a person who has had an accident and to a person without shelter. They concern the position of individuals in relation to a zoning of space: certain areas such as tracks should not be entered. They refer to the speed of movement of groups — a rapid gathering or dispersal of people is understood as being caused by dangerous events. And the single object relates to the fact that at stations objects occur in close proximity to people (pieces of luggage, dogs, children), but

oner's body "the bearer of the border, as it is inscribed with multiple coded boundaries of access" (Amoore 2006: 347f.).

can become dangerous if separated (imago of the bomb case). However, insofar as these patterns are imaginaries of danger that are regularly and extensively played out in popular culture (cf. Horn 2018; Koch/Nanz/Pause 2018), there can be no question of requiring AI technologies to recognize them here. It can be assumed that it is not only a matter of recognizing such first-order patterns of danger, but that, with reference to the two police functions mentioned above (tracking and retrograde evaluation), second-order patterns are expected here, too, i.e., correlations that only a mass analysis of material collected in this way can produce.

However, it is not only what figures as security patterns in the first place that points to the gap between the ambition and reality of AI systems; the production of these patterns also proves to be difficult. As with facial recognition technologies, the hope with AI systems of situation and behaviour recognition was also that they wouldn't only provide support for the work of station control, but would also help to reduce personnel in the security sector. The reality, however, was different. The introduction of such systems always makes the respective settings more complex overall: For instead of simplifying things, AI systems intermingle with other actors – people who manage and repair them, the technical and physical building infrastructure on site, software companies, institutional regulations and legal requirements, the people they are supposed to detect. They can therefore never represent simple systems of control, but must in turn be elaborately controlled, regulated, and monitored, which raises new questions and problems. The complexity they seek to reduce is thus continually increased by the AI systems in question. Even their purely technical functioning requires a high degree of customization and adaptation: the creation of training data (through the invention of 'scenes' and production of own video material), the calibration of the systems (for example, due to the changing light conditions in real space), the preparation, extraction and classification of features, the manual removal of random and systematic errors. Furthermore, an appropriate evaluation and analysis as well as reports are necessary, but also accompanying public discourses that prove such a test to be a success. This is not provided in the case of the second Südkreuz test, the (poor) results of which have not been made public so far;[7] instead, the project has been extended (Federal Police/Deutsche Bahn 2020).

Figure 8: Processing crowded public spaces. Taken from Golda et al. (2019: 1).

However, problems with the conceptualization and implementation of patterns can be observed at a more fundamental level. Motion and activity recognition is one of the strongly discussed fields in computer vision. Public places such as train stations pose particular difficulties, since they are crowded and present unstructured everyday situations with many people (fig. 8), which makes the recognition process complex and computationally intensive due to lighting conditions, multiple occlusions as well as rapid movement of many people. A pilot project in Mannheim is working with the Fraunhofer Institute of Optronics, System Technologies and Image Exploitation (IOSB) to develop a software based on artificial neural networks specifically for police situation assessment in such environments (cf. Golda/Cormier/Beyerer 2022).[8] In this

7 Even though the final report of the second test was not publicly available, it was available to the Federal Data Protection Commissioner who concluded that the results did not justify further "similarly elaborate tests" but that security should be increased by "other measures" (BfDI 2022: 74).

8 The "Mannheim Way" project (2018–2023), a cooperation between the City of Mannheim, Mannheim Police Headquarters, the responsible Ministry of the Interior, Digitalization and Migration Baden-Württemberg, and the IOSB, is testing video

process, the image data originating from the live feed of a static surveillance camera is detected for recognizable persons (fig. 9); to protect privacy and to avoid other forms of bias, these are converted into skeletal representations, which are then filtered for anomalies. These anomalies are classified in relation to defined activities (such as hitting or kicking). If there is an accumulation of such 'critical' activities within a certain period of time, a warning is generated. Otherwise, the data is deleted after a defined time (e.g., one minute).

Figure 9: Human Pose Estimation. Edited version of image taken from Golda/Cormier/Beyerer (2022: 1493).

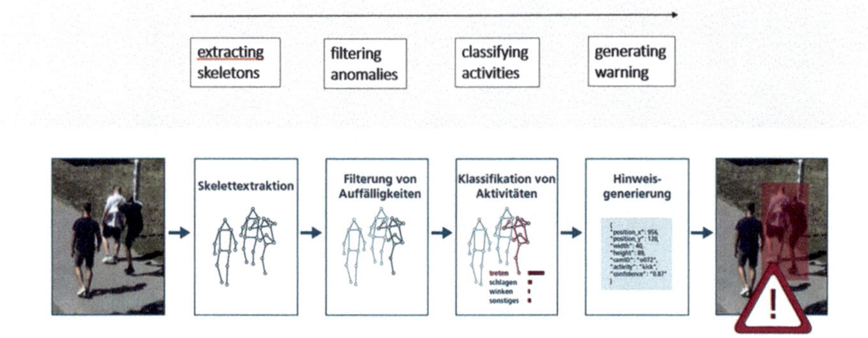

I would like to concentrate on one aspect here, namely the human pose estimation. Interestingly, this is a workaround made necessary by German data protection regulations: Namely, human pose estimation bypasses the process of identifying individuals in favor of anonymity conditions and the analysis of group-related behavioral constellations. This method is based on identifying and classifying joints of the human body. For this purpose, every joint (elbow, head, torso, etc.), also called "key point", assumed to describe the posture of a person is captured from a given video input in order to obtain a skeletal representation of the human body (Golda/Cormier/Beyerer 2022: 1494f.). A normal surveillance camera takes a new image every 33 milliseconds, the processing of which requires a correspondingly large amount of computing time, depending on the number of poses to be detected (ibid.: 1491). Especially in crowded

surveillance in urban areas (https://www.iosb.fraunhofer.de/en/projects-and-products/intelligent-video-surveillance.html) (31.03.23).

scenarios, such as a metropolitan train station (fig. 10), the bounding boxes in the background quickly become confused. Therefore, the live operation of such AI systems represents a compromise between the accuracy and the speed of detection, as in a real-world scenario "the surveillance footage requires immediate processing in order to provide human assistance on-site in a matter of minutes" (Cormier et al. 2022: 591).

Figure 10: Increasing occlusions in the background. Taken from Cormier et al. (2022: 597).

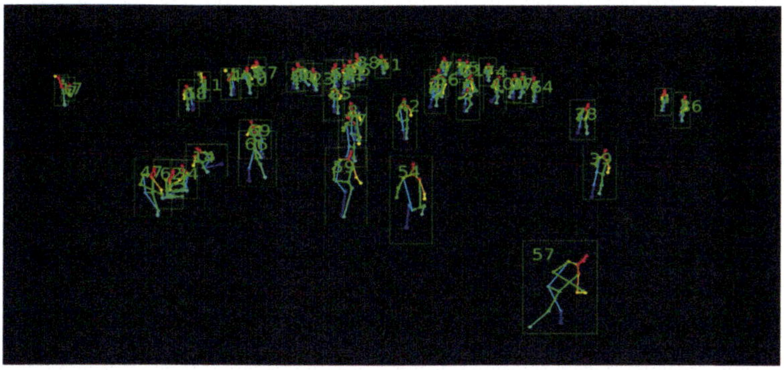

Such a compromise is of course understandable under real-world conditions, since an alarm system should be able to react quickly. However, it already pertains to the production of the training data. The quality of activity recognition or the respective pose estimation depends here, amongst other things, on the number of "keypoints" (body joints). So time is a significant factor here, as "annotating a single human body pose for activity recognition requires 40–60 seconds in complex sequences" (Cormier et al. 2021: 1649). That means, the higher the number of body joints, the more accurate the representation, but the longer the overall computing time. The crowdedness of real-world conditions require even further trade-offs: To build a data set for crowded scenarios with many people, one resorts to automated annotation (Cormier 2021: 36f.) and "data augmentation methods" in order to supplement hidden limbs by "synthetic training data" (Golda et al. 2019: 2), which reduces the authenticity of the data structures in favor of completing poses. AI systems are therefore already used for the production of training data. Of course, this specific constellation depends on the current state of implementation of AI systems; insofar as technological development is progressing, it can be assumed that the problem in question will be solved, for example, by an increase in computing power. What matters to me, however, is the recursive logic in operation here: AI systems produce certain problems, difficulties, and affordances (such as the time-consuming annotation of motion data), the solution of which again requires the use of AI systems. In this small example, we see patterns in several ways: We see them already in the concept of the pose as such, that refers to the position of a body with respect to its position and orientation in space. It is not without reason that Roland Barthes, in his analysis of photography, refers to the meaning of the pose as a form of still time, as "immobility" and pausing (Barthes 1981: 78). We also see patterns as related to the skeletal representation, which is already a "heavy abstraction" (Cormier et al. 2022: 591) compared to the original images; we see them in the classifications upon which activities are filtered (kicking and punching as specific crime patterns); and we see them related to what is called the 'overall picture' of the police situation, when activity recognition is merged with other data. However, what machines achieve here is precisely not a phenomenal Gestalt perception in the sense of Christian von Ehrenfels (1890), but an act of ultimate (binary) classification (for example, do we deal with a conspicuous activity or not?), based on statistical threshold values.

In security contexts, such classifications are in the service of forecasting and are intended to legitimize preventive action. In predictive policing, for example, a software like PRECOBS (Pre Crime Observation System), which was also tested in some German police stations, uses offense data from the recent past in order for police authorities to make predictions in which area repeat crimes (mainly residential burglaries) are most likely to occur within the next 72 hours (cf. Egbert/Leese 2021; Perry et al 2013; Ferguson 2017).[9] What I would like to emphasize in relation to predictive policing is the conservative and normative dimension of patterns, which I have elsewhere called the "*temporal vector* of patterns" (Schabacher forthcoming: 160; original emphasis). According to Mareile Kaufmann, Simon Egbert and Matthias Leese, the respective programs reinforce the "epistemological authority" of patterns in policing (2019: 684), however developing different "styles" of pattern identification (ibid.: 680) with own rationalities and conceptions of crime that in turn inform police work: "They [patterns, G.S.] give form to and formalize different

9 Website of LogObject Deutschland Gmbh 2021: https://logobject.com/en/solutions/precobs-predictive-policing/ (accessed March 31, 2023).

understandings about crime, which are in turn based on specific ideas of governing crime. This makes patterns political" (ibid.: 684). Four implications of this pattern politics are highlighted by the authors: Firstly, patterns emerge only where regularities exist: "Patterns can only capture offenses that follow rules" (ibid.). Thus, a solitary crime cannot occur. Secondly, the future is not extrapolated from live data, but from past data; patterns are therefore "conservative" (ibid.: 685). Thirdly, they exhibit a "self-reinforcing logic" (ibid.: 687), because assumptions about crime patterns feed back into policing cultures by establishing a direct link between assumed pattern and the efficiency of respective police action. Finally, patterns alter the general relationship between crime and norm: From this perspective, criminal behavior must be regular, otherwise it could not be captured by patterns (ibid.). Patterns thus normalize crime without stimulating reflection on motives and causes (ibid.), which is why predictive policing has discriminatory and stigmatizing effects.

The considerations of Kaufmann, Egbert and Leese highlight the political-cultural implications of reinforcing and habituating (crime) patterns based on regularity and repetition. Together with the described German tests on AI-based surveillance systems in public spaces, it can be shown how the regularity of patterns informs practices of predicting, for example, dangerous situations. Thus, it is first-order patterns that inform AI systems at the level of training data: This is equally true for the fact that, in the context of human pose estimation, rapid striking and standing close together are understood as expressions of aggression and fight, that a lying person or an unaccompanied suitcase at a train station are perceived as deviant, or that burglaries are more likely to be expected where burglaries have already occurred. Although in all three cases the patterns of danger are easily recognizable even without AI systems – they are culturally familiar 'templates' – this is rarely highlighted in the context of AI systems.

4. The work of patterns

What is at stake in the question of predictive analytics is made clear by Rainer Mühlhoff in his plea for "prediction privacy" (2023). Mühlhoff (2023: 3) refers to the regulation of the currently unregulated possibilities of economic actors to match data of individuals with anonymous mass data in such a way that individual predictions become possible. One thus encounters the becoming-environmental of surveillance discussed at the beginning of the previous section in a more generalized form: The prediction concerns information on all conceivable human categories (gender, ethnicity, purchasing behavior, age, health, sexual orientation, etc.) (cf. Hirschauer 2021) and is generated on the basis of individual data of the person concerned (usage, tracking, or activity data) in comparison to anonymous mass data. By operating in this way, AI systems do not foresee a future that exists independently of them, but rather they modulate the future according to their own specifications: "Algorithms 'manufacture' with their operations the future they anticipate" (Esposito 2022: 11). In doing so, AI systems are having a self-fulfilling performativity (Rona-Tas 2020; on the logic of prevention see Bröckling 2012), they structure and govern our (image of the) future and are thus "world-shaping instruments" (Lazzaro/Rizzi 2022: 16).

Accordingly, the phenomena discussed with respect to security research and predictive policing can also be formulated in more abstract terms: It is not only about the hope of preemption generally associated with AI systems, i.e., the proactive action that anticipates the future and prevents an undesirable outcome of things. It is also about a serious problem that all such automated detection systems have and that is intimately related to their pattern-based nature. For due to their training being based solely on things that already exist, predictions can only vary them virtuously; ultimately, however, a crime warning remains directed at such events that one already knows in principle, just as a purchase recommendation is oriented toward transactions that have already been made. This orientation towards the past can also be related to Wendy Chun's critique of the "homophily" of network research. For, as Chun argues, network analyses follow the paradigm of similarity; this has the effect of weakening the importance of difference in favor of self-similarity, which reinforces the segregation of societies (2021: 81–137). However, insofar as homophily itself is to be understood as a form of patterning in which the repetition of similarity is rewarded (not least because this is easier to calculate than difference), the kind of reference to the past that also played a role in crime predictions is evident here: "Because [...] predictions rely exclusively on past regularities, the future made present in the here and now is impoverished and reduced to a mere repetition of the possible, of what has already happened at least once" (Lazaro/Rizzi 2022: 13).

But how does this apply to the supposed creativity of AI systems? As we came to know large language models such as ChatGPT or DALL-E in the past year, AI systems are quite capable of producing aesthetic content. True to its name, a portmanteau of Dali and the Pixar garbage robot WALL·E, DALL-E creates images based on prompts. The underlying artificial neural network operates on billions of parameters trained with text-image pairs from the Internet so that it can convert text into pixels. Such a "'prompt design'" and the associated "'promptology'" have two sides (Bajohr 2022: 67): Although the AI remains without consciousness and thus "dumb" (ibid.: 66), what it produces is no longer mere syntax, but rather "dumb meaning". Thus, interaction with these systems becomes a feedback loop between artificial and human meaning: "Not only does the machine learn to correlate the semantics of words with those of the images we have given it, but we learn to anticipate the limitations of the system in our interaction with it" (ibid.: 67). In particular, styles are patterns that can be readily addressed by prompts. Thus, to ask DALL-E to 'paint' the WALL·E robot in a Paul Klee style leads to quite appropriate results, just as with the corresponding request to Yves Klein or Barbara Kruger, the typical blue or large-scale text-photo combinations are generated (fig. 11).

Figure 11: Pictures generated by DALL-E 2. Courtesy of the author.

Seen in this way, AI's repertoire is quite broad. As Roland Meyer correctly observes, in this context style is no longer a historical category; instead, styles are "typical visual patterns extracted from a latent space of possible images accessed through generative (and often iterative) search queries" (Meyer 2022: 107), i.e., monetizable "images about images, filtered through language" (ibid.: 108). Still, beyond the legitimate question of copyright violations, forgeries, and data hallucinations by such systems,[10] which, despite their training (reinforcement learning), nonetheless also produce much fake knowledge in flawless prose, the more important point is: The produced artifacts are "*statistical*

10 See, for example, the pending lawsuit on consumer harm caused by the practices of Open AI and Microsoft (Kang/Metz 2023).

art" (Pasquinelli 2019: 15; original emphasis). AI systems such as DALL-E or ChatGPT produce only the most probable, that is the statistically reasonable, answers on the basis of their (large amount of) training data.

AI systems are thus characterized by a limit that Matteo Pasquinelli very aptly calls "*undetection of the new*" (ibid.; original emphasis). At the core of machine learning lies the "inability to predict and recognise a new *unique anomaly*" (ibid.: 14; original emphasis), because every anomaly, even a social or political one, would be the creation of "a new code or rule" (ibid.: 16). And that is precisely what AI systems cannot do. Rather, they represent a constantly further "standardized world", which is why their most decisive effect on society consists in a "social *normalisation*" (ibid.: 17; original emphasis). Are the patterns of AI therefore de facto stereotypic stencils? In a certain way they are. They are, because AI is not able to produce negation, lack, or workaround. There is no place for surprise in the sense of revaluations. And they are, too, because there is never anything like 'the new': Every invention is a re-combination of existing entities or concepts.

5. Conclusion

To conclude, I would like to highlight four aspects that seem significant to me for the pattern regimes associated with AI technologies in general.

Firstly, there is the vision of simplification. With new technologies such as AI systems, environments become more complex because the entanglement of different actors becomes denser and less manageable. Promises of a fundamental simplification of communication, work or control through automated systems are thus de facto accompanied by the constant complication of the concrete constellations. In an ethnographic analysis of a township in eastern South Africa, Thomas Kirsch (2019) very convincingly shows how the introduction of security technologies leads to a recursive securitization – "*security needs to be secured*" (ibid.: 124; original emphasis). To continue, one can easily add such recursive structures for other contexts: security also requires maintenance, it requires energy, it requires trust etc. As AI systems are embedded within socio-technical-*discursive* infrastructures, they will never represent technical effects and solutions alone, but will concern the respective structures as a whole. Therefore, they can be seen as mediators of knowledge, of societal relations as well as of cultural and aesthetic perspectives.

Secondly, pattern recognition systems are part of a politics of rationalization and convenience. On the one hand, an important reason for the enforcement of AI systems is the optimization of operational processes and human resources through automation. For example, ChatGPT is claimed to be able to relieve clinical staff of burdensome documentation duties through its auto-completion capabilities. The consumer sector on the other hand focuses on convenience, allowing passengers to pass through gates without contact, to pay more quickly with face recognition payment systems (FayPace, Paybyface), and to complete homework with less effort due to using a large language model. From a cultural and media critical perspective, it should not only be noted here that it is questionable what the greater efficiency or freed-up time can be used for. In view of the capitalist logic driving these changes, it is to be assumed that no 'free space to do others things' will emerge, but only that new areas will become calculable for economic value creation. Rather, it must be emphasized that their application close to everyday life (smart home, smart driving, etc.) will lead to a familiarization with AI technologies that will make their presence fade into the background of functioning infrastructure, making them more and more invisible.

Thirdly, there is the scientific legitimation respectively authentication of AI technologies. Within the scientific-economic complex, the scientific use of AI technologies, for example in the medical field or in biology, legitimizes, ennobles even, their use in incomparably more critical areas such as security. Science thus contributes to the social acceptance of AI, without it having to be covered in detail by the findings obtained by AI systems. And although proprietary AI systems have a black-boxed status, interestingly enough, this lack of transparency that is based on corporate policies seems to even increase rather than decrease the public belief in capabilities of these systems. The approach of Explainable AI, which aims at elaborating methods to make the functioning of artificial neural networks (more) transparent, must be seen as an attempt to deal with this problem (cf. Samek et al. 2019); this is equally true for efforts to clarify the different uses and horizons of crucial terms (e.g., 'autonomy' or 'agents') (Powers/Ganascia 2020), the emphasis on the need for decidedly political action (McQuillan 2019), and the call to pay more attention to the production of large data sets as well as to the movement of data through the sciences (Leonelli 2020).

Finally, the pattern reference of AI technologies, that is, their recourse to stencils (templates) in the sense of feature spaces and classification schemes, can be seen as their normative dimension, which is discursively invisibilized.

In that AI technologies constantly find new correlations seemingly 'on their own' (what I have called second-order patterns), the operations of stenciling, gridding, and classification that are at the outset and indispensable for training AIs remain strangely unobserved. This is why correlation can so easily be mistaken for causation, even though we know it is not the same thing (cf. Pasquinelli 2019: 14), and why differences of degree can be interpreted as differences of kind (cf. Mackenzie 2017: 149). This blending of two understandings of patterns – the repetitive stencil and the statistical correlation, the ornamental and the numerical patterns – is what I take to be representative of the politics of patterns of AI technologies.

List of references

Amoore, Louise (2006): "Biometric Borders: Governing Mobilities in the War on Terror." In: Political Geography 25, pp. 336–351.

Andrejevic, Mark (2020): Automated Media, New York and London: Routledge.

Andrejevic, Mark (2021): "Granular Biopower: Touchlessness, Mass-Recognition and Milieu Modulation in Pandemic Times." In: Gay Hawkins/Ned Rossiter (eds.), Contagion Design: Labour, Economy, Habit, Data, London: Open University Press, pp. 148–160.

Apprich, Clemens (2018): "Data Paranoia: How to Make Sense of Pattern Discrimination." In: Clemens Apprich/Wendy Hui Kyong Chun/Florian Cramer/Hito Steyerl (auths.): Pattern Discrimination, Minneapolis and London: The University of Minnesota Press, and Lüneburg: meson press, pp. 99–123.

Bajohr, Hannes (2023): "Dumb Meaning: Machine Learning and Artificial Semantics." In: IMAGE. The Interdisciplinary Journal of Image Sciences 37/1, pp. 58–70.

Barthes, Roland (1981): Camera Lucida. Reflections on Photography, New York: Hill & Wang.

Benjamin, Ruha (2019): Race After Technology: Abolitionist Tools for the New Jim Code, Cambridge and Melford: Polity.

BfDI (Der Bundesbeauftragte für den Datenschutz und die Informationsfreiheit) (2022): Tätigkeitsbericht 2021. 30. Tätigkeitsbericht für den Datenschutz und die Informationsfreiheit, Bonn: BfDI.

Bröckling, Ulrich (2012): "Dispositive der Vorbeugung: Gefahrenabwehr, Resilienz, Precaution." In: Christopher Daase/Philipp Offermann/Valentin

Rauer (eds.): Sicherheitskultur. Soziale und politische Praktiken der Gefahrenabwehr, Frankfurt a. M.: Campus, pp. 93–108.
Buolamwini, Joy/Gebru, Timnit (2018): "Gender Shades: Intersectional Accuracy Disparities in Commercial Gender Classification." In: Proceedings of Machine Learning Research 81, pp. 1–15.
Chun, Wendy Hui Kyong (2021): Discriminating Data: Correlation, Neighborhoods, and the New Politics of Recognition, Cambridge and London: The MIT Press.
Cormier, Mickael (2021): "A Data Annotation Process for Human Activity Recognition in Public Places." In: Jürgen Beyerer/Tim Zander (eds.), Proceedings of the 2020 Joint Workshop of Fraunhofer IOSB and Institute for Anthropomatics, Vision and Fusion Laboratory, Karlsruhe, Germany, pp. 33–51.
Cormier, Mickael/Clepel, Aris/Specker, Andreas/Beyerer, Jürgen (2022): "Where are we with Human Pose Estimation in Real-World Surveillance?" In: IEEE/CVF Winter Conference on Applications of Computer Vision Workshops (WACVW), Waikoloa, HI, USA, pp. 591–601.
Cormier, Mickael/Röpkel, Fabian/Golda, Thomas/Beyerer, Jürgen (2021): "Interactive Labeling for Human Pose Estimation in Surveillance Videos." In: Proceedings of the IEEE/CVF International Conference on Computer Vision (ICCV) Workshops, Online, pp. 1649–1658.
Crawford, Kate (2021): Atlas of AI: Power, Politics, and the Planetary Costs of Artificial Intelligence, New Haven, CT and London: Yale University Press.
Crevier, Daniel (1993): AI: The Tumultuous History of the Search for Artificial Intelligence, New York: BasicBooks.
Daston, Lorraine (2022): Rules. A Short History of What We Live By, Princeton and Oxford: Princeton University Press.
Day, Lewis F. (1887): The Anatomy of Pattern, London: Batsford.
Deleuze, Gilles/Guattari, Félix (1987): A Thousand Plateaus. Capitalism and Schizophrenia, Minneapolis and London: University of Minnesota Press.
Egbert, Simon/Leese, Matthias (2021): Criminal Futures. Predictive Policing and Everyday Police Work, London and New York: Routledge.
Ehrenfels, Christian von (1890): "Über 'Gestaltqualitäten'." In: Vierteljahrsschrift für wissenschaftliche Philosophie 14, pp. 249–292.
Esposito, Elena (2022): "The Future of Prediction: From Statistical Uncertainty to Algorithmic Forecasts." In: Artificial Communication (https://artificialcommunication.mitpress.mit.edu/pub/m8xpxiru).

Eysenck, Michael W./Keane, Mark T. (2003): Cognitive Psychology: A Student's Handbook, 4th ed., Hove, Philadelphia, and New York: Taylor & Francis.

Federal Ministry of the Interior and Community (2017): "Projekt Sicherheitsbahnhof Berlin Südkreuz geht an den Start.", April 11, 2017 (https://www.bmi.bund.de/SharedDocs/pressemitteilungen/DE/2017/04/sicherheitsbahnhof.html).

Federal Police (2019): "Test intelligenter Videoanalyse-Technik.", June 7, 2019 (http://www.bundespolizei.de/Web/DE/04Aktuelles/01Meldungen/2019/06/190607_videoanalyse.html). Since this press release is no longer available, please refer to the following for a reprint of the text, Tempelhof-Schöneberg Zeitung (2019): "Bahnhof Südkreuz wird Testfeld.", June 9, 2019 (http://www.tempelhof-schoeneberg-zeitung.de/bahnhof-suedkreuz-wird-testfeld-fuer-videoanalyse-technik/).

Federal Police/Deutsche Bahn (2020): "The Federal Government and Deutsche Bahn Agree on Additional Measures to Step Up Security in Railway Stations." December 13, 2020 (https://www.bmi.bund.de/SharedDocs/pressemitteilungen/EN/2020/12/sicherheit-bahnhoefe-en.html).

Ferguson, Andrew G. (2017): The Rise of Big Data Policing: Surveillance, Race, and the Future of Law Enforcement, New York: New York University Press.

Ganascia, Jean-Gabriel (2010): "Epistemology of AI Revisited in the Light of the Philosophy of Information." In: Knowledge, Technology & Policy 23/1, pp. 57–73.

Golda, Thomas/Cormier, Mickael/Beyerer, Jürgen (2022): "Intelligente Bild- und Videoauswertung für die Sicherheit." In: Dieter Wehe/Helmut Siller (eds.), Handbuch Polizeimanagement. Polizeipolitik – Polizeiwissenschaft – Polizeipraxis. 2nd rev. and expanded ed., Wiesbaden: Springer, pp. 1487–1507.

Golda, Thomas/Kalb, Tobias/Schumann, Arne/Beyerer, Jürgen (2019): "Human Pose Estimation for Real-World Crowded Scenarios." In: 2019 16th IEEE International Conference on Advanced Video and Signal Based Surveillance (AVSS), Taipei, Taiwan, pp. 1–8.

Gombrich, Ernst H. (1984): The Sense of Order: A Study in the Psychology of Decorative Art, 2nd ed., Ithaca and New York: Cornell University Press.

Groys, Boris (1992): Über das Neue. Versuch einer Kulturökonomie, München and Wien: Hanser.

Groys, Boris (2000): "Editorial." In: RES: Anthropology and Aesthetics 38, pp. 5–17.

Grünbaum, Branko/Shephard, G. C. (2016 [1987]): Tilings and Patterns, 2nd ed., Garden City and New York: Dover Publications.

Haeckel, Ernst (1904): Kunstformen der Natur. Supplement-Heft. Allgemeine Erläuterung und systematische Übersicht, Leipzig and Wien: Bibliographisches Institut.

Hirschauer, Stefan (2021): "Menschen unterscheiden. Grundlinien einer Theorie der Humandifferenzierung." In: Zeitschrift für Soziologie 50, pp. 155–174.

Horn, Eva (2018): The Future as Catastrophe. Imagining Disaster in the Modern Age, New York: Columbia University Press.

Kang, Cecilia/Metz, Cade (2023): "F.T.C. Opens Investigation Into ChatGPT Maker Over Technology's Potential Harms.", July 13, 2023 (https://www.nytimes.com/2023/07/13/technology/chatgptinvestigation-ftc-openai.html).

Kaufmann, Mareile/Egbert, Simon/Leese, Matthias (2019): "Predictive Policing and the Politics of Patterns." In: British Journal of Criminology 59/3, pp. 674–692.

Kirsch, Thomas G. (2019): "Securing Security. Recursive Security Assemblages in South Africa." In: Setha Low/Mark Maguire (eds.), Spaces of Security. Ethnographies of Securityscapes, Surveillance, and Control, New York: New York University Press, pp. 122–140.

Koch, Lars/Nanz, Tobias/Pause, Johannes (2018): "Imagined Scenarios of Disruption. A Concept." In: Lars Koch/Tobias Nanz/Johannes Pause (eds.), Disruption in the Arts: Textual, Visual, and Performative Strategies for Analyzing Societal Self-Descriptions, Boston: de Gruyter, pp. 63–81.

Koffka, K. (1935): Principles of Gestalt Psychology, London: Keagon, Trench, Trubner and New York: Harcourt, Brace and Company.

Krünitz, Johann Georg (1805): "Muster." In: Johann Georg Krünitz (auth.), Oeconomische Encyclopädie oder allgemeines System der Land-, Haus- und Staats-Wirthschaft in alphabetischer Ordnung, Bd. 99, Berlin: Pauli, pp. 219–225.

Lazaro, Christophe/Rizzi, Marco (2022): "Predictive Analytics and Governance: a new Sociotechnical Imaginary for Uncertain Futures." In: International Journal of Law in Context, pp. 1–21.

Leonelli, Sabina (2020): "Learning from Data Journeys." In: Sabina Leonelli/Niccolò Tempini (eds.), Data Journeys in the Sciences, Cham: Springer Open, pp. 1–23.

Mackenzie, Adrian (2017): Machine Learners. Archaeology of a Data Practice, Cambridge and London: The MIT Press.

Mandelbrot, Benoit B. (1983): The Fractal Geometry of Nature. Updated and Augmented, San Franscisco: Freeman and Company.

McQuillan, Dan (2019): "The Political Affinities of AI." In: Andreas Sudmann (ed.), The Democratization of Artificial Intelligence. Net Politics in the Era of Learning Algorithms, Bielefeld: transcript, pp. 163–173.

Meyer, Roland (2023): "The New Value of the Archive: AI Image Generation and the Visual Economy of 'Style'." In: IMAGE. The Interdisciplinary Journal of Image Sciences 37/1, pp. 100–111.

Mühlhoff, Rainer (2020): "Human-Aided Artificial Intelligence: Or, How to Run Large Computations in Human Brains? Toward a Media Sociology of Machine Learning." In: New Media & Society 22/10, pp. 1868–1884.

Mühlhoff, Rainer (2023): "Predictive Privacy: Collective Data Protection in the Context of Artificial Intelligence and Big Data." In: Big Data & Society 10/1 (https://doi.org/10.1177/20539517231166886).

Nassehi, Armin (2019): Muster. Theorie der digitalen Gesellschaft, München: Beck.

Noble, Safiya Umoja (2018): Algorithms of Oppression: How Search Engines Reinforce Racism, New York: New York University Press.

OED Online (2023a): "medium, n. and adj." In: Oxford University Press (https://www.oed.com/view/Entry/115772).

OED Online (2023b): "pattern, n. and adj." In: Oxford University Press (https://www.oed.com/view/Entry/138977).

Pasquinelli, Matteo (2019): "How a Machine Learns and Fails." In: spheres: Journal for Digital Cultures 5 (https://doi.org/10.25969/mediarep/13490).

Perry, Walter L./McInnis, Brian/Price, Carter C./Smith, Susan/Hollywood, John S. (2013): Predictive Policing: The Role of Crime Forecasting in Law Enforcement Operations, Santa Monica: RAND Corporation.

Powers, Thomas M./Ganascia, Jean-Gabriel (2020): "The Ethics of the Ethics of AI." In: Markus D. Dubber/Frank Pasquale/Sunit Das (eds.), The Oxford Handbook of Ethics of AI, Oxford: Oxford Univerity Press, pp. 26–51.

Reckwitz, Andreas (2020): The Society of Singularities, Cambridge and Melford: polity.

Rickert, Heinrich (1926): Kulturwissenschaft und Naturwissenschaft, 7th rev. and expanded ed, Tübingen: Mohr.

Rona-Tas, Akos (2020): "Predicting the Future: Art and Algorithms." In: Socio-Economic Review 18/3, pp. 893–911.

Russell, Stuart/Norvig, Peter (2022): Artificial Intelligence: A Modern Approach, 4th ed., Harlow: Pearson.

Samek, Wojciech/Montavon, Grégoire/Vedaldi, Andrea/Hansen, Lars Kai/Müller, Klaus-Robert (eds.) (2019): Explainable AI: Interpreting, Explaining and Visualizing Deep Learning, Cham: Springer.

Sammut, Claude/Webb, Geoffrey I. (eds.) (2011): Encyclopedia of Machine Learning, New York: Springer.

Schabacher, Gabriele (2021): "Infrastrukturen und Verfahren der Humandifferenzierung. Medienkulturwissenschaftliche Perspektiven." In: Dilek Dizdar/Stefan Hirschauer/Johannes Paulmann/Gabriele Schabacher (eds.), Humandifferenzierung. Disziplinäre Perspektiven und empirische Sondierungen, Weilerswist: Velbrück, pp. 287–313.

Schabacher, Gabriele (forthcoming): "In Control of Algorithms: Human-Machine Relations at the Train Station." In: Christiane Voss/Lorenz Engell (eds.), Anthropologies of Entanglements – Media and Modes of Existence, London: Bloomsbury, pp. 151–168.

Schabacher, Gabriele/Spallinger, Sophie (forthcoming): "Tests als Medien der Gewöhnung: Pilotversuche am Bahnhof." In: Zeitschrift für Medienwissenschaft 29.

Schneider, Birgit (2007): Textiles Prozessieren. Eine Mediengeschichte der Lochkartenweberei, Zürich and Berlin: Diaphanes.

Seitter, Walter (2002): Physik der Medien. Materialien, Apparate, Präsentierungen, Weimar: Verlag der Geisteswissenschaften.

Stanley, Jay (2019): The Dawn of Robot Surveillance: AI, Video Analytics, and Privacy, New York: American Civil Liberties Union (https://www.aclu.org/wp-content/uploads/legal-documents/061119-robot_surveillance.pdf).

Stewart, Ian (2001): What Shape is a Snowflake?, New York: Freeman and Company.

Steyerl, Hito (2018): "A Sea of Data: Pattern Recognition and Corporate Animism (Forked Version)." In: Clemens Apprich/Chun, Wendy Hui Kyong/Cramer, Florian/Steyerl, Hito (auths.): Pattern Discrimination, Minneapolis and London: The University of Minnesota Press, and Lüneburg: meson press, pp. 1–22.

Sudmann, Andreas (2019): "On the Media-Political Dimension of Artificial Intelligence. Deep Learning as a Black Box and OpenAI." In: Andreas Sudmann (ed.), The Democratization of Artificial Intelligence. Net Politics in the Era of Learning Algorithms, Bielefeld: Transcript, pp. 223–244.

Weigel, Sigrid (2017): "Punkt-Punkt-Komma-Strich. Urzeichen und ars combinatoria des menschlichen Gesichts." In: Sigrid Weigel (ed.), Das Gesicht. Bilder – Medien – Formate, Göttingen: Wallstein, pp. 124–129.

Images

Fig. 1: Ernst Haeckel (1904): Kunstformen der Natur, Leipzig/Wien: Bibliographisches Institut, Plate 84 (Diatomea).

Fig. 2: Mandelbrot, Benoit B. (1983): The Fractal Geometry of Nature. Updated and Augmented, San Franscisco: Freeman and Company, Table 44, p. 44.

Fig. 3: Ancient tilings from Portugal. After Simoes 1969. Branko Grünbaum/Shephard, G. C. (2016 [1987]): Tilings and Patterns. Second Edition, Garden City andNew York: Dover Publications, p. 7.

Fig. 4: Baby Cap. Riego de la Branchardière (1848): The Knitting Book, 3rd ed., London: Simpkin, Marshall, and Co, pp. 44–45.

Fig. 5: Lewis F. Day (1887): The Anatomy of Pattern, London: Batsford, plate 3.

Fig. 6: Jacquard loom with clapper tacker. Musée d'art et d'industrie de Saint-Étienne, France. photographie: Hélène Rival, November 27, 2012.

Fig. 7: Brassaï, Graffiti from the Series III: "Naissance du visage" (left and middle) and IV: "Masques et visages" (right). Untitled. Gelatin silver prints. c. 1930. Print run c. 1950. MACBA Collection. MACBA Foundation. © Estate Brassaï, 2023.

Fig. 8: Golda, Thomas/Kalb, Tobias/Schumann, Arne/Beyerer, Jürgen (2019): "Human Pose Estimation for Real-World Crowded Scenarios." In: 2019 16th IEEE International Conference on Advanced Video and Signal Based Surveillance (AVSS), Taipei, Taiwan, p. 1.

Fig. 9: Golda, Thomas/Cormier, Mickael/Beyerer, Jügen (2022): "Intelligente Bild- und Videoauswertung für die Sicherheit." In: Dieter Wehe/Helmut Siller (ed.), Handbuch Polizeimanagement. Polizeipolitik – Polizeiwissenschaft – Polizeipraxis. 2nd rev. and expanded ed., Wiesbaden: Springer, p. 1493. Edited version of the image.

Fig. 10: Cormier, Mickael/Clepel, Aris/Specker, Andreas/Beyerer, Jürgen (2022): "Where are we with Human Pose Estimation in Real-World Surveillance?" In: IEEE/CVF Winter Conference on Applications of Computer Vision Workshops (WACVW), Waikoloa, HI, USA, p. 597.

Fig. 11: Pictures generated with DALL-E 2 (https://labs.openai.com), Gabriele Schabacher 2022.

Artificial Intelligence in medicine

Potential applications and barriers to deployment

Urvi Sonawane, Matthieu Komorowski

1. Introduction

The application of Artificial Intelligence (AI) to healthcare has gained tremendous momentum in the last decade, offering the potential to streamline patient clinical encounters and improve patient experience, augment clinical decision making, deliver personalised assessments, and reduce healthcare expenditures (Khanna et al. 2022). However, despite these promises, there remains a vast gap between the large number of 'proof of concept' studies published (AI models with restricted clinical application and limited validation) and the relatively few validated and certified AI tools currently deployed in healthcare settings (Esmaeilzadeh 2020; van de Sande et al. 2021; Gómez-González et al. 2020). The reasons behind this lag are complex, multifaceted and vary across settings and healthcare systems, but broadly include technical, ethical, legal, and human factors (Gerke/Minssen/Cohen 2020).

In this chapter, we will delve deeper into the current and potential applications of AI in medicine, exploring the many ways in which this technology can be utilised to improve patient experience and outcomes and/or healthcare effectiveness. Then, we will examine the major barriers preventing deployment and widespread use of these technologies in healthcare settings.

2. Survey of current AI applications in medicine

Clinical encounters can broadly be classified into three categories, these being primary care (usually a patient's first point of contact, e.g., general practice, community pharmacy or dental services), secondary care (planned or elective care – usually in a hospital, urgent and emergency care or mental health care) and tertiary care (highly specialised treatment), along with community health services (see fig. 1 & 2). Because each of these domains presents challenges, bottlenecks and process inefficiencies, AI applications are being developed on all levels of this 'healthcare ecosystem'.

In primary care, AI solutions have been proposed for a number of applications which can be classified into three categories: 1) clinical decision making and care management, 2) predictive modelling and proactive detection of health conditions and 3) administrative tasks (Mistry 2019).

Figure 1: Overview of the healthcare ecosystem; original figure by NHS Digital (2022).

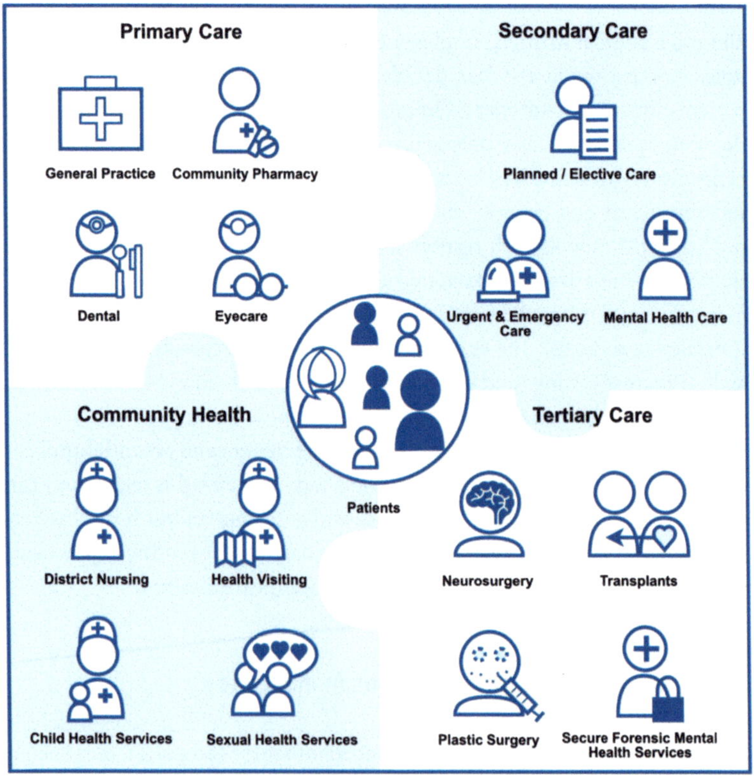

Figure 2: Summary of a unique clinical encounter. AI applications are being developed at all levels of the healthcare ecosystem and targeting all steps of the clinical encounter; adapted from Groenewegen et al. (2014).

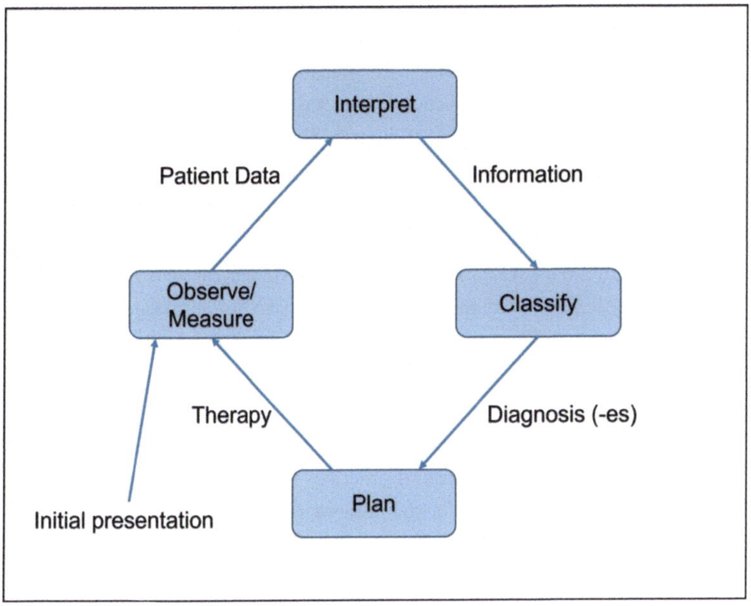

One example of how clinical decision-making and care management have been influenced by AI in primary care settings is 'Doctor AI', developed by Choi and colleagues (Choi et al. 2016). The predictive model is based on a recurrent neural network and was trained on data from over 260,000 patients over the course of eight years, with the aim to predict diagnoses and medication requirements for the subsequent patient visit (ibid.). With this large dataset, the model achieved 79% recall for diagnosis prediction, which is comparatively higher than other baseline models such as logistic regression or multilayer perceptron (ibid.). In the primary care setting, where integration of specialties and holistic patient care is at the forefront, 'Doctor AI' could contribute to planning subsequent appointment discussion points, thereby assisting primary care clinicians when treating patients that have multiple comorbidities.

Predictive modelling and proactive detection of health conditions has been deployed in diagnosing skin conditions since a seminal publication in *Nature* in 2017 (Esteva et al. 2017; cf. Liu et al. 2020). Around one in four patients seek out their general practitioner due to skin problems every year (Schofield et al. 2011), and there is an increased demand for a dermatologist review (Eedy 2016). This model used 16,114 anonymised cases from 17 sites to distinguish between 26 skin conditions commonly seen in primary care (Liu et al. 2020). 963 cases were used to validate the system, with the model performing just as well as six board-certified dermatologists and better than six primary care physicians and six nurse practitioners (ibid.). Whilst this is no permanent solution for the increased burden on dermatology in secondary care, it provides an aid for primary care physicians currently facing the impact of strained secondary care.

Administrative tasks have been reported to account for over 50% of general practice time in the UK (Clay/Stern 2015). Furthermore, in the US, one study reported primary care physicians spending nearly two hours on electronic health record tasks for every hour of patient care (Arndt et al. 2017). This indicates an administrative burden on primary care that is not limited to one country. Considering this, Willis and colleagues concluded that there was a potential to 'completely or mostly' automate 44% of administrative tasks carried out by three urban and three rural general practices in England (Willis et al. 2020). This shows massive potential for machine learning to be integrated into primary care. One new development by Microsoft in collaboration with Nuance Communications Inc. is to use conversational AI to provide clinical documentation that writes itself during a clinician-patient encounter (Langston 2019).

In hospital care, AI applications have been developed across the whole patient pathway, from admission prediction, patient triaging, early diagnosis, decision treatment support and outcome prediction. A large research effort is also focusing on auxiliary tasks such as drug discovery, clinical trial enrolment or administrative tasks including appointment scheduling and medical data management.

Progress has been particularly abundant in the field of radiology. As of February 2023, the United States Food and Drug Administration (FDA) has approved 521 machine learning-enabled medical devices, with 71% of them related to radiology (U.S. Food & Drug Administration n.d.). These devices use AI algorithms to analyse images for diagnostics – particularly detecting tumours and identifying patterns in X-rays, CT-scans, MRIs, or tissue samples (Vora et al. 2019). Through the use of AI in medical image analysis, radiologists may potentially provide faster and more accurate diagnoses, which could lead to better patient outcomes. For example, a company developed an AI solution which automatically detects and alerts clinicians of the presence of large vessel occlusions in the brains of patients suspected to suffer from strokes, with a high sensitivity and specificity, in a real-world prospective setting (Vitellas et al. 2022).

Personalised medicine (also known as precision medicine) is based on the belief that treating, monitoring and preventing diseases must be tailored towards an individual's specific biochemical, physiological, environmental and behavioural profile (Goetz/Schork 2018). The aim is to provide tailored medical care specific to individual patients instead of a broad 'one-size-fits-all' approach, typically provided by expert guidelines (Ruiz-Rodriguez et al. 2022). For example, the management of severe infections in the hospital is dictated by international guidelines such as the "Surviving Sepsis Campaign" (Evans et al. 2021). However, many of the recommendations in such guidelines are based on weak evidence and specific, personalised treatments are not available (Vincent/van der Poll/Marshall 2022). The most likely explanation for this is that sepsis represents a highly heterogeneous patient population, and it is very challenging to identify patients who are more likely to benefit from a specific intervention, for example one targeting components of an immune system (ibid.).

In turn, the concept of data-driven, personalised medicine is becoming increasingly popular, particularly after the COVID-19 pandemic's strain on healthcare provision (Vicente/Ballensiefen/Jönsson 2020). Predictive modelling, which involves using AI algorithms which do not only identify patients at risk for progression of certain diseases, but also predict their responses to treatments, is a particularly promising area of research (Makino et al. 2019; Xu et al. 2021). By accurately predicting a patient's disease progression, healthcare professionals can administer more intensive treatments earlier on, in order to limit long-term disease complications (Makino et al. 2019). This leads to a combination of better patient outcomes coupled with cost cutting, through the reduction of the use of more complex treatments indicated at later stages of disease progression (ibid.). One example for this is taken from a promising study by Makino et al. (2019) where a predictive model was constructed using medical records from over 64,500 diabetic patients to predict diabetic kidney disease progression. The authors suggest that the model can predict diabetic kidney disease progression with 71% accuracy and may reduce the use of haemodialysis, which is known to be a costly intervention in diabetic patients (ibid.; Kent et al. 2015).

AI is also routinely used in natural language processing for popular speech recognition softwares such as 'Google Assistant' and 'Siri' (Google n.d.; Apple

Inc. n.d.). This could also be applied to natural language processing of electronic medical records. Using AI to compile and analyse healthcare records from different staff members could reveal new patterns otherwise almost impossible to spot, let alone diagnose, by human eyes (Mintz/Brodie 2019).

The development of new drugs is imperative for addressing evolving health challenges, such as antibiotic resistance, and AI has the potential to accelerate this process through identifying new drug targets (David et al. 2021). An example of this was shown by Zoffman and colleagues, who used machine learning to search through available antibiotic compounds, eliminate known substances from past projects, and prioritise substances based on factors such as a potency, novelty and availability (Zoffman et al. 2019). This approach can lead to the enhanced discovery of new drugs, particularly in the primary screening stage, as well as the narrowing down and prediction of specific modes of action (ibid.). Drug discovery and development is a complex and expensive process that involves rigorous testing and regulations to ensure safety and efficacy (Chan et al. 2019). One challenge in drug development is ascertaining the toxicology profile of the compound, which can be time-consuming and expensive (Blomme/Will 2016). However, AI systems such as 'DeepTox', which have shown promising accuracy in predicting the toxicology profile of compounds (Mayr et al. 2016), can help to reduce the uncertainty inherent in those processes.

The use of AI in surgical procedures has the potential to significantly improve patient outcome by enhancing precision and accuracy in surgical techniques, with some already being approved by the FDA (Bhandari/Zeffiro/Reddiboina 2020). For instance, AI can be used to identify kidney tumours from bulk CT, allowing surgeons to plan their approach before the procedure commences, or to practice surgical technique in low-risk surgeries (ibid.). However, a systematic review suggested that research surrounding AI in surgery (robot-assisted surgery in particular) is not yet of sufficient quality to safely rely on, primarily due to its limited dataset size (Moglia et al. 2021).

Apart from physical health, there is also growing interest in the application of AI in mental health. Virtual reality and gaming technology, particularly, could help patients with conditions such as depression, bipolar disorder, or chronic pain. By transporting patients into immersive virtual environments, these technologies can provide a safe space for patients to receive psychological therapy and acquire coping mechanisms for their conditions (Hatta et al. 2022; Goudman et al. 2022). While these sessions are currently conducted in the presence of professional staff, the potential for remote sessions should be explored, particularly following the COVID-19 pandemic. This would provide patients with greater flexibility and accessibility to mental health care, particularly for those living in remote or impoverished areas. Machine learning software also analyses patient responses and feedback during the sessions and learns to adapt more effectively to individual patients, as every manifestation of mental health conditions is unique.

3. Limited deployment of AI tools

Despite the extensive list of promising AI applications we detailed above, real world evidence of benefits is lacking for most applications and the validation of AI tools in relevant clinical settings against patient experience and outcomes remains a major challenge.

For example, a rapid search in google scholar for the keywords 'sepsis' and 'prediction' yields over 800,000 results. Comparatively, a 2020 systematic review of the literature focusing on AI identified only 28 published papers, which include mere 3 prospective trials (of which only one was randomised and involved only 142 patients) (Fleuren et al. 2020). Although this number has marginally increased since then with recent publications (e.g., Adams et al. 2022) considering the overall burden of sepsis in the world and the correlated scientific interest generated by sepsis predictions models, the evidence-based benefit of this technology appears worryingly thin.

In a 2021 systematic review of AI applications in the intensive care unit, van de Sande and colleagues produced an insightful summary plot (see fig. 3). While there is an increasing number of AI prototypes and early models being developed and trialled, there seems to be a disproportionate disparity when it comes to translating these AI models from production to clinical evaluation. Consequently, the wide gap between the development and clinical implementation of AI tools in intensive care persists, thus limiting the potential benefits that these technologies were intended to achieve (van de Sande et al. 2021).

A group of experts associated with the Joint Research Centre of the European Commission came to a similar conclusion when they reviewed and classified the application of AI in healthcare in terms of current and near-future applications and ethical/social impact (Gómez-González 2020). A novel scale was created to qualify how 'available' healthcare applications were to the public, ranging from 'TAL 0-Unknown status, not considered feasible according to references' to 'TAL 9-Available for the public'. From their systematic search of AI and AI-mediated technologies, most technologies with a positive social impact were found to have a rating of 'TAL-4-Results of academic/partial projects disclosed', 'TAL-5-Early design of product disclosed' or 'TAL-6-Operational prototype/'first case' disclosed'. This shows that there is still room to drive AI and AI-mediated technologies into the band of TAL-7 to TAL-9, or that there is a fundamental block that needs to be addressed in order to allow more technologies to reach public availability (ibid.).

In the following section, we will explore some of the potential reasons for this phenomenon.

Figure 3: Number of studies published in Artificial Intelligence in the intensive care unit, according to their level of readiness and year of publication (total number of studies = 494); original figure under CC-BY-NC license by van de Sande et al. (2021).

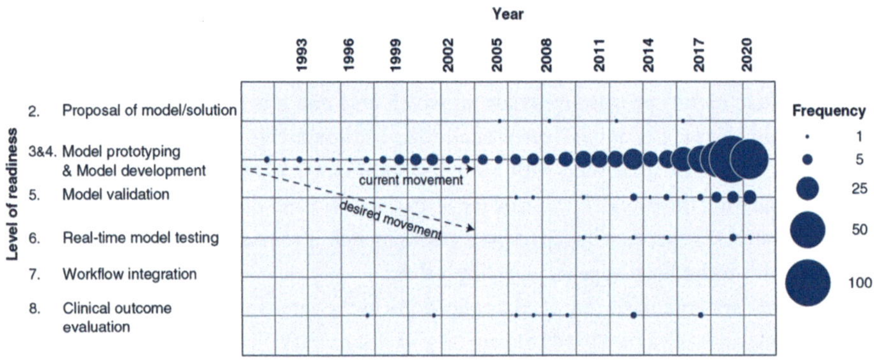

4. Challenges to validation and deployment of AI tools in medicine

The process of developing, testing and deploying AI tools in healthcare at scale involves three major steps (see fig. 4: 1) enabling the data, 2) model development and 3) model deployment. Challenges and hurdles are present at each step of this pipeline/process (Mamdani/Slutsky 2021).

The deployment of AI in medicine falls short in comparison to the sheer number of new machine learning inventions proposed in the field of research. This discrepancy can be attributed to several challenges, the first being the lack of model performance in new clinical settings. It is expected for machine learning models to lose some performance due to the differences between real life clinical settings versus the 'developmental environment'. AI models are often trained with simulated data ortightly controlled parameters, and so models transitioning from the developmental environment to complex real-life clinical settings may face significant differences (Topol 2019). However, this lack of adaptability leads to skepticism about their reliability in crucial clinical judgements that must be accurate. For example, the UK NEWS score was shown to perform poorly in predicting prognosis (AUC 0,6) in a cohort of COVID-19 patients, thus leading to researchers recommending the use of UK NEWS scores as adjuncts to clinical judgment rather than replacements (Colombo et al. 2021).

Figure 4: Overview of the pipeline for developing, testing, and deploying AI tools in healthcare at scale. This involves three major steps: 1) enabling the data, 2) model development and 3) model deployment; adapted from Mamdani/Slutsky (2021).

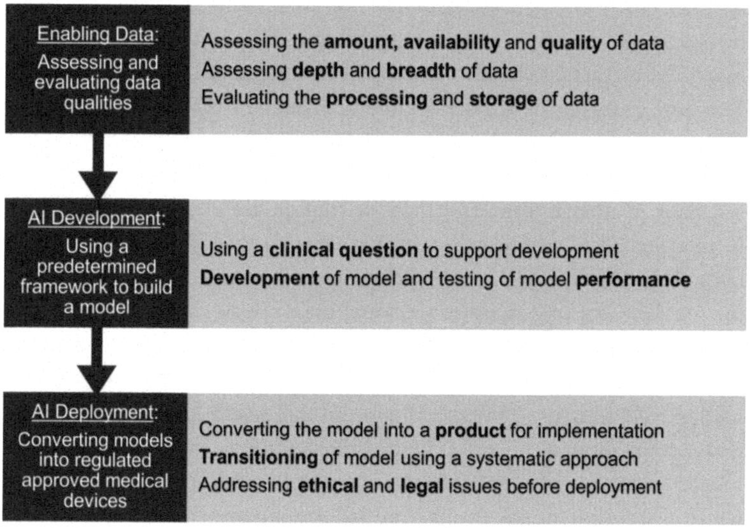

Another hurdle is the scarcity of available data (Ibrahim et al. 2021). AI algorithms require vast amounts of high-quality data to train and test their models. However, obtaining such data is challenging due to patient confidentiality and consent, data sensitivity and lack of cohesive data sharing between the hospitals (Atkin et al. 2021; Kaplan 2016). Electronic health records are often stored in slightly different variations of the same parameter, which creates difficulties in aggregating data and conducting large-scale studies (Holmgren/Adler-

Milstein/McCullough 2018; Dhruva et al. 2020). This can lead to issues with reproducibility and scalability of AI models, as well as difficulties in comparing the performance of different models across different datasets. This limitation means that even the successful algorithms are less suited to be rolled out on a large-scale healthcare service or even across a country (Liang et al. 2022).

Moreover, AI systems are notoriously difficult to integrate within and between systems (this is true within and outside of healthcare) (Baxter/Lee 2021). Currently, most medical AI systems connected to patient data are developed by academic institutions and are not easily usable by external institutions due to the profound discrepancies in IT systems and database structures. Hospitals and small-scale clinics use personalised and/or purpose-built databases to store patient information. Although this makes it easy to navigate through the local area with unique patient demographics, it makes it challenging to adapt the AI code from external institutions to local concept identifiers (Kasparick et al. 2019). Furthermore, hospitals and healthcare systems are often constrained by budget and resource limitations, making it difficult to invest in the necessary infrastructure required to support AI integration (Liang et al. 2022). Therefore, ensuring the AI system is interoperable with other systems, and that data can be shared between different stakeholders in a secure and controlled manner, is challenging.

Many AI models (especially those relying on deep learning) are difficult to interpret and comprehend, which makes it challenging for patients to trust them (Amann et al. 2020). Additionally, patients may not consent to the machine learning software accessing their private data and feed it into ever-changing algorithms due to data security concerns (Atkin et al. 2021). To address this information governance issue, the General Data Protection Regulation (GDPR) in the European Union and the Health Insurance Portability and Accountability Act (HIPAA) in the United States subjects the obtained AI medical data to strict regulatory and compliance scrutiny. These regulations, which also govern the storage, sharing and use of patient data, can be difficult to navigate through in the context of AI (Liang et al. 2022). All these factors compound into a giant hurdle that has to be overcome.

Another difficulty is the acceptance of this new technology by clinicians. The lack of explainability at this stage makes it challenging to encourage healthcare staff to trust early models. It is possible to observe a gap between expected effect and observed effect even in simple and seemingly innocuous interventions, such as a 'pop-up alert' for acute kidney injury upon the opening of a patient's electronic health record – which shockingly led to a sharp increase in patient mortality (Wilson et al. 2021). Clinicians' bias with use of technology and evolving (but not confirmed) evidence may be contributing factors. Furthermore, the use of this model for very sick septic patients in the ICU may compound to their lack of trust.

There is no established gold-standard process to demonstrate patient benefit from AI solutions and indeed, there are no recognised best practices for evaluating the efficacy, reliability and safety of commercially available algorithms (Wu et al. 2021). What level of evidence can be accepted by patients, clinicians and regulators? Is retrospective evidence sufficient? Are developers required to conduct multiple randomised trials comparing the standard of care to care supported by their AI solution? Assessment frameworks for the clinical validation of AI have been both proposed (Tsopra et al. 2021; Hawkins et al. 2021; Kickbusch et al. 2020) and surveyed (de Hond et al. 2022), but developing a common set of guidelines for AI model development and implementation remains challenging.

Even those AI applications that have managed to overcome this giant hurdle have issues that need to be considered. The lack of standardisation between AI studies approved for hospital use by regulators (such as the FDA or MHRA) makes it difficult to compare results, mainly due to the varied level of the implementation of studies across different areas of healthcare (Pashkov/Harkusha/Harkusha 2020).

Furthermore, a number of additional human factors must be considered. Healthcare professionals may have a limited or developing understanding and familiarisation of AI tools and would therefore naturally be skeptical of its potential (Gama et al. 2022). This skepticism can make it difficult to integrate AI-based tools into their workflow and practice (Amann et al. 2020). AI tools undoubtedly would also initially add significant expenditure on the already stretched financial healthcare landscape, particularly in the post pandemic period (Kickbusch et al. 2020). One concern of healthcare providers is the legal implications in clinical practice. Healthcare providers may be held liable for potential or actual harm that is caused by AI systems, particularly if they delayed or failed to properly assess or monitor AI's performance. In an era of already burnt-out healthcare staff, the additional responsibility of overseeing the performance of an AI system is unappealing (Gooding/Kariotis 2021). Furthermore, developers of AI systems would also be more cautious when establishing/introducing the software in a position of responsibility given this legal liability (Luxton 2014). The hesitance from both sides is a contributing factor to the lack of implementation of AI software in mainstream healthcare.

5. Ethical considerations

The ethical considerations of AI are widely debated, and these concerns are not limited to healthcare. However, certain ethical arguments are particularly pertinent when considering the introduction of AI into mainstream medicine.

AI must provide a real benefit to patients and improve health outcomes and its use must be justified based on patient benefit (Hamet/Tremblay 2017). AI is increasingly being seen as the future of everyday life and financial gain from this cannot be ignored. Deviation from practical patient benefit is certainly possible amidst the desire for investment. Therefore, improving health outcomes should be at the core of AI development in healthcare, which can be done by working in conjunction with patients and healthcare staff.

The presence of discriminative biases in healthcare is undeniable (Ibrahim et al. 2021; Norori et al. 2021). The implication of this, however, could be amplified by AI systems. If they are designed to recognise patterns, these may also perpetuate the existing discrimination in healthcare, leading to further inequality in treatment and health outcomes in patient populations that already experience prejudice and discrimination (Ibrahim et al. 2021; Fletcher/Nakeshima/Olubeko 2021). This ultimately hinders progress towards achieving the desired healthcare equality. For example, an algorithm developed by Gijberts and colleagues using data derived from almost exclusively Caucasian people performed poorly when attempting to predict cardiovascular risk for patients of other ethnicities (e.g., African American and Hispanic ethnicities) (Gijsberts 2015).

AI should be accessible to all patients and should not widen existing health disparities. There is also a potential for AI systems to provide ambiguous or unhelpful answers in critical healthcare situations (Topol 2019). This could again lead to a lack of trust, especially if this happens at the start of implementing the software. It is crucial that the results of this are audited regularly, and the opinions of healthcare staff using the software should be monitored through focus groups and questionnaires to ensure that trust in the software is maintained (Vela et al. 2022).

Table 1: Summary of the main challenges involved in developing and deploying AI tools in medicine.

Step in integrating AI tools in medicine	Challenges involved
Enabling data	· **Data availability and quality:** effective AI algorithms require large volumes of quality data to refine their models. However, obtaining such data can be difficult, especially when it comes to sensitive medical information (Ibrahim et al. 2021). · **Interoperability and data sharing:** healthcare providers must also ensure that AI systems are interoperable with other systems and that data can be shared between different stakeholders in a standardised, secure and controlled manner. Currently, most systems developed ad-hoc by academic institutions are not usable in external institutions due to profound differences in IT systems and database structures (Baxter/Lee 2021; Kasparick et al. 2019).
Developing AI models	· **Familiarity with alternate clinical settings:** ML models generally perform sub optimally when deployed in settings dissimilar to those in development. The need for generalisability of AI tools is critical and history is replete with examples of applications that fell short in this regard. For example, the UK NEWS score was shown to perform poorly (AUC 0,6) in a cohort of COVID-19 patients (Colombo et al. 2021). · **Lack of standardisation:** there is a discrepancy in how AI is implemented and used in medicine, which can make it difficult to compare results across different studies and applications (Pashkov/Harkusha/Harkusha 2020; Gama et al. 2022).

Step in integrating AI tools in medicine	Challenges involved
Deploying AI models	· **Bias and discrimination:** AI systems may perpetuate and amplify existing biases in healthcare, leading to unequal treatment and outcomes (Norori et al. 2021). · **Privacy and confidentiality:** data storage, security and protection within AI systems must be compliant with all regulations (e.g., GDPR) (NHS England 2023). · **Responsibility and accountability:** responsibility must be taken for decisions and actions taken by AI systems, particularly in cases where they may cause harm (Gupta/Kamboj/Bag 2021). · **Explainability and transparency:** AI systems may produce results that are difficult to interpret. It's crucial to make sure that the methods and decisions of AI systems are transparent and can be audited (Amann et al. 2020). · **Clinical validity:** evidence based medicine and latest guidelines must be regularly incorporated into AI systems to ensure validity (Crossnohere et al. 2022). · **Clinical utility:** AI must provide a real benefit to patients and improve health outcomes and its use must be justified on the basis of patient benefit (NHS England 2023). · **Equity and access:** AI should be accessible to all patients and should not widen existing health disparities (Gómez-González 2020). · **Regulation and compliance:** AI in medicine is subject to strict regulation and compliance requirements, such as GDPR (EU) and HIPAA (USA), that govern the storage and use of patient data (Crossnohere et al. 2022). · **Legal concerns and liability:** healthcare providers may be held liable for harm caused by AI systems, particularly if they failed to properly assess or monitor the AI's performance (Gupta/Kamboj/Bag 2021). · **Limited understanding of AI by healthcare providers:** healthcare professionals may have a limited understanding of AI, which can make it difficult to integrate AI-based tools into their workflow and practice (Amann et al. 2020).

Privacy and confidentiality are critical factors that are essential to maintain. Concerns over the risk of patient re-identification have profoundly limited the development of large, publicly available datasets for research. For example, we evaluated what data sources had been used in machine learning models for sepsis resuscitation in the ICU and found that nearly two thirds relied on the same dataset (the MIMIC database) (Johnson et al. 2016). We argue that the benefit of open data sharing outweighs the risks. Indeed, a recent analysis of potential reidentifications of patients in publicly available datasets confirmed that the risk was extremely low (Seastedt et al. 2022). The authors argued that

> the cost – measured in terms of access to future medical innovations and clinical software – of slowing ML progress is too great to limit sharing data through large publicly available databases for concerns of imperfect data anonymization (ibid.).

Data security and patient privacy must be preserved at the phase of model deployment and real-time use, which is a legal requirement and a key aspect of regulatory approval (NHS England 2023).

The security and protection of data is expected by the patient population, yet we may not really know if AI models will be successful in this until they are fully implemented in clinical practice. As per GDPR and NHS medical ethics principles, the patients should be explicitly informed about the use of AI in their care and should also have the autonomy for decision making. If they decided to opt-out of its use, a suitable alternative to the role of the AI software in their care should be offered to all (ibid.).

In situations where AI is involved, accountability and responsibility must be established at all times, including crucial life-or-death decision making. There is also a lack of clarity on which areas of decision making are legally accountable to AI and therefore it is important to identify a clear line of responsibility, including shared responsibility for the decisions and actions taken by AI systems (ibid; Gupta/Kamboj/Bag 2021).

The use of AI in medicine must be supported by clinical evidence and validated through rigorous testing to ensure its accuracy and reliability. Each geographical area has different guidelines supported by different bodies of evidence to suit the varied case mix, and AI systems may be produced in areas with different guidelines. Therefore, it is important to constantly update the treatment algorithms so they comply with constantly evolving medical research and clinical evidence (Crossnohere et al. 2022).

The ethical consideration of financial gain also leads to the point of financial disparities between patients. As mentioned in the first section, the need for AI integration towards 'personalised medicine' would go a long way to make significant savings both by avoiding ineffective treatment costs and better prognosis/quality of life (ibid.).

In conclusion, AI in healthcare holds significant potential to revolutionise the way healthcare operates, from administrative tasks, diagnostics, drug development to surgery. However, despite the many avenues of research that have been and will be explored, there is currently a bottleneck when it comes to the deployment and widespread use of these technologies. This is due to a multitude of factors, including data availability and standardisation, privacy and ethical concerns, clinician and patient skepticism, clinical utility and legal regulations. To overcome these challenges, a collaborative and multidisciplinary approach involving regulatory bodies, healthcare professionals, government entities and patient committees is necessary. This collaboration can produce a clear and regulated framework that will allow innovative and life-changing AI projects to be seamlessly integrated into mainstream healthcare practices.

List of references

Adams, Roy/Henry, Katherine E./Sridharan, Anirudh/Soleimani, Hossein/Zhan, Andong/Rawat, Nishi/Johnson, Lauren/et al. (2022): "Prospective, Multi-Site Study of Patient Outcomes After Implementation of the TREWS Machine Learning-Based Early Warning System for Sepsis." In: Nature Medicine 28/7, pp. 1455–1460.

Amann, Julia/Blasimme, Alessandro/Vayena, Effy/Frey, Dietmar/Madai, Vince I. (2020): "Explainability for Artificial Intelligence in Healthcare: A Multidisciplinary Perspective." In: BMC Medical Informatics and Decision Making 20/310 (https://doi.org/10.1186/s12911-020-01332-6).

Apple Inc. (n.d.): "Siri – Apple (UK).", accessed March 17, 2023 (https://www.apple.com/uk/siri/).

Arndt, Brian G./Beasley, John W./Watkinson, Michelle D./Temte, Jonathan L./Tuan, Wen Jan/Sinsky, Christine A./Gilchrist, Valerie J. (2017): "Tethered to the EHR: Primary Care Physician Workload Assessment Using EHR Event

Log Data and Time-Motion Observations." In: Annals of Family Medicine 15/5, pp. 419–426.

Atkin, Catherine/Crosby, B./Dunn, K./Price, G./Marston E./Crawford C./O'Hara, M./et al. (2021): "Perceptions of Anonymised Data Use and Awareness of the NHS Data Opt-out Amongst Patients, Carers, and Healthcare Staff." In: Research Involvement and Engagement 7/40 (https://doi.org/10.1186/s40900-021-00281-2).

Baxter, Sally L./Lee, Aaron Y. (2021): "Gaps in Standards for Integrating Artificial Intelligence Technologies into Ophthalmic Practice." In: Current Opinion in Ophthalmology 32/5, pp. 431–438.

Bhandari, Mahendra/Zeffiro, Trevor/Reddiboina, Madhu (2020): "Artificial Intelligence and Robotic Surgery: Current Perspective and Future Directions." In: Current Opinion in Urology 30/1, pp. 48–54.

Blomme, Eric A. G./Will Yvonne (2016): "Toxicology Strategies for Drug Discovery: Present and Future." In: Chemical Research in Toxicology 29/4, pp. 473–504.

Chan, H. C. Stephen/Shan, Hanbin/Dahoun, Thamani/Vogel, Horst/Yuan, Shuguang (2019): "Advancing Drug Discovery via Artificial Intelligence." In: Trends in Pharmacological Sciences (Regular ed.) 40/10, p. 801.

Choi, Edward/Bahadori, Mohammad Taha/Schuetz Andy/Stewart, Walter F./Sun Jimeng (2016): "Doctor AI: Predicting Clinical Events via Recurrent Neural Networks." In: JMLR Workshop and Conference Proceedings 56, pp. 301–318.

Clay, Henry/Stern, Rick (2015): "Making Time in General Practice: Freeing GP Capacity by Reducing Bureaucracy and Avoidable Consultations, Managing the Interface with Hospitals, and Exploring New Ways of Working.", October 2015 (https://thehealthcreationalliance.org/wp-content/uploads/2018/11/Making-Time-in-General-Practice-FULL-REPORT-06-10-15.pdf).

Colombo, Christopher J./Colombo, Rhonda E./Maves, Ryan C./Branche, Angela R./Cohen, Stuart H./Elie, Marie-Carmelle/George, Sarah L./et al. (2021): "Performance Analysis of the National Early Warning Score and Modified Early Warning Score in the Adaptive COVID-19 Treatment Trial Cohort." In: Critical Care Explorations 3/7, e0474.

Crossnohere, Norah L./Elsaid, Mohamed/Paskett, Jonathan/Bose-Brill, Seuli/Bridges, John F. P. (2022): "Guidelines for Artificial Intelligence in Medicine: Literature Review and Content Analysis of Frameworks." In: Journal of Medical Internet Research 24/8, e36823.

David, Liliana/Brata, Anca Monica/Mogosan, Christina/Pop, Christina/Czako, Zoltan/Muresan, Lucian/Ismaiel, Abdulrahman/et al. (2021): "Artificial Intelligence and Antibiotic Discovery." In: Antibiotics (Basel) 10/11, p. 1376.

de Hond, Anne A. H./Leeuwenberg, Artuur, M./Hooft, Lotty/Kant, Ilse M. J./Nijman, Steven W. J./van Os, Hendrikus J. A./Aardoom, Jiksa J./et al. (2022): "Guidelines and Quality Criteria for Artificial Intelligence-Based Prediction Models in Healthcare: A Scoping Review." In: NPJ Digital Medicine 5/2 (https://doi.org/10.1038/s41746-021-00549-7).

Dhruva, Sanket S./Ross, Joseph S./Akar, Joseph G./Caldwell, Brittany/Childers, Karla/Chow, Wing/Ciaccio, Laura/et al. (2020): "Aggregating Multiple Real-World Data Sources Using a Patient-Centered Health Data-Sharing Platform." In: NPJ Digital Medicine 3/60.

Eedy, David (2016): "Dermatology: A Specialty in Crisis." In: Clinical Medicine 16/1, p. 61.

Esmaeilzadeh, Pouyan (2020): "Use of AI-Based Tools for Healthcare Purposes: A Survey Study from Consumers' Perspectives." In: BMC Medical Informatics and Decision Making 20/170 (https://doi.org/10.1186/s12911-020-01191-1).

Esteva, Andre/Kuprel, Brett/Novoa, Roberto A./Ko, Justin/Swetter, Susan M./Blau, Helen M./Thrun, Sebastian (2017): "Dermatologist-Level Classification of Skin Cancer with Deep Neural Networks." In: Nature 542/7639, pp. 115–118.

Evans, Laura/Rhodes, Andrew/Alhazzani, Waleed/Antonelli, Massimo/Coopersmith, Craig M./French, Craig/Machado, Flávia R./et al. (2021): "Surviving Sepsis Campaign: International Guidelines for Management of Sepsis and Septic Shock 2021." In: Intensive Care Medicine 47/11, pp. 1181–1247.

Fletcher, Richard Ribón/Nakeshimana, Audace/Olubeko, Olusubomi (2021): "Addressing Fairness, Bias, and Appropriate Use of Artificial Intelligence and Machine Learning in Global Health." In: Frontiers in Artificial Intelligence 3 (https://doi.org/10.3389/frai.2020.561802).

Fleuren, Lucas M./Klausch, Thomas L. T./Zwager, Charlotte L./Schoonmade, Linda J./Guo, Tingjie/Roggeveen, Luca F./Swart, Eleonora/et al. (2020): "Machine Learning for the Prediction of Sepsis: a Systematic Review and Meta-Analysis of Diagnostic Test Accuracy." In: Intensive Care Medicine 46/3, pp. 383–400.

Gama, Fábio/Tyskbo, Daniel/Nygren, Jens/Barlow, James/Reed, Julie/Svedberg, Petra (2022): "Implementation Frameworks for Artificial

Intelligence Translation into Health Care Practice: Scoping Review." In: Journal of Medical Internet Research 24/1, e32215.

Gerke, Sara/Minssen, Timo/Cohen, Glenn (2020): "Ethical and Legal Challenges of Artificial Intelligence-Driven Healthcare." In: Bohr, Adam/ Memarzadeh Kaveh (eds.), Artificial Intelligence in Healthcare, London, San Diego, Cambridge and Oxford: Academic Press, pp. 295–336.

Gijsberts, Crystal M./Groenewegen, Karlijn A./Hoefer, Imo E./Eijkemans, Marinus J. C./Asselbergs, Folkert W./Anderson, Todd J./Britton, Annie R./et al. (2015): "Race/Ethnic Differences in the Associations of the Framingham Risk Factors with Carotid IMT and Cardiovascular Events." In: PLoS ONE 10/7, e0132321.

Goetz, Laura H./Schork, Nicholas J. (2018): "Personalized Medicine: Motivation, Challenges, and Progress." In: Fertility and Sterility 109/6, pp. 952–963.

Gómez-González, Emilio/Gomez, Emilia/Márquez-Rivas, Javier/Guerrero-Claro, Manuel/Fernández-Lizaranzu, Isabel/Relimpio-López, María Isabel/Dorado, Manuel E./Mayorga-Buiza, María José/Izquierdo-Ayuso, Guillermo/Capitán-Morales, Luis (2020): Artificial Intelligence in Medicine and Healthcare: A Review and Classification of Current and Near-Future Applications and Their Ethical and Social Impact, arXiv Preprint (https://doi.org/10.48550/arXiv.2001.09778).

Gooding, Piers/Kariotis, Timothy (2021): "Ethics and Law in Research on Algorithmic and Data-Driven Technology in Mental Health Care: Scoping Review." In: JMIR Mental Health 8/6, e24668.

Google (n.d.): "Google Assistant, Your Own Personal Google.", accessed March 17, 2023 (https://assistant.google.com/).

Goudman, Lisa/Jansen, Julie/Billot, Maxime/Vets, Nieke/Smedt, Ann De/ Roulaud, Manuel/Rigoard, Philippe/Moens, Maarten (2022): "Virtual Reality Applications in Chronic Pain Management: Systematic Review and Meta-analysis." In: JMIR Serious Games 10/2, e34402.

Groenewegen, André/Tofighy, Azita/Ryvlin, Philippe/Steinhoff, Bernhard J./ Dedeken, Peter (2014): "Measures for Improving Treatment Outcomes for Patients with Epilepsy – Results from a Large Multinational Patient-Physician Survey." In: Epilepsy & Behavior 34, pp. 58–67.

Gupta, Shivam/Kamboj, Shampy/Bag, Surajit (2021): "Role of Risks in the Development of Responsible Artificial Intelligence in the Digital Healthcare Domain." In: Information Systems Frontiers (https://doi.org/10.1007/s10796-021-10174-0).

Hamet, Pavel/Tremblay, Johanne (2017): "Artificial Intelligence in Medicine." In: Metabolism, Clinical and Experimental 69, pp. S36–S40.

Hatta, Muhammad Hizri/Sidi, Hatta/Sharip, Shalisah/Das, Srijit/Saini, Suriati Mohamed (2022): "The Role of Virtual Reality as a Psychological Intervention for Mental Health Disturbances During the COVID-19 Pandemic: A Narrative Review." In: International Journal of Environmental Research and Public Health 19/4, p. 2390.

Hawkins, Richard/Paterson, Colin/Picardi, Chiara/Jia, Yan/Calinescu, Radu/Habli, Ibrahim (2021): Guidance on the Assurance of Machine Learning in Autonomous Systems (AMLAS), arXiv Preprint (https://doi.org/10.48550/arxiv.2102.01564).

Holmgren, A. Jay./Adler-Milstein, Julia/McCullough, Jeffrey (2018): "Are All Certified EHRs Created Equal? Assessing the Relationship Between EHR Vendor and Hospital Meaningful Use Performance." In: Journal of the American Medical Informatics Association 25/6, pp. 654–660.

Ibrahim, Hussein/Liu, Xiaoxuan/Zariffa, Nevine/Morris, Andrew D./Denniston, Alastair K. (2021): "Health Data Poverty: An Assailable Barrier to Equitable Digital Health Care." In: The Lancet Digital Health 3/4, e260–e265.

Johnson, Aistair E. W./Pollard, Tom J./Shen, Lu/Lehman, Li-wei H./Feng, Mengling/Ghassemi, Mohammad/Moody, Benjamin/Szolovits, Peter/Celi, Leo Anthony/Mark, Roger G. (2016): "MIMIC-III, a Freely Accessible Critical Care Database." In: Scientific Data 3/160035 (https://doi.org/10.1038/sdata.2016.35).

Kaplan, Bonnie (2016): "How Should Health Data Be Used?" In: Cambridge Quarterly of Healthcare Ethics 25/2, pp. 312–329.

Kasparick, Martin/Andersen, Björn/Franke, Stefan/Rockstroh, Max/Golatowski, Frank/Timmermann, Dirk/Ingenerf, Josef/Neumuth, Thomas (2019): "Enabling Artificial Intelligence in High Acuity Medical Environments." In: Minimally Invasive Therapy and Allied Technologies 28/2, pp. 120–126.

Kent, Seamus/Schlackow, Iryna/Lozano-Kühne, Jingky/Reith, Christina/Emberson, Jonathan/Haynes, Richard/Gray, Alastair/et al. (2015): "What Is the Impact of Chronic Kidney Disease Stage and Cardiovascular Disease on the Annual Cost of Hospital Care in Moderate-to-Severe Kidney Disease?" In: BMC Nephrology 16/65 (https://doi.org/10.1186/s12882-015-0054-0).

Khanna, Narendra N./Maindarkar, Mahesh A./Viswanathan, Vijay/Fernandes, Jose Fernandes E./Paul, Sudip/Bhagawati, Mrinalini/Ahluwalia, Puneet/et al. (2022): "Economics of Artificial Intelligence in Healthcare: Diagnosis vs. Treatment." In: Healthcare (Basel) 10/12, p. 2493.

Kickbusch, Ilona/Leung, Gabriel M./Bhutta, Zulfiqar A./Matsoso, Malebona Precious/Ihekweazu, Chikwe/Abbasi, Kamran (2020): "COVID-19: How a Virus Is Turning the World Upside Down." In: BMJ 369, m1336.

Langston, Jennifer (2019): "Microsoft and Nuance Join Forces in Quest to Help Doctors Turn Their Focus Back to Patients.", October 17, 2019 (https://news.microsoft.com/source/features/ai/nuance-exam-room-of-the-future/).

Liang, Likeng/Hu, Jifa/Sun, Gang/Hong, Na/Wu, Ge/He, Yuejun/Li, Yong/et al. (2022): "Artificial Intelligence-Based Pharmacovigilance in the Setting of Limited Resources." Drug Safety 45/5, pp. 511–519.

Liu, Yuan/Jain, Ayush/Eng, Clara/Way, David H./Lee Kang/Bui, Peggy/Kanada, Kimberly/et al. (2020): "A Deep Learning System for Differential Diagnosis of Skin Diseases." In: Nature Medicine 26/6, pp. 900–908.

Luxton, David D. (2014): "Recommendations for the Ethical Use and Design of Artificial Intelligent Care Providers." In: Artificial Intelligence in Medicine 62/1, pp. 1–10.

Makino, Masaki/Yoshimoto, Ryo/Ono, Masaki/Itoko, Toshinari/Katsuki, Takayuki/Koseki, Akira/Kudo, Michiharu/et al. (2019): "Artificial Intelligence Predicts the Progression of Diabetic Kidney Disease Using Big Data Machine Learning." In: Scientific Reports 9, 11862.

Mamdani, Muhammad/Slutsky, Arthur S. (2021): "Artificial Intelligence in Intensive Care Medicine." Intensive Care Medicine 47/2, pp. 147–149.

Mayr, Andreas/Klambauer, Günter/Unterthiner, Thomas/Hochreiter, Sepp (2016): "Toxicity Prediction Using Deep Learning." In: Frontiers in Environmental Science 3 (https://doi.org/10.3389/fenvs.2015.00080).

Mintz, Yoav/Brodie, Ronit (2019): "Introduction to Artificial Intelligence in Medicine." In: Minimally Invasive Therapy and Allied Technologies 28/2, pp. 73–81.

Mistry, Pritesh (2019): "Artificial Intelligence in Primary Care." In: British Journal of General Practice 69/686, pp. 422–423.

Moglia, Andrea/Georgiou, Konstantinos/Georgiou, Evangelos/Satava, Richard M./Cuschieri, Alfred (2021): "A Systematic Review on Artificial Intelligence in Robot-assisted Surgery." In: International Journal of Surgery 95, p. 106151.

NHS Digital (2022): "The Healthcare Ecosystem.", last edited October 26, 2022 (https://digital.nhs.uk/developer/guides-and-documentation/introduction-to-healthcare-technology/the-healthcare-ecosystem).

NHS England (2023): "Artificial Intelligence- NHS Transformation Directorate.", February 10, 2023 (https://transform.england.nhs.uk/information-governance/guidance/artificial-intelligence).

Norori, Natalia/Hu, Qiyang/Aellen, Florence Marcelle/Faraci, Francesca Dalia/Tzovara, Athina (2021): "Addressing Bias in Big Data and AI for Health Care: A Call for Open Science." In: Patterns (N.Y.) 2/10, 100347.

Pashkov, Vitlaii M./Harkusha, Andrii O./Harkusha, Yevheniia O. (2020): "Artificial Intelligence in Medical Practice: Regulative Issues and Perspectives." In: Wiadomości Lekarskie 73/12, pp. 2722–2727.

Ruiz-Rodriguez, Juan Carlos/Plata-Menchaca, Erika P./Chiscano-Camón, Luis/Ruiz-Sanmartin, Adolfo/Pérez-Carrasco, Marcos/Palmada, Clara/Ribas, Vicent/Martínez-Gallo, Mónica/et al. (2022): "Precision Medicine in Sepsis and Septic Shock: From Omics to Clinical Tools." In: World Journal of Critical Care Medicine 11/1, pp. 1–21.

Schofield, Julia K./Fleming, D./Grindlay, D./Williams, H. (2011): "Skin Conditions are the Commonest New Reason People Present to General Practitioners in England and Wales." In: British Journal of Dermatology 165/5, pp. 1044–1050.

Seastedt, Kenneth P./Schwab, Patrick/O'Brien, Zach/Wakida, Edith/Herrera, Karen/Marcelo, Portia Grace F./Agha-Mir-Salim, Louis/Frigola, Xavier Borrat/Ndulue, Emily Boardman/Marcelo, Alvin/et al. (2022): "Global Healthcare Fairness: We Should Be Sharing More, Not Less, Data." In: PLOS Digital Health 1/10, e0000102.

Topol, Eric J. (2019): "High-Performance Medicine: The Convergence of Human and Artificial Intelligence." In: Nature Medicine 25/1, pp. 44–56.

Tsopra, Rosy/Fernandez, Xose/Luchinat, Claudio/Alberghina, Lilia/Lehrach, Hans/Vanoni, Marco/Dreher, Felix/et al. (2021): "A Framework for Validating AI in Precision Medicine: Considerations from the European ITFoC Consortium." In: BMC Medical Informatics and Decision Making 21/274 (https://doi.org/10.1186/s12911-021-01634-3).

U.S. Food & Drug Administration (n.d.): "Artificial Intelligence and Machine Learning (AI/ML)-Enabled Medical Devices.", accessed March 17, 2023 (https://www.fda.gov/medical-devices/software-medical-device-samd/artificial-intelligence-and-machine-learning-aiml-enabled-medical-devices).

van de Sande, Davy/van Genderen, Michel E./Huiskens, Joost/Gommers, Diederick/van Bommel, Jasper (2021): "Moving From Bytes to Bedside: A Systematic Review on the Use of Artificial Intelligence in the Intensive Care Unit." In: Intensive Care Medicine 47/7, pp. 750–760.

Vela, Monica B./Erondu, Amarachi I./Smith, Nichole A./Peek, Monica E./Woodruff, James N./Chin, Marshall H. (2022): "Eliminating Explicit and Implicit Biases in Health Care: Evidence and Research Needs." In: Annual Review of Public Health 43/1, pp. 477–501.

Vicente, Astrid M./Ballensiefen, Wolfgang/Jönsson, Jan-Ingvar (2020): "How Personalised Medicine Will Transform Healthcare by 2030: The ICPerMed Vision." In: Journal of Translational Medicine 18/1, p. 180.

Vincent, Jean-Louis/van der Poll, Tom/Marshall, John C. (2022): "The End of 'One Size Fits All' Sepsis Therapies: Toward an Individualized Approach." In: Biomedicines 10/9 p. 2260.

Vitellas, Carol A./Mannix, Nicholas C./Nimjee, Shahid M./Shujaat, Mohammad T./Heaton, Sharon/Lee, Vivien H. (2022): "Abstract 130: Real World Experience With Viz.AI Automated Large Vessel Occlusion Detection." In: Stroke 53/Suppl_1, A130.

Vora, Amisha/Malik, Paras/Pathania, Monika/Rathaur, Vyas Kumar (2019): "Overview of Artificial Intelligence in Medicine." In: Journal of Family Medicine and Primary Care 8/7, pp. 2328–2331.

Willis, Matthew/Duckworth, Paul/Coulter, Angela/Meyer, Eric T./Osborne, Michael (2020): "Qualitative and Quantitative Approach to Assess the Potential for Automating Administrative Tasks in General Practice." In: BMJ open 10, e032412.

Wilson, F. Perry/Martin, Melissa/Yamamoto, Yu/Partridge, Caitlin/Moreira, Erica/Arora, Tanima/Biswas, Aditya/Feldman, Harold/Garg, Amit X./Greenberg, Jason H. (2021): "Electronic Health Record Alerts for Acute Kidney Injury: Multicenter, Randomized Clinical Trial." In: BMJ 372, m4786.

Wu, Eric/Wu, Kevin/Daneshjou, Roxana/Ouyang, David/Ho, Daniel E./Zou, James (2021): "How Medical AI Devices Are Evaluated: Limitations and Recommendations from an Analysis of FDA Approvals." In: Nature Medicine 27/4, pp. 582–584.

Xu, Ying/Su, Guan-Hua/Ma, Ding/Xiao, Yi/Shao, Zhi-Ming/Jiang, Yi-Zhou (2021): "Technological Advances in Cancer Immunity: From Immunogenomics to Single-Cell Analysis and Artificial Intelligence." In: Signal Transduction and Targeted Therapy 6/312 (https://doi.org/10.1038/s41392-021-00729-7).

Zoffman, Sannah/Vercruysse, Maarten/Benmansour, Fethallah/Maunz, Andreas/Wolf, Luise/Marti, Rita Blum/Heckel, Tobias/et al. (2019): "Machine Learning-Powered Antibiotics Phenotypic Drug Discovery." In: Scientific Reports 9, 5013 (https://doi.org/10.1038/s41598-019-39387-9).

Subsymbolic, hybrid and explainable AI

What can it change in medical imaging?

Isabelle Bloch

1. Introduction

While symbolic methods and statistical machine learning methods for artificial intelligence (AI) have been developing rather independently for decades, with alternated predominance of one or the other across time, a current trend is to merge both types of approaches. Examples include neuro-symbolic approaches (see e.g., De Raedt et al. 2020; d'Avila Garcez/Lamb 2023; Garnelo/Shanahan 2019; Kautz 2022; Marcus 2020), among others. However, in this paper, hybrid artificial intelligence is intended in a broader sense, as the combination of several AI methods, whatever their type.[1] These methods may belong to the domains of abstract knowledge representation and formal reasoning, based on logic, structural representation (such as graphs and hypergraphs, ontologies, concept lattices, etc.), machine learning, etc. Additionally, imprecision in data, knowledge and reasoning can benefit from the fuzzy sets theory.

Such combinations of approaches take inspiration from cognitive functions. Roughly speaking, according to Kahneman (2012), who distinguished two systems for thinking named system 1 and system 2, we may consider, from a (strongly simplified) AI point of view, modeling system 1 (rapid, intuitive) by deep learning and system 2 (slower, more controlled, logical) by symbolic reasoning. Developing neuro-symbolic approaches is a new trend to combine the two systems (see e.g., Kautz 2022). But again, more theories will be committed in our view of hybrid AI, in particular for image understanding.

The aim of this paper is not to propose new methods for hybrid AI, but rather, as a position paper, to highlight how this way of thinking and design-

1 We should note here that AI is already the umbrella term for very different methods, and that many AI methods or systems are actually by essence hybrid.

ing AI systems offers opportunities towards explainability in the field of explainable AI (XAI) and as a mean to maintain the link between knowledge and data. In that domain, too, the two main branches are developed quite independently, with early work (e.g., Peirce at the end of the 19th century) focusing on logical reasoning based on abduction on the one hand, versus recent methods focusing on features or data most involved in a decision on the other hand (to name but a few). In the first paradigm, knowledge is represented by symbols in a given logic and the reasoning power of this logic then plays a major role. Reasoning is based on axioms, theories and inference rules, leading to provable, non-refutable conclusions. In the second paradigm, where data and experience play the major role, statistical guarantees can be achieved, but conclusions are potentially refutable. As an example, fuzzy sets can cope with both approaches and establish links between them.

These ideas are illustrated in the field of image understanding and formulated as a spatial reasoning problem (section 2). Examples of combinations of different AI methods are given, both for knowledge and data representation, in section 3, and for reasoning in section 4. These methods find concrete applications in several domains such as medical imaging (only briefly mentioned in this paper). The question of explanations is addressed in section 5. Finally a short discussion on open research directions concludes the paper (section 6).

This paper is an extension of Bloch (2022), and focuses on the explainability aspects as well as the usefulness of hybrid AI and XAI for medical image understanding, in particular in pediatrics. The example of pediatric imaging is relevant here for illustrating the main topics developed in this paper, because of the challenging issues it raises (few data, very specific images, anatomy and pathologies, etc.). In addition, as mentioned in the next section, it is important with regards to the availability of domain knowledge and the usefulness of developing tools for explainable image understanding. This paper does not contain technical details – those can be found in the listed references.

2. Image understanding and spatial reasoning

Image understanding, at the simplest level, refers to the problem of recognizing an object or structure, or several objects in an image, which can either be real, as an observation of a part of the real world, or synthetic. But this may not be sufficient and more generally, relations between these objects should be considered towards a global recognition of the scene and a higher level interpretation, beyond individual objects. Furthermore, the recognition of an individual object can benefit from the recognition of others.

The question of semantics is central, since it is not directly in the image, but should be inferred based on visual features. We advocate that knowledge should be involved in this process. Indeed, while purely data driven approaches have proven powerful in image and computer vision problems, with sometimes impressive results, they still require a good accessibility to numerous and annotated data, where annotations bring the semantic information. This is not always possible and induces high costs (in terms of both human interactions and computation). Knowledge and models have then an important role to play. Image understanding is formulated as a spatial reasoning problem, combining representations of data and knowledge, pertaining to both objects and relations between objects (in particular spatial relations), as well as reasoning on them.

Let us take the example of pediatric medical imaging. In this domain, data may be scarce and present a high variability. Data are also very heterogeneous when they come from multicentric studies, with different hospitals, different imaging machines, different protocols and acquisition parameters. This makes the appearance of the same tissues, organs or pathologies vary a lot from one image to the other. This problem is sometimes addressed by transferring a model learned on adult images to children images. However, there is a huge domain gap, since the relative sizes of body parts, organs and pathologies vary considerably (in particular depending on the development stage of the children). Pathologies of children may differ from those observed in adults, the acquisitions should be as short as possible on children, thus inducing differences in image appearance. The contrast between tissues can also be quite different, even with the same acquisition protocol. Control cases and images of healthy children are even more rare, in particular due to ethical reasons. All this makes the problem particularly difficult. On the other hand, anatomical and medical knowledge is important, and was gathered over centuries. Using it is undoubtedly helpful.

Spatial reasoning has been largely developed in symbolic AI, based mostly on logic and benefitting from the reasoning apparatus of this logic (Aiello/Pratt-Hartmann/Benthem 2007). It has been much less developed for image understanding, where purely symbolic approaches are limited to account for numerical information. This again votes for hybrid approaches. Spatial reasoning evolved from purely qualitative and symbolic approaches, to more and more hybrid methods involving methods from mathematical morphology, fuzzy sets, graphs, machine learning, etc. to gain in expressivity (sometimes at the price of increased complexity). As an example, let us mention region connection calculus (RCC), that was first proposed in logical frameworks (first order, modal) and then augmented with fuzzy sets to handle imprecision, with mathematical morphology, lattice-based reasoning, etc. (Aiello/Pratt-Hartmann/Benthem 2007; Aiguier/Bloch 2019; Bloch 2021b; Landini et al. 2019; Randell/Cui/Cohn 1992; Schockaert et al. 2008; Schockaert/De Cock/Kerre 2009). The main ingredients in spatial reasoning include knowledge representation, imprecision representation and management, fusion of heterogeneous information (whether it is knowledge or data), reasoning and decision making. Approaches for spatial reasoning take a lot of inspiration from work in philosophy, linguistics, human perception, cognition, neuroimaging, art, etc. (see e.g., a related discussion for the case of spatial distances in Bloch 2003).

Models for image understanding are particularly useful to represent, in a formal way, knowledge (about the domain, the scene content and in particular its structure), image information (type of acquisition, geometry, characteristics of signal and noise, etc.), the potential imperfections of knowledge and data (imprecision, uncertainty, incompleteness, etc.), as well as the combination of knowledge and image information. These models are then included in algorithms to guide image understanding in concrete applications. Conversely, models can be built from data, to infer knowledge, or to provide a digital twin of a patient as a 3D model, useful to plan a surgery or a therapy, as well as to explain the plan (e.g., to other surgeons, to the patients and their parents in the case of pediatrics).

An important issue is the semantic gap (Smeulders et al. 2000), with the following question: how to link visual percepts from the images to symbolic descriptions? In artificial intelligence, this is close to the notions known as the anchoring or symbol grounding problem (Coradeschi/Saffiotti 1999; Harnad 1990). Solving the semantic gap issue has bidirectional consequences: on the one hand, it allows moving from a concept to its instantiation in the image (or feature) space, as a guide during spatial reasoning. On the other hand, it is part of the explainability, since it links results inferred from the image to concepts related to prior knowledge. For instance, anatomical knowledge says that the heart is between the lungs. Since the heart might be difficult to recognize directly in a medical image (e.g., a non-enhanced CT image), we may rely on its relative position with respect to the lungs (which are easier to detect in such images) to perform the task. This is an example where the recognition of an object benefits from the recognition of other objects, as mentioned at the beginning of this section. Conversely, we can explain the recognition of an image region as the heart because it is between the lungs (see section 5).

3. Information and knowledge representation

Representations of spatial entities can take various forms, either in the spatial domain (region, key points, bounding box, etc.), or abstractly, as in region connection calculus (RCC), as formulas in a given logic. Semi-quantitative (or semi-qualitative) representations as fuzzy sets (in either domain) constitute a good midway and can accommodate both numerical and symbolic representations (Zadeh 1965). Representations as numbers, imprecise numbers, intervals, distributions and linguistic values can all find a unifying framework with fuzzy sets. In this framework, different types of imperfections can be easily modeled, such as imprecision or blurriness on the boundaries of an object, on its location, shape or appearance, ambiguity, partial lack of information, etc. These imperfections can have varied sources, starting with the observed phenomenon, the sensors and the associated image reconstruction algorithms, and can also result from image processing steps such as filtering, registration and segmentation.

Spatial reasoning involves models of spatial entities, but also spatial relations between these entities. Here, the advantages of fuzzy representations become even more significant. This was already stated in the 1970s (Freeman 1975), but formal mathematical models were developed only later (see the review in Bloch 2005). The objective is to account for the intrinsic imprecision of concepts such as "close to", "to the left of" and "between", which are nevertheless perfectly understandable by humans in a given context and to account for the imprecision of the objects (even for a conceptually well-defined relation). In our previous work, we have designed mathematical models of several relations (set theoretical, topological, distances, directional relations and more complex relations such as between, along, parallel, etc.) by combining formalisms from mathematical morphology and fuzzy sets. They are detailed in Bloch and Ralescu (2023), chapter 6, and in the references cited therein. From a mathematical point of view, the common underlying structure is the one of complete lattices that allows instantiating the definitions, with the very same formalism in different frameworks: sets, fuzzy sets, graphs and hypergraphs, formal concept lattices, conceptual graphs, ontologies, etc., that can all be endowed with a lattice structure with appropriate partial orders. This becomes particularly useful when defining spatial relations based on mathematical morphology, a theory where deterministic operators are usually defined in a lattice. Our main idea was to design structuring elements, defined as fuzzy sets in the spatial domain, that provide the semantics of the spatial relation. Then applying a fuzzy morphological dilation of a reference object (whether fuzzy or not) using this structuring element provides the region of space where the considered relation is satisfied. The membership value of a point to the resulting fuzzy set is then interpreted as the degree to which the relation of this point to the reference object is satisfied. This approach can be applied to several classes of spatial relations: topological, distances, relative direction and more complex ones such as along, parallel, between, etc. (see e.g., Bloch 2021a; Bloch/Ralescu 2023 and the references therein). It applies to objects defined as sets or fuzzy sets in the spatial domain, but also those defined more abstractly as logical formulas, vertices of a (hyper-)graph, concepts, etc.

Note that most of the frameworks mentioned above carry structural information, useful for instance when representing the spatial arrangement of objects in a scene and in an image. To take a simple example, a graph can represent this structure, where vertices correspond to objects (e.g., anatomical structures in medical images) and edges correspond to relations between objects (e.g., contrast between two structures in a given imaging modality, relative position between objects, etc.), this graph being enhanced with the fuzzy representations of objects and their properties, as well as relations. For instance, the representation of a spatial relation can be abstract, as extracted from an ontology for example, or linked to the concrete domain of an image (degree of satisfaction of the relation, region of space where the relation to some object is satisfied, etc.), using linguistic variables, as explained next. Other structured representations of knowledge (including spatial knowledge) may rely on grammars, decision trees, relational algebras, or on temporal or spatial configurations and graphical models. They can also benefit from a fuzzy modeling layer, helping them cope with imprecision.

The relevance of fuzzy sets for knowledge representation, combined with other representations, lies in their ability to capture linguistic as well as quantitative knowledge and information. A useful notion is the one of linguistic variable (Zadeh 1975), where symbolic values, defined at an ontological level, have semantics defined by membership functions on a concrete domain at the image or features level. The membership functions and their parameters can be handcrafted, according to some expert knowledge on the application domain. They can also be learned, for instance from annotated data (Atif et al. 2007). The advantage of such representations is that linguistic characterizations may be less specific than numerical ones (and therefore need less information). Their two levels (syntactic and semantic) allow on the one hand for approximate modeling of vague concepts, and reasoning on them, and on the other hand constitute an efficient way to solve the semantic gap issue (see section 2) by providing semantics in concrete domains, according to each specific context. Linguistic variables, maintaining the consistency between concepts and data, therefore play an important role for explainability. Similarly, the goals of an image understanding problem can be expressed in an imprecise way, and again, translating vague concepts into useful representations and algorithms benefits from fuzzy modeling, in particular when using linguistic variables.

4. Reasoning

Based on the previous representations, the reasoning part takes various forms, separately or in combination, again in the spirit of hybrid AI. It is important to mention a few, mostly from previous work, which led to applications in medical imaging, in particular for brain structure recognition:[2] matching between a model and an image based on graph representations (Aldea/Bloch 2010; Cesar et al. 2005; Fasquel/Delanoue 2019; Perchant/Bloch 2002); sequential spatial reasoning mimicking the usual cognitive process where one may focus on an object that is easy to detect and to recognize, and then move progressively to more and more difficult objects by exploring the space based on the spatial relations with respect to previously recognized objects (Bloch/Géraud/Maître 2003; Colliot/Camara/Bloch 2006; Delmonte et al. 2019; Fouquier/Atif/Bloch 2012); exploration of the whole space and reducing progressively the potential region for each object, again mimicking a type of cognitive process, for instance by expressing the task as a constraint satisfaction problem (Deruyver/Hodé 1997; Nempont/Atif/Bloch 2013), logical reasoning based on abduction, to find the best explanations to the observations according to the available knowledge (Yang/Atif/Bloch 2015) and logical reasoning driven by an ontology (Hudelot/Atif/Bloch 2008).

2 These are only examples and similar approaches have been developed in other application domains, such as satellite imaging, video, music representations, etc.

In all these methods, an important feature is the combination of several approaches within the framework of hybrid AI, with the aim of explainability. Abstract knowledge representation and formal reasoning (typically using logics) are appropriate to build a knowledge base representing prior information (on anatomy for the considered examples) and to reason on it – the expressivity and the reasoning power depending on the chosen logic. Structural representations (graphs and hypergraphs, ontologies, conceptual graphs, concept lattices, etc.) are frameworks to convert expert knowledge on the spatial organization of objects (e.g., organs in medical imaging) into operational computational models. As mentioned in section 3, converting knowledge into meaningful representations and algorithms highly benefits from fuzzy modeling, in particular linguistic variables used to fill the semantic gap. This is indeed key to explainability. These models are then associated with structural representations to enrich them. For instance, fuzzy models of object features (shape, appearance) and of spatial relations can be attributes of vertices or edges of graphs, associated with concept descriptions in ontologies or conceptual graphs, providing semantics for these concepts, and considering them properties in fuzzy extensions of concept lattices, or providing semantics of logical formulas.

Usually several pieces of knowledge are involved together in the reasoning process. The advantages of fuzzy sets lie in the variety of combination operators, offering a lot of flexibility in their choice, that can be adapted to any situation at hand, and which may deal with heterogeneous information (Dubois/Prade 1985; Yager 1991). A classification of these operators was proposed by Bloch (1996), with respect to their behavior (in terms of conjunctive, disjunctive, compromise (Dubois/Prade 1985), the possible control of this behavior, their properties and their decisiveness.

Now, considering the recent huge developments in machine learning, and in particular deep learning, a recent trend is to combine such approaches with knowledge driven methods. This can be done at several levels (see e.g., Xie et al. 2021): to enhance the input (e.g., by including in the input of a neural network as a result of some image processing method as in Couteaux et al. 2019), as regularization terms in the loss function (e.g., to force the satisfaction of some relations), or to focus attention on specific patches based on geometric or topological information (e.g., vessel tree, see Virzi et al. 2018), or as post-processing to improve results (e.g., Chopin et al. 2022). Conversely, in some situations, the neural networks can use implicit spatial relations to solve a task such as object segmentation and recognition, as soon as the concerned objects are within the receptive field (Riva et al. 2022). Again, one of the advantages of such hybrid approaches is to improve interpretability and explainability. This is particularly important in medical imaging for increasing the confidence the user may have in an approach based on deep learning, consequently also increasing the adoption of such techniques.

Finally, the result of an image understanding system can be expressed in various forms (sets of (fuzzy) objects representing recognized structures, classes (of objects or pathologies for instance), properties of objects or structures and the relations they share, linguistic descriptions providing in a given vocabulary sentences describing the content of the image, etc.), finding yet again a unifying representation framework in fuzzy sets. The next step is then to provide explanations to these results.

5. Explanations

A first way to provide explanations is to rely on abductive reasoning in some logic.[3] Mathematical morphology is a useful theory for abductive reasoning and various logics (Aiguier et al. 2018; Bloch 2006; Bloch et al. 2018). An example is the use of erosion or derived operators to provide explanations for observations according to a knowledge base by applying these operators to a set of models for logical formulas or to a concept lattice. For instance, from a knowledge base on anatomy, expressed in some logics, and from segmentation and recognition results, higher level interpretations of an image can be derived using such a method of abductive reasoning (Atif/Hudelot/Bloch 2014; Yang/Atif/Bloch 2015). Then the image understanding problem itself is formulated as an explanatory process. The logic is endowed with fuzzy semantics, used to cope with imprecise statements in the knowledge base, such as "the lateral ventricles are dark in T1 weighted magnetic resonance images, the caudate nuclei are external to the lateral ventricles and close to them". Observation is the image and results from segmentation and recognition procedures. Hence, there is an interpretation on two levels: first at the object level, using the approaches presented in the previous sections involving fuzzy representations and structural models, and secondly globally, at the scene level. The advantages of using abstract formulation in a logic is that this second, higher level, interpretation can

3 Note that this is very natural, and explored since the antiquity, while it is much more difficult with machine learning that performs mostly inductions.

take intelligible forms, such as "this image presents an enhanced tumor, which is subcortical and has a small deforming impact on the other structures".

The language in which the knowledge is expressed should be defined according to the granularity level expected of the interpretation and based on whom the description is dedicated to (the explainee). For instance, the description of the content of a pathological brain image will depend on whether the explainee is anyone (without assuming any particular expertise), the patient, or a medical expert who wants to make a decision guided by this description and aims to interact with other experts. Other important questions are related to what should be explained. For instance, a medical expert needs mostly explanations of a result rather than explanations of every step of the algorithm as well as explanations of the links between the results, the data, and the available knowledge. More importantly, explanations are required when the results are unexpected. This is related to the question of when an explanation is needed and refers to the idea of contrastive explanations (why is the result A, when B was expected?).

To go further, another level of explanation is to identify which part of the knowledge base has actually been involved in the reasoning process or is relevant in the object or scene description. An implicit method to do so was mentioned above (Riva et al. 2022). More explicit methods are also very relevant for providing meaningful explanations to users. Fuzzy sets are then useful for establishing a link between the results derived from the image and concepts expressed in the knowledge base, as mentioned at the end of section 2. A simple example is to assess to which degree a spatial relation is satisfied between the resulting objects. Then explanations such as "this object is the left caudate nucleus because it is close to the left ventricle and to the left of it" are easy to derive. For instance, a given spatial relation between two identified objects can be computed, as a number or as a distribution, and then compared to the fuzzy model of this relation (Bloch/Atif 2016). An approach based on fuzzy frequent itemset mining has also been proposed (Pierrard/Poli/Hudelot 2021). Considering the example of structure recognition based on spatial reasoning, explanations become natural by identifying the spatial relations that actually play a role in the recognition. Furthermore, we can make use of hedges and quantifiers to find out whether "most" of the relations in a given set are indeed satisfied by a result, or involved in the image understanding process.

In all that precedes, hybrid AI and the combination of several approaches are at the core of:

- knowledge representation (object properties and relations between objects),
- associating attribute definition with structural frameworks such as graphs hypergraphs or other computational models representing the structure (in the sense of spatial organization) of a scene,
- semantics of logics,
- semantic gap solving,
- spatial reasoning for image understanding, computing similarities between a model and a result,
- providing descriptions of an image in a given language, providing cues for explainability.

They are the main medium to travel from knowledge to data and conversely explain results obtained from data according to the available knowledge.

6. Discussion

To go further in the field of hybrid AI and XAI for image understanding, principles expressed and discussed more generally in AI could be instantiated in this particular domain of application and pave the way for new research directions.

This starts with the definition of interpretability and explainability. An interesting distinction is proposed by Denis and Varenne (2022), where interpretability is defined as the composition of elements that are meaningful for humans, while explanation is strongly related to causality, and understanding is linked to unifying diversity under a common principle (this is may be somewhat different when interpreting an individual image as in medical imaging). In the works summarized in this paper, fuzzy sets are an example that can be used to make explicit the components of knowledge and image information that are involved in a reasoning process. This is done in a semi-qualitative way, close to human understanding, and therefore directly useful to provide explanations.

Seeing explanations as causality has been widely addressed, in particular by Halpern and Pearl (Halpern/Pearl 2005a; Halpern/Pearl 2005b) and by Miller (Miller 2019; Miller 2021), where structural models play a major role. Links with argumentation frameworks (Munro et al. 2022) and extensions of contrastive explanations for fuzzy sets (Bloch/Lesot 2022) have recently been proposed. Notions such as contrast and relevance are put to the fore, and would be also important to consider in image understanding. For instance, explaining why a certain decision was proposed by an algorithm, and not another, is a way to make explanations more convincing. A simple way to do so based on the methods presented here would be to compare resulting image descriptions with different models or decisions, and to identify which components in the knowledge or in the reasoning was responsible for a particular decision proposal. This would be particularly interesting in medical imaging, where explanations are mostly required when the result provided by an algorithm differs from the expected one. This deserves further investigation. The level of explanation should depend on the explainee, as mentioned above, and a deeper study of this aspect could take inspiration from the work on intelligibility by Coste-Marquis and Marquis (2020) (for instance based on projections on a given vocabulary). This goes with the idea of a human-centered evaluation of AI systems.

It has been advocated by Marcus (2020: 1) that new research should aim at developing "a hybrid, knowledge driven, reasoning based approach, centered around cognitive models, that could provide the substrate for a richer, more robust AI than is currently possible." This is exactly what research in image understanding based on hybrid AI is trying to do, but still at a modest level. The question of bias is related to the one of robustness. Statistical biases, on the one hand, are usually quite well identified in medical imaging. They may come from the limited data, from the under-representativity of parts of a population, from the specificities of the study (which intrinsically limit the population) and of the imaging center to the evolution of the data and the update of the algorithms, etc. This raises difficulties to adapt a method to a different population for instance. One may also wonder whether learning methods implicitly use information that can be relevant or that can be biased (which is then not explicitly identified). On the other hand, cognitive biases (such as confirmation, framing, complacency biases) may be more difficult to assess. An interesting direction of research is to investigate how hybrid AI can cope with these questions.

Finally, it would be interesting to investigate more deeply to which extent hybrid AI and XAI could help answering questions related to ethics, for instance in radiology, where these questions are often raised.

Acknowledgements

The author would like to thank all her co-authors and emphasize that the ideas summarized in this paper benefitted from many joint works with PhD candidates, post-doctoral researchers, colleagues in universities, research centers in several countries, as well as university hospitals and industrial partners. This work was partly supported by the author's chair in Artificial Intelligence (Sorbonne Université and SCAI). A part of the work was performed while the author was with LTCI, Télécom Paris, Institut Polytechnique de Paris.

List of references

Aiello, Marco/Pratt-Hartmann, Ian/Benthem, Johan (eds.) (2007): Handbook of Spatial Logics, Dordrecht: Springer.

Aiguier, Marc/Atif, Jamal/Bloch, Isabelle/Pino Pérez, Ramón (2018): "Explanatory Relations in Arbitrary Logics Based on Satisfaction Systems, Cutting, and Retraction." In: International Journal of Approximate Reasoning 102, pp. 1–20.

Aiguier, Marc/Bloch, Isabelle (2019): "Logical Dual Concepts Based on Mathematical Morphology in Stratified Institutions." In: Journal of Applied Non-Classical Logics 29/4, pp. 392–429.

Aldea, Emanuel/Bloch, Isabelle (2010): "Toward a Better Integration of Spatial Relations in Learning with Graphical Models." In: Fabrice Guillet/Gilbert Ritschard/Djamel Abdelkader Zighed/Henri Briand (eds.), Advances in Knowledge Discovery and Management, Berlin and Heidelberg: Springer, pp. 77–94.

Atif, Jamal/Hudelot, Céline/Bloch, Isabelle (2014): "Explanatory Reasoning for Image Understanding Using Formal Concept Analysis and Description Logics." In: IEEE Transactions on Systems, Man and Cybernetics: Systems 44/5, pp. 552–570.

Atif, Jamal/Hudelot, Céline/Fouquier, Geoffroy/Bloch, Isabelle/Angelini, Elsa (2007): "From Generic Knowledge to Specific Reasoning for Medical Image Interpretation using Graph-based Representations." In: International Joint Conference on Artificial Intelligence IJCAI'07, Hyderabad, India, pp. 224–229.

Avila Garcez, Artur d'/Lamb, Luís C. (2023): "Neurosymbolic AI: the 3rd wave." In: Artificial Intelligence Review (https://doi.org/10.1007/s10462-023-10448-w).

Bloch, Isabelle (1996): "Information Combination Operators for Data Fusion: A Comparative Review with Classification." In: IEEE Transactions on Systems, Man, and Cybernetics 26/1, pp. 52–67.

Bloch, Isabelle (2003): "On Fuzzy Spatial Distances." In: Peter W. Hawkes (ed.), Advances in Imaging and Electron Physics, Amsterdam: Elsevier, pp. 51–122.

Bloch, Isabelle (2005): "Fuzzy Spatial Relationships for Image Processing and Interpretation: A Review." In: Image and Vision Computing 23/2, pp. 89–110.

Bloch, Isabelle (2006): "Spatial Reasoning under Imprecision using Fuzzy Set Theory, Formal Logics and Mathematical Morphology." In: International Journal of Approximate Reasoning 41/2, pp. 77–95.

Bloch, Isabelle (2021a): "Mathematical Morphology and Spatial Reasoning: Fuzzy and Bipolar Setting." In: TWMS Journal of Pure and Applied Mathematics, special issue 12/1, pp. 104–125.

Bloch, Isabelle (2021b): "Modeling Imprecise and Bipolar Algebraic and Topological Relations Using Morphological Dilations." In: Mathematical Morphology – Theory and Applications 5/1, pp. 1–20.

Bloch, Isabelle (2022): "Hybrid Artificial Intelligence for Knowledge Representation and Model-Based Medical Image Understanding – Towards Explainability." In: Discrete Geometry and Mathematical Morphology, Second International Joint Conference, DGMM 2022, Straßbourg, France, pp. 17–25.

Bloch, Isabelle/Atif, Jamal (2016): "Defining and Computing Hausdorff Distances Between Distributions on the Real Line and on the Circle: Link Between Optimal Transport and Morphological Dilations." In: Mathematical Morphology: Theory and Applications 1/1, pp. 79–99.

Bloch, Isabelle/Géraud, Thierry/Maître, Henri (2003): "Representation and Fusion of Heterogeneous Fuzzy Information in the 3D Space for Model-Based Structural Recognition – Application to 3D Brain Imaging." In: Artificial Intelligence 148, pp. 141–175.

Bloch, Isabelle/Lang, Jérôme/Pino Pérez, Ramón/Uzcátegui, Carlos (2018): Morphology for Knowledge Dynamics: Revision, Fusion, Abduction, arXiv Preprint (https://arxiv.org/abs/1802.05142).

Bloch, Isabelle/Lesot, Marie-Jeanne (2022): "Towards a Formulation of Fuzzy Contrastive Explanations." In: IEEE International Conference on Fuzzy Systems (FUZZ-IEEE), Padua, Italy, pp. 1–8.

Bloch, Isabelle/Ralescu, Anca (2023): Fuzzy Sets Methods in Image Processing and Understanding: Medical Imaging Applications, Cham: Springer Nature.

Cesar, Roberto M. Jr./Bengoetxea, Endika/Bloch, Isabelle/Larranaga, Pedro (2005): "Inexact Graph Matching for Model-Based Recognition: Evaluation and Comparison of Optimization Algorithms." In: Pattern Recognition 38/11, pp. 2099–2113.

Chopin, Jérémy/Fasquel, Jean-Baptiste/Mouchère, Harold/Dahyot, Rozenn/Bloch, Isabelle (2022): "Improving Semantic Segmentation with Graph-Based Structural Knowledge." In: International Conference on Pattern Recognition and Artificial Intelligence (ICPRAI 2022), Paris, France, pp. 173–184.

Colliot, Olivier/Camara, Oscar/Bloch, Isabelle (2006): "Integration of Fuzzy Spatial Relations in Deformable Models – Application to Brain MRI Segmentation." In: Pattern Recognition 39/8, pp. 1401–1414.

Coradeschi, Silvia/Saffiotti, Alessandro (1999): "Anchoring Symbols to Vision Data by Fuzzy Logic." In: European Conference on Symbolic and Quantitative Approaches with Uncertainty (ECSQARU'99), London, UK, pp. 104–115.

Coste-Marquis, Sylvie/Marquis, Pierre (2020): "From Explanations to Intelligible Explanations." In: 1st International Workshop on Explainable Logic-Based Knowledge Representation (XLoKR'20), 2020, Rhodes, Greece (https://univ-artois.hal.science/hal-03300380/document).

Couteaux, Vincent/Si-Mohamed, Salim/Nempont, Olivier/Lefevre, Thierry/Popoff, Alexandre/Pizaine, Guillaume/Villain, Nicolas/Bloch, Isabelle/Cotten, Anne/Boussel, Loic (2019): "Automatic Knee Meniscus Tear Detection and Orientation Classification with Mask-RCNN." In: Diagnostic and Interventional Imaging 100/4, pp. 235–242.

De Raedt, Luc/Dumančić, Sebastijan/Manhaeve, Robin/Marra, Giuseppe (2020): "From Statistical Relational to Neuro-Symbolic Artificial Intelligence." In: Proceedings of the Twenty-Ninth International Joint Conference on Artificial Intelligence (IJCAI-20), Yokohama, Japan, pp. 4943–4950.

Delmonte, Alessandro/Mercier, Corentin/Pallud, Johan/Bloch, Isabelle/Gori, Pietro (2019): "White Matter Multi-Resolution Segmentation Using Fuzzy Set Theory." In: IEEE International Symposium on Biomedical Imaging (ISBI), Venice, Italy, pp. 459–462.

Denis, Cristophe/Varenne, Franck (2022): "Interprétabilité et explicabilité de phénomènes prédits par de l'apprentissage machine." In: Revue Ouverte d'Intelligence Artificielle 3/3-4, pp. 287–310.

Deruyver, Aline/Hodé, Yann (1997): "Constraint Satisfaction Problem with Bilevel Constraint: Application to Interpretation of Over-Segmented Images." In: Artificial Intelligence 93/1-2, pp. 321–335.

Dubois, Didier/Prade, Henri (1985): "A Review of Fuzzy Set Aggregation Connectives." In: Information Sciences 36/1-2, pp. 85–121.

Fasquel, Jean-Baptiste/Delanoue, Nicolas (2019): "A Graph-Based Image Interpretation Method Using A Priori Qualitative Inclusion and Photometric Relationships." In: IEEE Transactions on Pattern Analysis and Machine Intelligence 41/5, pp. 1043–1055.

Fouquier, Geoffroy/Atif, Jamal/Bloch, Isabelle (2012): "Sequential Model-Based Segmentation and Recognition of Image Structures Driven by Visual Features and Spatial Relations." In: Computer Vision and Image Understanding 116/1, pp. 146–165.

Freeman, John (1975): "The Modelling of Spatial Relations." In: Computer Graphics and Image Processing 4/2, pp. 156–171.

Garnelo, Marta/Shanahan, Murray (2019): "Reconciling Deep Learning with Symbolic Artificial Intelligence: Representing Objects and Relations." In: Current Opinion in Behavioral Sciences 29, pp. 17–23.

Halpern, Joseph Y./Pearl, Judea (2005a): "Causes and Explanations: A Structural-Model Approach. Part I: Causes." In: The British Journal for the Philosophy of Science 56/4, pp. 843–887.

Halpern, Joseph Y./Pearl, Judea (2005b): "Causes and Explanations: A Structural-Model Approach. Part II: Explanations." In: The British Journal for the Philosophy of Science 56/4, pp. 889–911.

Harnad, Stevan (1990): "The symbol grounding problem." In: Physica 42/1-3, pp. 335–346.

Hudelot, Céline/Atif, Jamal/Bloch, Isabelle (2008): "Fuzzy Spatial Relation Ontology for Image Interpretation." In: Fuzzy Sets and Systems 159, pp. 1929–1951.

Kahneman, Daniel (2012): Thinking, Fast and Slow, New York: Penguin.

Kautz, Henry A. (2022): "The third AI summer: AAAI Robert S. Engelmore Memorial Lecture." In: AI Magazine 43/1, pp. 93–104.

Landini, Gabriel/Galton, Antony/Randell, David/Fouad, Shereen (2019): "Novel Applications of Discrete Mereotopology to Mathematical Morphology." In: Signal Processing: Image Communication 76, pp. 109–117.

Marcus, Gary (2020): The Next Decade in AI: Four Steps Towards Robust Artificial Intelligence, arXiv Preprint (https://arxiv.org/abs/2002.06177).

Miller, Tim (2019): "Explanation in Artificial Intelligence: Insights from the Social Sciences." In: Artificial Intelligence 267, pp. 1–38.

Miller, Tim (2021): "Contrastive Explanation: A Structural-Model Approach." In: The Knowledge Engineering Review 36, e14.

Munro, Yann/Bloch, Isabelle/Chetouani, Mohamed/Lesot, Marie-Jeanne/Pelachaud, Catherine (2022): "Argumentation and Causal Models in Human-Machine Interaction: A Round Trip." In: 8th International Workshop on Artificial Intelligence and Cognition, Örebro, Sweden (https://hal.sorbonne-universite.fr/hal-03739310).

Nempont, Olivier/Atif, Jamal/Bloch, Isabelle (2013): "A Constraint Propagation Approach to Structural Model-Based Image Segmentation and Recognition." In: Information Sciences 246, pp. 1–27.

Perchant, Aymeric/Bloch, Isabelle (2002): "Fuzzy Morphisms between Graphs." In: Fuzzy Sets and Systems 128/2, pp. 149–168.

Pierrard, Régis/Poli, Jean-Philippe/Hudelot, Céline (2021): "Spatial Relation Learning for Explainable Image Classification and Annotation in Critical Applications." In: Artificial Intelligence 292, 103434.

Randell, David A./Cui, Zhan/Cohn, Anthony G. (1992): "A Spatial Logic based on Regions and Connection." In: Principles of Knowledge Representation and Reasoning. Proceedings of the Third International Conference, Cambridge, MA, USA, pp. 165–176.

Riva, Mateus/Gori, Pietro/Yger, Florian/Bloch, Isabelle (2022): "Is the U-Net Directional-Relationship Aware?" In: 2022 IEEE International Conference on Image Processing (ICIP), Bordeaux, France, pp. 3391–3395.

Schockaert, Steven/De Cock, Martine/Cornelis, Chris/Kerre, Etienne E. (2008): "Fuzzy Region Connection Calculus: Representing Vague Topological Information." In: International Journal of Approximate Reasoning 48/1, pp. 314–331.

Schockaert, Steven/De Cock, Martine/Kerre, Etienne E. (2009): "Spatial Reasoning in a Fuzzy Region Connection Calculus." In: Artificial Intelligence 173/2, pp. 258–298.

Smeulders, Arnold/Worring, Marcel/Santini, Simone/Gupta, Amarnath/Jain, Ramesh (2000): "Content-based Image Retrieval at the End of the Early Years." In: IEEE Transactions on Pattern Analysis and Machine Intelligence 22/12, pp. 1349–1380.

Virzi, Alessio/Gori, Pietro/Muller, Cécile Olivia/Mille, Eva/Peyrot, Quoc/Berteloot, Laureline/Boddaert, Nathalie/Sarnacki, Sabine/Bloch, Isabelle (2018): "Segmentation of Pelvic Vessels in Pediatric MRI Using a Patch-based Deep Learning Approach." In: PIPPI MICCAI Workshop, Granada, Spain, pp. 97–106.

Xie, Xiaozheng/Niu, Jianwei/Liu, Xuefeng/Chen, Zhengsu/Tang, Shaojie/Yu, Shui (2021): "A Survey on Incorporating Domain Knowledge into Deep Learning for Medical Image Analysis." In: Medical Image Analysis 69, 101985.

Yager, Ronald R. (1991): "Connectives and Quantifiers in Fuzzy Sets." In: Fuzzy Sets and Systems 40, pp. 39–75.

Yang, Y./Atif, Jamal/Bloch, Isabelle (2015): "Abductive Reasoning Using Tableau Methods for High-level Image Interpretation." In: KI 2015: Advances in Artificial Intelligence, 38th Annual German Conference on AI, Dresden, Germany, pp. 356–365.

Zadeh, Lotfi A. (1965): "Fuzzy Sets." In: Information and Control 8/3, pp. 338–353.

Zadeh, Lotfi A. (1975): "The Concept of a Linguistic Variable and its Application to Approximate Reasoning." In: Information Sciences 8/3, pp. 199–249.

AI-based approaches in Cultural Heritage

Investigating archaeological landscapes in Scandinavian forestland

Giacomo Landeschi

1. Computational methods in archaeology

There is a long tradition of using computers and computational methods in archaeology. In 2023, the Computer and Quantitative methods in Archaeology Conference (CAA) turned 50, with the first meeting originally hosted in Birmingham back in 1973 (Djindjian et al. 2015). As a spatial discipline, archaeology relies on quantitative and statistical methods to investigate and detect patterns connectable to the presence of past humans in a given landscape. There is a constant need of measuring spatial distributions of artifacts, monuments and settlements in a multi-scalar and multi-temporal perspective. Quantification has rapidly become a standard procedure for generating deeper insights into the human past and for this scope the introduction of computational methods marked a tremendous advance in archaeological practice. Geographical Information Systems (GIS) are considered one of the first products to be introduced in archaeology for the purpose of managing spatial datasets related to archaeological excavations, field surveys or landscape investigations. Most of GIS-based analysis was aimed at the detection of new archaeological material in areas not previously investigated, but soon, the importance of these computational methods for generating more complex, explanatory models capable of providing archaeologists with interpretative tools for generating a better understanding of past human activities, became clear. In this context, predictive modelling was introduced as a methodological framework for forecasting archaeological presence in specific portions of landscapes (Kohler/Parker 1986; Wescott/Brandon 2000; Verhagen 2007). It comes as no surprise that the use of GIS among different institutions both in the public and private sector became increasingly popular with many projects starting with the purpose of managing cultural heritage in a more effective way. The very idea of developing a statistical/inferential method to detect, in a semi-automatic way, significant numbers of archaeological material did represent an important game changer in the discipline, enabling archaeologists to re-interpret past landscapes in a totally different manner. Beside predictive modelling, among the most popular GIS applications in landscape archaeology it is worth to mention the use of Least-Cost-Path (LCP) analysis for examining best-suited routes in a landscape that is likely to have been crossed by people in the past based on the analysis of factors that could have either facilitated or prevented human movement, such as slope, natural barriers and distance to be crossed (Herzog 2014). Another very widespread application is viewshed analysis, enabling archaeologists to determine locations in the landscape that were more visually exposed or secluded while considering a number of observation points used to perform the calculation (Wheatley 1995). More recently, thanks to the dramatic advances in hardware and software performance, more sophisticated and efficient tools have been introduced in support of archaeological research. 3D-based technology has marked a significant advance in the area of site documentation and museum communication and dissemination (Barcelo et al. 2000). Apart from traditional laser-scanning techniques, there are now image-based modelling techniques, enabling specialists to rely on relatively low-cost solutions to acquire and document archaeological features and monuments in 3D (Dell'Unto 2014). On a similar way, the advances in Unmanned Aircraft System (UAS) technology led to the definition of innovative pipelines for the data capture and the documentation of large portions of an archaeological landscape, making it possible to investigate archaeological features in a multi-scalar way, increasing the level of spatial definition to a detail that is unparalleled by any of the existing satellite sensors commonly used in landscape archaeology (Adamopoulos/Rinaudo 2020). Among the most notable innovations that impacted the discipline in the last ten years, it is important to mention Artificial Intelligence (AI) and its contributions to the analysis of 'big data' that is now produced on a daily basis as a result of the introduction of more advanced sensing technology and sophisticated methods of data collection. Before examining in detail the impact AI had and is having on archaeological data analysis, the next sections will briefly introduce two technologies that are particularly relevant for the setup of the described work pipeline, namely Remote Sensing and LiDAR, which are related to the techniques and the sensors specifically employed for data acquisition.

2. Remote sensing

Numerous studies have extensively documented the use of satellite remote sensing in archaeology (Campana/Forte 2006; Parcak 2009; Lasaponara/Masini 2012). These studies have specifically examined a wide range of geographical regions and time periods, providing valuable insights into the application of this technology in diverse contexts. Satellite remote sensing relies on sensors that can capture and analyze radiating energy across various wavelengths in the electromagnetic spectrum. These sensors can convert this energy into new information regarding the physical and chemical attributes of the specific area on the Earth's surface that is being examined. Archaeological use of satellite multispectral images can be traced back to the 1970s when the initial satellite missions were launched by NASA and the Landsat program was initiated (Giardino 2012). These early endeavours marked the beginning of employing satellite multispectral images for archaeological purposes. Right from the start, it became evident that this form of remote sensing would have a profound impact on archaeology. It provided specialists with the ability to survey expansive areas of land, enabling them to identify numerous ground anomalies. During the initial stages of using multispectral images, the spatial resolution was relatively low. As a result, the primary focus at that time was on identifying paleo-environmental elements and small-scale field systems (Rainey et al. 1976). This emphasis allowed archaeologists to develop a more comprehensive understanding of how landscapes were utilized and exploited during prehistoric and historical periods. By studying these features, researchers could gain valuable insights into the human interactions and activities that shaped the landscape in the past. Satellite remote sensing also plays a crucial role when examining landscapes that can be described as challenging from a logistical standpoint. Within the field of archaeology, there exist numerous geographical regions that have restricted accessibility due to environmental obstacles or administrative/political circumstances. An illustrative example of this is the exploration and identification of Mayan cities in Central America, where the dense and expansive rainforest poses a significant challenge to traditional on-site research methods (Saturno et al. 2007). An additional issue arises in conflict and war zones, where conducting archaeological investigations on the ground becomes either impossible or, if attempted, can only take place after heritage sites have suffered damage and looting. In this respect, Campana et al. (2022) showcased how remote sensing played a vital role in assessing the extent of war damages inflicted on the ancient city of Niniveh following the occupation and destructive actions carried out by ISIS. On a similar note, in the context of the EAMENA project, which focuses on safeguarding endangered heritage sites, a new open-access database was established. The main objective of this initiative was to provide archaeologists and cultural heritage experts with access to satellite imagery from regions in the Middle East and North Africa that have been impacted by war and looting (Bewley et al. 2016). This database allows users to visualize and analyze the imagery for research and preservation purposes. In summary, over the past two decades, satellite multispectral images have had a significant influence on landscape archaeology. There is now a widespread agreement on the importance of utilizing such datasets for investigating archaeological sites from various scales and temporal perspectives. The introduction of high-resolution sensors capable of producing satellite images with a spatial resolution of up to 30 cm has been a true game changer in this field. This significant advancement in landscape archaeology has permitted archaeologists to utilize multispectral information when investigating individual monuments or sites in a manner that was unimaginable during the early stages of satellite remote sensing. Similarly to geophysical prospecting techniques, it is crucial to emphasize the importance of conducting ground-truthing when interpreting satellite imagery. This process involves verifying the actual presence of archaeological material on the ground, which serves to validate the performance and accuracy of the sensor used in the investigation.

3. LiDAR

LiDAR, one of the latest technologies introduced in landscape archaeology, has undeniably had a significant impact on site detection, particularly in areas characterized by dense forest coverage. The acronym LiDAR stands for Light Detection and Ranging, which involves the use of a sensor that emits a laser beam towards a target surface. The receiver measures the time it takes for the laser beam to return, enabling the calculation of the distance between the sensor and the target. This data allows for the derivation of precise 3D coordinate values for each measured point. Through the application of specific filtering algorithms, the resulting point cloud from LiDAR data can be classified based on their positions on the land surface. This classification enables the differentiation of points belonging to the ground surface from those associated with vegetation elements. The ability of LiDAR to penetrate dense vegetation and detect ground anomalies makes it an ideal solution for investigating areas with extensive vegetation cover. This capability surpasses the limitations of other sources, such as satellite multispectral images, which may not be able to detect such ground-level details. Indeed, LiDAR has marked a significant transformation in various research scenarios, ranging from the tropical landscapes of Central America to the forests of Northern Europe. A notable example is the systematic investigation of the Mayan site of Caracol in Belize, where the utilization of airborne LiDAR enabled archaeologists to detect and map extensive sections of an ancient city, including structures, causeways, and agricultural terraces, unveiling the complex nature of the site (Chase et al. 2011). Similarly, in a completely different context, this technology has enabled archaeologists to reexamine the archaeological landscape surrounding Stonehenge in Southern England. Through LiDAR, they were able to map a substantial number of features, such as field systems, burial mounds, and ancient river courses, in a manner that surpassed the limitations of solely analyzing aerial photographs (Bewley/Crutchley/Shell 2005). While the conventional method for data acquisition involves the use of aircrafts, such as small planes or helicopters, in the last few years a new generation of drones equipped with LiDAR sensors has emerged. This development has resulted in a significant enhancement in the point density of the acquired surface data and has made lower-cost solutions available for individual data acquisitions. Traditionally, data collection was limited to professional commissioned flights conducted with aircrafts, but the advent of LiDAR-equipped drones has revolutionized this process (Casana et al. 2021). Regarding data output, the point cloud obtained from LiDAR acquisition is commonly filtered to extract points classified as 'terrain'. These filtered points are then utilized to generate a Digital Terrain Model (DTM). The DTM is typically represented as a raster grid, where each grid cell corresponds to an elevation value, providing a detailed representation of the terrain. DTMs can be further processed and converted into thematic maps, where ground anomalies can be emphasized using specialized algorithms. One notable application, as further described in the next sessions, is the integration of LiDAR-derived raster images with Artificial Intelligence (AI) techniques. By training AI models on known features within a dataset, this approach enables the semi-automatic extraction of similar features from the larger landscape. Archaeologists can benefit from this method as it facilitates the faster and more efficient detection of numerous archaeological features, aiding in their research efforts (Küçükdemirci et al. 2022).

4. Artificial Intelligence and archaeology

Artificial Intelligence (AI) has emerged as a powerful tool in archaeology, revolutionizing various aspects of research and analysis. AI techniques, such as machine and deep learning for computer vision tasks, are being applied to archaeological data to assist in tasks such as feature detection, classification, data interpretation, and predictive modelling. One of the significant contributions of AI in archaeology is in the field of image analysis. AI algorithms can be trained to recognize and identify archaeological features, artifacts and patterns in large datasets of images, including satellite imagery, aerial photographs and ground-based photographs. This enables archaeologists to automate the process of feature identification, saving time and effort in data analysis. AI also plays a crucial role in data processing and analysis. By utilizing machine and deep learning algorithms, large archaeological datasets can be analyzed to identify patterns, correlations and trends that may not be easily discernible by human researchers. This allows for more comprehensive and efficient data analysis, leading to new insights and interpretations.

Artificial Intelligence (AI) has been introduced in the archaeological discourse as early as the 1980s, with the purpose of supporting expert systems for the definition of heuristic frameworks in the analysis of the archaeological record based on a joint effort involving domain specialists (archaeologists), software engineers and computer scientists (Wilcock 1985).

Baker (1987) instead seems to use the definition of 'expert systems' synonymously with 'AI', pointing out the problematic nature of these computational tools and its applicability in the archaeological domain. Patel and Stutt (1989) identify different application areas for AI/expert system technology, highlighting the urgency for archaeologists to get confronted with significant amounts of data. Archaeological reasoning being an important field in the application of AI, the authors introduce KIVA, a programming language capable of simulating reasoning in connection with archaeological data, providing different interpretations based on the combination of data and context conditions where artifacts and single findings have been collected.

More recently, the use of AI-based applications had a dramatic increase in archaeological practice, becoming a de-facto standard in many sub-fields of the discipline. As Mantovan and Nanni (2020) show, the research areas include (but are not limited to) musealization, artifact and ecofact analysis, landscape interpretation, ancient building monitoring and underwater archaeology. Image recognition has been employed for automatic detection and comparison of categories of pictures belonging to different museum collections from all over the world with the aim to describe objects from the same cultural/historical context in order to facilitate findability and accessibility of material that would be otherwise difficult to retrieve (Wilbrink et al. 2023). Similarly, AI-based approaches including machine learning and deep learning have been used to develop supportive tools for archaeologists in the field to allow a quick and efficient recognition of ancient pottery classes based on the examination of images taken from sherds and other fragmentary material that is typically found in the archaeological stratigraphy (Gualandi/Gattiglia/Anichini 2021; Anichini et al. 2021). Concerning the study and the analysis of ancient buildings, significant results have been obtained in the analysis of the Forbidden City in China by introducing advanced point cloud classification tools thanks to the introduction of more refined algorithms such as PointNet++ which enabled users to improve the accuracy of the 3D point segmentation, reducing the number of data sample to be collected (Hu et al. 2022). Still, it is in archaeological remote sensing that most of current AI-based approaches are employed with image classification and object detection being the main functionalities applicable to investigate an archaeological landscape by examining the presence of ancient features and any transformation occurring in the natural environment. Karamitrou et al. (2022) recently explored the possibility of using Google Earth's freely available satellite high-resolution images to test deep learning networks for the automatic detection of archaeological features in very diverse geographical areas distributed worldwide. The application of AI in the analysis of satellite multispectral images has proven significant results also on relatively low spatial resolution datasets such as Corona, enabling specialists to refine the quality of data interpretation due to an improved performance of the classification tools, with a lower number of false positives obtained (Soroush et al. 2020). Orengo and Garcia-Molsosa (2019) further improved the capabilities of small finds detection in UAS-derived images by introducing a machine learning approach that allows archaeologists to easily spot small pottery shards scattered over a field surface and to obtain a better performance than the one obtained by on-site visual inspection. In underwater archaeology, machine learning approaches have been recently explored for detecting shipwrecks and other categories of submerged sites based on the processing of datasets of images derived from Autonomous Underwater Vehicles (AUVs) acquisition, in which data augmentation was applied in order to increase the number of samples for the training dataset, due to the relative scarcity of submerged sites available (Nayak et al. 2021).

Another important field of application for AI is geophysical prospections, a very effective tool for the detection of buried structures and to collect subsoil information. A typical dataset produced during a Ground Penetrating Radar (GPR) survey consists of a very large number of images where it is possible to extract in an automatic way information that is useful for the archaeological interpretation. In this sense, CNN-based approaches have proven to be very effective for the automated interpretation of these datasets (Küçükdemirci/Sarris 2020; Küçükdemirci/Sarris 2022).

5. Investigating archaeological features in a forestland

The combination of non-destructive methods presents a vast range of case studies that can be explored and examined. In the context of Scandinavia and specifically Sweden, the utilization of AI-based techniques plays a crucial role in identifying and studying archaeological features within the landscape. Particularly in Sweden, the use of image recognition methods holds great potential in the analysis of LiDAR datasets. This is because LiDAR enables the observation of archaeological features on the ground surface, even in areas covered by vegetation where traditional satellite or aerial multispectral imagery fails to provide adequate information. By employing AI-driven approaches, the analysis of LiDAR data can yield highly effective results in the detection and analysis of archaeological traces in these challenging environments. So far, only a few studies have tried to investigate archaeological traces hidden in the woodland, consisting of several categories of sites including burial areas, settlements and productive areas such as kilns or mints. Recently, Lindholm et al. (2021) demonstrated the pivotal role of boreal forest land by providing its ancient inhabitants with important sources of the economy of Scandinavian regions from the Roman Iron Age (1st to 4th century CE) to the later Middle Ages (1050 to 1520 CE). Such research now allows archaeologists to challenge the current view of Scandinavian forest land as a marginal space and to investigate more thoroughly vast portions of landscape where traditional forms of survey have long been discarded due to a significant imbalance between benefits and costs. To fill this gap, researchers at Lund University have recently tried to introduce innovative approaches to the study of forestland regions by relying on integrated methods including AI, ML, LiDAR and GIS. The main purpose for this project is to understand diachronic transformations that occurred in the landscape of the Scania region (Southern Sweden), witnessing the change in destination from agricultural fields into woodland areas. In this context, there is a significant variety of archaeological features that lie beneath the dense canopy coverage and that consists of artifacts connected to the ancient agricultural exploitation of the landscape. These consist of stone walls, boundaries, clearance cairns, terraces and Celtic fields and all of these features that can only be detected by examining the LiDAR-derived imagery where the ground-related information is visualized in the form of a Digital Terrain Model (DTM). Indeed, differences in the elevation values observed in DTMs are important markers of the presence of buried structures or features whose appearance is marked by patterns of discontinuity in the topography of the area under scrutiny. As for this project, the main focus was the analysis of the so-called clearance cairns, human-made piles of stones that were created in ancient times as a result of clearing space for agriculture in selected portions of land. This category of finds is very widespread all over Scandinavia and represents one of the most common archaeological features identifiable in Swedish forests. Their shape is quite regular (2–6 meters in diameter and 0,2–0,5 meters high) and is characterised by a moss or grass turf coverage (Lagerås/Bartholin 2003).

Figure 1: Clearance cairn located in the study area of Söderhånsen National Park. Typically, prehistoric or medieval ones can be recognized either by its diameter (between 2 and 6 meters), or the reduced size of the stones and the presence of moss partially covering it. Image courtesy of the author.

Figure 2: Case study area in the national park of Söderhånsen, central Scania (Southern Sweden). Red-marked features indicate areas of possible clearance cairns as a result of CNN data processing. Image courtesy of the author.

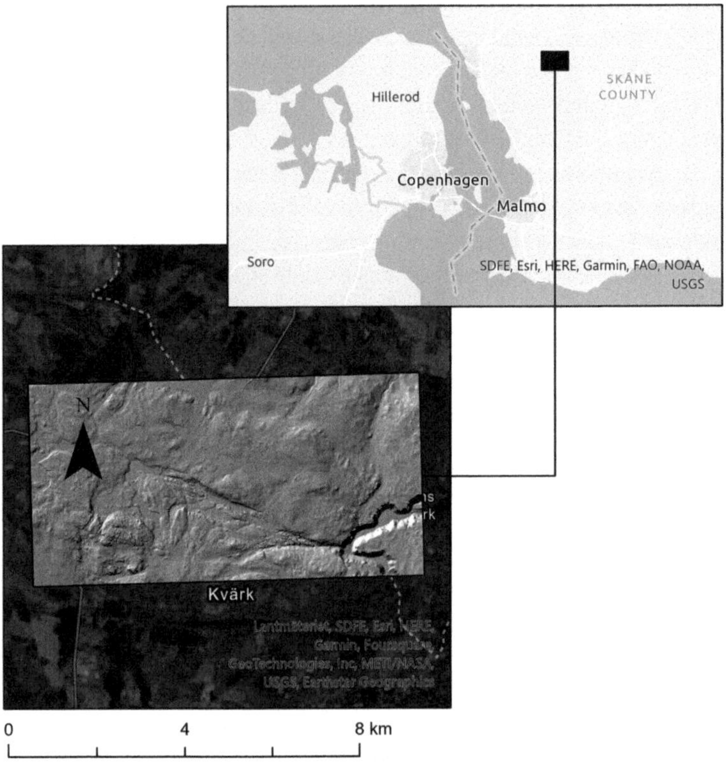

Concerning their chronology, the oldest clearance cairns date back to the Bronze Age (9th to 6th centuries BCE) and their presence indicates an area that used to be agricultural land and whose boundaries were often defined by straight lines made of clearance stones too. Identifying those features can mark an important advance in the study of forestland and provide a significant contribution to the management of cultural heritage and forest resources along with a powerful instrument for planning new development. So far, only a relatively small number of clearance cairns has been identified and reported on the Swedish National Heritage Board (RAÄ, https://app.raa.se/open/fornsok/), with most of them still to be identified.

In this respect, advances in remote sensing techniques, including the use of LiDAR-derived sources combined with AI and GIS can now dramatically contribute to a more effective identification of these archaeological traces. To prove this working hypothesis, a test case area was selected in the Söderånsen National park, located in central Scania, where a very vast portion of land is now covered by protected forest land (fig. 2). In this area, a number of clearance cairns was previously identified and reported in the RAÄ registry. Still, by examining a LiDAR-derived DTM it is possible to observe an even larger number of ground anomalies in areas not previously documented and that can be possibly interpreted as ancient clearance cairns. Having the geometrical reference provided by the known previously identified clearance cairns allowed to obtain a training and a comparison dataset to be used for testing the prediction of the AI network.

6. Methodology

As previously stated, the workflow for the identification, classification and interpretation of clearance cairns in the study areas is based on the integration of different acquiring techniques and data processing methods. At the core of the system, a spatial geodatabase was set up to collect, store and process all the datasets related to the landscape of Söderåsen National Park. LiDAR-derived raster DTMs were chosen as a primary source for performing the AI-based spatial analysis. This source is freely made available for researchers through the Swedish Cadastral Agency web portal (Lantmäteriet, https://www.lantmateriet.se) and comes in the form of a vector 3D point cloud with an average spatial density of 0,75 points per square meter. These data are then processed and converted into raster DTMs with a spatial resolution of 0,5. These raster images are the result of GIS-based filtering operations that allow users to remove any vegetation point and to obtain a 'clean' model of the terrain made by ground surface points. These points are eventually used to derive a Triangulated Irregular Network (TIN) model that will be in turn transformed into a raster DTM by applying specific interpolating algorithms. As a final step of this process, a slope and a hillshade map are generated in order to enhance the visibility of the archaeological features that need to be spotted (fig. 3).

Figure 3: Portion of the study area covered by woodland, as appears in a RGB aerial image (A). From the LiDAR-derived DTM, slope (B) and hillshade (C) algorithms were applied to enhance the visibility of clearance cairns (that show up in a pretty circular shape and are evenly distributed throughout the selected areas). Image courtesy of the author.

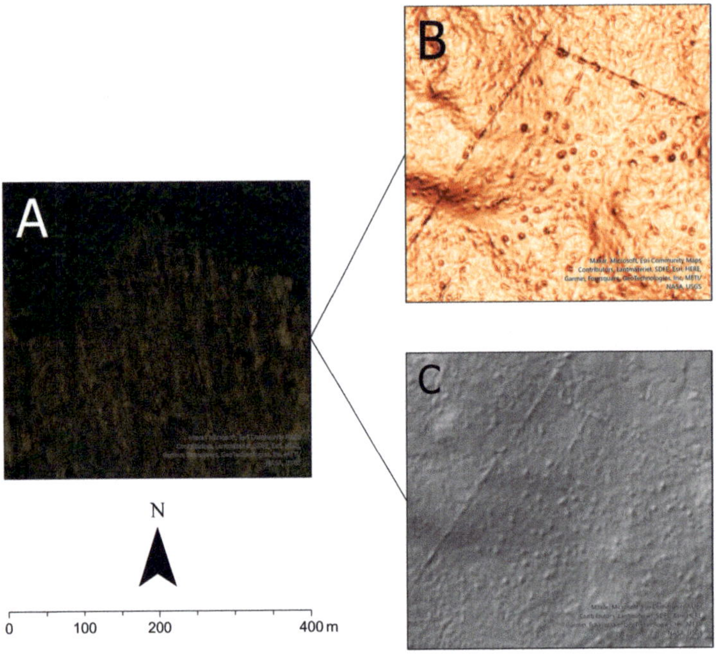

More GIS-based operations are performed to extract tiles that must have included known clearance cairns in order to create a valid training dataset to feed the network. Typical metadata configuration for each tile was characterized by an uncompressed .tiff file with a depth of 8 bit.

As thoroughly described by Küçükdemirci et al. (2022), the present research utilizes a U-net, a U-shaped convolutional neural network (CNN) in order to identify, detect and segment the data and extract the mentioned archaeological characteristics of clearance cairns from the LiDAR dataset. This modified CNN architecture goes beyond conventional approaches by enabling pixel-level localization, classification as well as learning from limited training samples, which offers significant advantages.

The initial findings of this study are presented based on a limited amount of labeled data. At the beginning, 290 images containing cairns, each measuring 64x64 pixels, were labeled. Subsequently, the training dataset was expanded to include a total of 1054 images through the application of data augmentation techniques such as varying shear range, zoom range, flipping, and rotation ranges. However, the training metrics did not yield satisfactory results, possibly due to the extensive distortion in the training image datasets, causing them to deviate significantly from their original forms. Consequently, a decision was made to enhance the data augmentation solely by incorporating vertical and horizontal flipping. This led to a dataset consisting of 627 images, which were randomly divided into a training set of 501 images and a validation set of 126 images (ibid.).

7. Preliminary results

As a result of a preliminary investigation of the selected area (fig. 1), measuring 9984x4992 meters, the following findings are presented. As figure 4 shows, there is an apparent matching between areas predicted as likely to have clearance cairns with those ones reported in the Swedish national heritage registry where actually these features were located. The red pixels on the image represent ground anomalies, potentially indicating clearance cairns, which were detected using the proposed CNN model. Despite using a limited amount of labeled training data during this phase of the study, the outcomes are promising and showcase the model's effectiveness in identifying previously unknown or undocumented archaeological features, as evidenced in this portion of the sample image.

Figure 4: A portion of Söderhånsen National Park, where the areas previously known and reported on the Swedish National Heritage Board website as 'fossil fields' are marked in blue. As a result of the CNN data processing, several ground anomalies are detected in the hillshade map used to feed the network (red pixels). Image courtesy of the author.

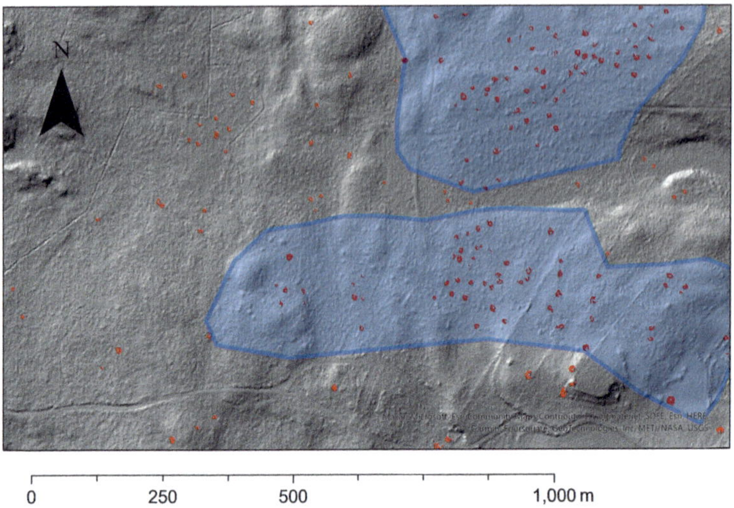

8. Ground truthing

The external validation survey was conducted in a specific area within the network's predicted region, within the Söderånsen National Park, which revealed the presence of numerous anomalies both inside and outside the boundaries defined by the RAÄ surveyed areas. Through field surveying and by comparing the GPS position of the observed clearance cairns with the location of the red-marked ground anomalies detected by the network, the results indicate an approximate average 74 percent success rate in accurately predicting clearance cairns (fig. 5). This percentage derives from the examination of 3 separate cluster areas with a concentration of ground anomalies with a matching ratio of 7/9, 8/13, and 10/12 good predictions corresponding to 77, 61, and 83 percent of relative success rate.

However, it has become evident that the terrain morphology and vegetation type introduced background noise into the quality of the LiDAR data. The presence of bedrock outcrops in the landscape created uneven areas, which can negatively impact the visual interpretation of the data and the model's effectiveness in detecting cairns, potentially resulting in false positive identifications. From a methodological perspective, another limitation in the field of data collection is the weakness of the GPS signal, due to the tree canopy coverage, which makes it difficult to properly use any differential or single-antenna GPS, thereby reducing the instrument accuracy to a few meters.

Figure 5: Ground truthing was performed to validate the model prediction on new external, independent data collected in the field (asterisk points). The selected areas were not previously reported as fossil fields, probably due to a lack of surveying coverage. Interestingly, as figures A and B show, there is a good matching (around 74 percent) between clearance cairns observations and the model prediction characterized by red pixels. Image courtesy of the author.

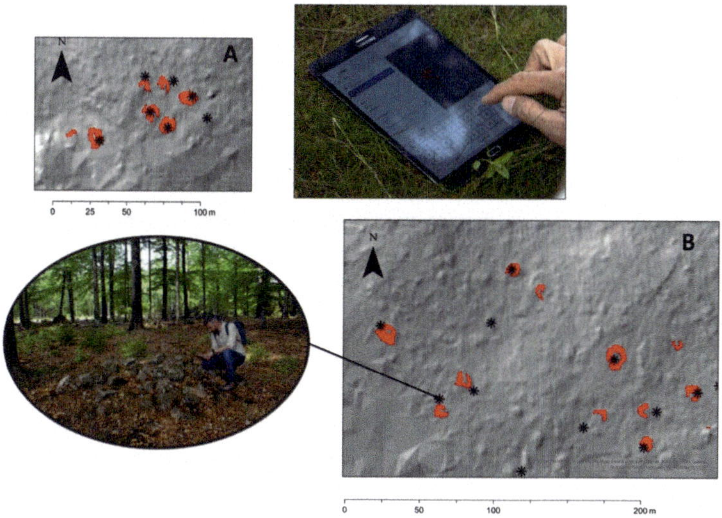

9. Conclusion

Despite being at a very preliminary stage, the project conducted so far provided very encouraging results in terms of prediction accuracy. Based on the field surveying assessment, most of the predicted ground anomalies have been identified on the ground and interpreted as actual clearance cairns. This pilot project marks a significant advance in the use of AI-based approaches for the study of archaeological landscapes and the identification of spatial patterns related to past land exploitation and human activity. Nevertheless, from an interpretative perspective, it is important to outline the need for more solid and accurate information to be used as a training dataset. Taking clearance cairns for instance, the features, as they appear in DTM-derived slope or hillshade maps, can be easily misinterpreted, if not misexamined, in relation to the surrounding context.

While our primary focus was on clearance cairns, it is apparent that this approach holds promise for providing fresh insights into the examination of complex agricultural systems from the past. It also offers a means to gain a deeper understanding of various types of farming landscapes in Scandinavia. In this sense, this contribution has sought to demonstrate the feasibility of utilizing a tool to semi-automatically detect archaeological features in challenging and peripheral areas where traditional survey methods are impractical.

As for the future, we are developing an alternative network for multiclass segmentation. This network will employ annotated data associated with different ground anomalies linked to agricultural activities, such as linear boundary walls and Celtic fields. Due to the combination of more features with different geometries, we believe this new approach can provide more accurate information about the presence of areas of past agricultural activity, reducing the risk of misinterpretation. Nonetheless, to construct an effective model that can assist archaeologists, heritage specialists and developers in addressing the challenge of archaeological predictability and expanding our knowledge of landscape transformations, we need to incorporate even more parameters. These parameters include geology, geomorphology, hydrological conditions and historical maps. Adopting a multi-scalar and multi-temporal perspective will enable us to comprehend human interactions with the environment and landscape.

List of references

Adamopoulos, Efstathios/Rinaudo, Fulvio (2020): "UAS-based Archaeological Remote Sensing: Review, Meta-analysis and State-of-the-art." In: Drones 4/3, 46.

Anichini, Francesca/Dershowitz, Nachum/Dubbini, Nevio/Gattiglia, Garbiele/Itkin, Barak/Wolf, Lior (2021): "The Automatic Recognition of Ceramics from Only One Photo: The ArchAIDE app." In: Journal of Archaeological Science: Reports 36, 102788.

Baker, Katherine G. (1987): "Towards an Archaeological Methodology for Expert Systems." In: Computer and Quantitative Methods in Archaeology 393, pp. 229–236.

Bewley, Robert H./Crutchley, Simon P./Shell, Colin A. (2005): "New Light on an Ancient Landscape: Lidar Survey in the Stonehenge World Heritage Site." In: Antiquity 79/305, pp. 636–647.

Bewley, Robert/Wilson, Andrew/Kennedy, David/Mattingly, David/Banks, Rebecca/Bishop, Michael/Bradbury, Jennie/et al. (2016): "Endangered archaeology in the Middle East and North Africa: Introducing the EAMENA project." In: Proceedings of the 43rd Annual Conference on Computer Applications and Quantitative Methods in Archaeology (CAA2015), Siena, Italy, pp. 919–932.

Campana, Stefano/Forte, Maurizio (eds.) (2006): From Space to Place: 2nd International Conference on Remote Sensing in Archaeology. Proceedings of the 2nd International Workshop, Rome, Italy.

Campana, Stefano/Sordini, Matteo/Berlioz, Stefania/Vidale, Massimo/Al-Lyla, Rowaed/al-Araj, Ammar Abbo/Bianchi, Alessandro (2022): "Remote Sensing and Ground Survey of Archaeological Damage and Destruction at Nineveh During the ISIS Occupation." In: Antiquity 96/386, pp. 436–454.

Chase, Arlen F./Chase, Diane Z./Weishampel, John F./Drake, Jason B./Shrestha, Ramesh L./Slatton, K.Clint/Awe, Jaime J./Carter, William E. (2011): "Airborne LiDAR, Archaeology, and the Ancient Maya Landscape at Caracol, Belize." In: Journal of Archaeological Science 38/2, pp. 387–398.

Djindjian, François (2015): "Computers and Mathematics in Archaeology, Anatomy of an Ineluctable Success." In: 21st Century Archeology. Concepts, Methods and Tools. Proceedings of the 42nd Annual Conference on Computer Applications and Quantitative Methods in Archaeology (CAA2014), Paris, France, pp. 1–6.

Giardino, Marco J. (2012): "NASA Remote Sensing and Archaeology." In: Rosa Lasaponara/Nicola Masini (eds.), Satellite Remote Sensing. A New Tool for Archaeology, Dordrecht: Springer, pp. 157–176.

Gualandi, Maria Letizia/Gattiglia, Gabriele/Anichini, Francesca (2021): "An Open System for Collection and Automatic Recognition of Pottery Through Neural Network Algorithms." In: Heritage 4/1, pp. 140–159.

Herzog, Irmela (2014): "A review of case studies in archaeological least-cost analysis." In: Archeologia e Calcolatori 25, pp. 223–239.

Hu, Chunmei/Zhang, Yunhui/Xia, Guofang/Liu, Xi/Ma, Xinjian (2022): "Automatic Classification of Ancient Building Components Based on PointNet++." In: 2022 International Conference on Artificial Intelligence, Information Processing and Cloud Computing (AIIPCC), Kunming, China, pp. 99–103.

Karamitrou, Alexandra/Sturt, Fraser/Bogiatzis, Petros/Beresford-Jones, David (2022): "Towards the Use of Artificial Intelligence Deep Learning Networks for Detection of Archaeological Sites." In: Surface Topography: Metrology and Properties 10/4, 044001.

Kohler, Timothy A./Parker, Sandra C. (1986): "Predictive Models for Archaeological Resource Location." In: Advances in Archaeological Method and Theory 9, pp. 397–452.

Küçükdemirci, Melda/Landeschi, Giacomo/Ohlsson, Mattias/Dell'Unto, Nicolo (2022): "Investigating Ancient Agricultural Field Systems in Sweden From Airborne LIDAR Data by Using Convolutional Neural Network." In: Archaeological Prospection (https://doi.org/10.1002/arp.1886).

Küçükdemirci, Melda/Sarris, Apostolos (2020): "Deep Learning Based Automated Analysis of Archaeo-geophysical Images." In: Archaeological Prospection 27/2, pp. 107–118.

Küçükdemirci, Melda/Sarris, Apostolos (2022): "GPR Data Processing and Interpretation Based on Artificial Intelligence Approaches: Future Perspectives for Archaeological Prospection." In: Remote Sensing 14/14, 3377.

Lagerås, Per/Bartholin, Thomas (2003): "Fire and Stone Clearance in Iron Age Agriculture: New Insights Inferred from the Analysis of Terrestrial Macroscopic Charcoal in Clearance Cairns in Hamneda, Southern Sweden." In: Vegetation History and Archaeobotany 12, pp. 83–92.

Lasaponara, Rosa/Masini, Nicola (eds.) (2012): Satellite Remote Sensing: A New Tool for Archaeology, Dordrecht: Springer.

Lindholm, Karl-Johan/Ersmark, Erik/Hennius, Andreas/Lindgren, Sakarias/Loftsgarden, Kjetil/Svensson, Eva (2021): "Contesting Marginality: the Boreal Forest of Middle Scandinavia and the Worlds Outside." In: The Medieval Globe 7/1, pp. 9–34.

Mantovan, Lorenzo/Nanni, Loris (2020): "The Computerization of Archaeology: Survey on Artificial Intelligence Techniques." In: SN Computer Science 1, 267.

Nayak, Nandeeka/Nara, Makoto/Gambin, Timmy/Wood, Zoë/Clark, Christopher M. (2021): "Machine learning Techniques for AUV Side-scan Sonar Data Feature Extraction as Applied to Intelligent Search for Underwater Archaeological Sites." In: Field and Service Robotics: Results of the 12th International Conference (FSR 2019), Tokyo, Japan, pp. 219–233.

Orengo, Hector A./Garcia-Molsosa, Arnau (2019): "A Brave New World for Archaeological Survey: Automated Machine Learning-based Potsherd Detection Using High-resolution Drone Imagery." In: Journal of Archaeological Science 112, 105013.

Parcak, Sarah H. (2009): Satellite Remote Sensing for Archaeology, London and New York: Routledge.

Patel, Jitu/Stutt, Arthur (1989): "Beyond Classification: the Use of Artificial Intelligence Techniques for the Interpretation of Archaeological Data." In: Computer Applications and Quantitative Methods in Archaeology 1989 (CAA89), Tübingen, Germany, pp. 338–347.

Rainey, Froelich (1976): "Archaeology: New Tools for an Old Art: Thermoluminescence, Magnetometers, Satellite Photography, and Other Techniques and Instruments Aid the Archaeologist." In: IEEE Spectrum 13/9, pp. 39–43.

Saturno, William/Sever, Thomas L./Irwin, Daniel E./Howell, Burgess F./Garrison, Thomas G. (2007): "Putting Us on the Map: Remote Sensing Investigation of the Ancient Maya Landscape." In: James Wiseman/Farouk El-Baz (eds.), Remote Sensing in Archaeology, New York: Springer, pp. 137–160.

Soroush, Mehrnoush/Mehrtash, Alireza/Khazraee, Emad/Ur, Jason A. (2020): "Deep Learning in Archaeological Remote Sensing: Automated Qanat Detection in the Kurdistan Region of Iraq." In: Remote Sensing 12/3, 500.

Verhagen, Philip (2007): Case Studies in Archaeological Predictive Modelling, Leiden: Leiden University Press.

Wescott, Konnie L./Brandon, R. Joe (eds.) (2000): Practical Applications of GIS for Archaeologists: A Predictive Modelling Toolkit, London and Philadelphia, PA: Taylor & Francis.

Wheatley, David (1995): "Cumulative Viewshed Analysis: a GIS-based Method for Investigating Intervisibility, and Its Archaeological Application." In: Gary Lock/Zoran/Stančič (eds.), Archaeology and Geographical Information Systems: A European Perspective, London and Bristol, PA: Taylor & Francis, pp. 171–185.

Wilbrink, Heleen/Roberson, Joshua A. (2023): "Cleo: The Artificial Intelligence Egyptology Platform." In: Rita Lucarelli/Joshua A. Roberson/Steve Vinson

(eds.), Ancient Egypt, New Technology. The Present and Future of Computer Visualization, Virtual Reality and Other Digital Humanities in Egyptology, Leiden: Brill, pp. 592–604.
Wilcock, John (1985): "A Review of Expert Systems: Their Shortcomings and Possible Applications in Archaeology." In: Computer Applications in Archaeology 13, pp. 139–144.

Interfaces of AI

Two examples from popular media culture and their analytical value for studying AI in the sciences

Sabine Wirth

1. Introduction: Perspectives from critical interface studies

Deep learning algorithms are currently introducing new forms of agency into many different fields at the same time: from various scientific disciplines like archaeology, art history or medical diagnostics to public sectors such as transportation or security and surveillance to popular media culture – forms of machine learning-based pattern recognition and generation are expected to affect many areas of private and professional life.[1] As can also be observed from the history of other media such as photography or personal computers, this "democratization"[2] of AI technologies leads to the strange circumstance that the same basic technologies (e.g., ML-based pattern recognition and generation) are applied to achieve completely different tasks in different areas.[3] Despite their universal appeal, these technologies inscribe themselves in very disparate ways in different fields of application.

1 As Pasquinelli and Joler (2021) describe it: "In this sense, pattern recognition has truly become a new cultural technique that is used in various fields." (1268)

2 For the ambivalent use of the term "democratization" regarding AI technologies see Sudmann 2019b: 11.

3 Adrian Mackenzie (2017) describes for instance how an image recognition system (kittydar) trained on cat images from Social Media and the Web could be applied to very different areas of use: "Based on how kittydar locates cats, we can begin to imagine similar pattern recognition techniques in use in self-driving cars (Thrun et al. 2006), border control facial recognition systems, military robots, or wherever something seen implies something to do." (19)

Within the field of media studies, the ubiquity of AI technologies has led to a variety of publications in recent years that can roughly be sorted into three main categories.[4] First, there are publications that address the development of AI technologies from a media-theoretical or philosophical point of view by discussing for instance theories of the artificial (e.g. Negrotti 2000), the history and foundations of pattern recognition (e.g. Apprich 2018), human-machine relations (e.g. Kasprowicz 2022), the role of *aisthesis* in machine learning (e.g. Krämer 2022) or the general question of creativity (e.g. Mersch 2019; 2022) and forms of intelligence that might differ from an anthropocentric understanding of it.[5] Secondly, there are publications that investigate specific media environments where AI is currently introducing new forms of agency, temporalities, decision making processes, politics and/or new aesthetics (e.g. Sprenger/Engemann 2015; Beverungen 2019; Manovich 2019; Sudmann 2019a; Ashri 2020; Karnouskos 2020; Sprenger 2020). Third, there are publications that deal with the question how different media are creating and shaping cultural imaginations and narratives of AI, which in turn can influence the actual development of AI tools (e.g. Bucher 2017; Kazansky/Milan 2021; Schulz 2022).

In all these publications there is a growing awareness that ML-based technologies are transforming media cultures in such a comprehensive way that we are already dealing with "media cultures of artificial intelligence" (Ernst et al. 2019: 19). This transformational development allows us to reexamine existing methods and approaches of media and culture studies like discourse analysis and media history or theory as well as integrate them into interdisciplinary research fields like software studies, platform studies or critical data studies. Especially in the field of critical data studies there is a growing amount of research that focuses on ML-induced bias and discrimination (e.g. Chun 2021; Apprich 2018; Kember 2013), dispositives of classification (e.g. Bechmann/Bowker 2019), questions of infrastructure, platformization and AI industries (e.g. Luchs/Apprich/Broersma 2023), and/or the material costs

4 This categorization is by no means able to encompass the many facets of research on AI-based technologies in the field of media and culture studies, but simply serves as an orientation for the purpose of this paper.

5 The question of a non-anthropocentric understanding of the agency of AI technologies can be traced back to similar discussions about the computer as a medium in the 1990s asking for a non-anthropocentric understanding of human-computer interaction, see Krämer 1997.

and planetary consequences of AI (e.g. Crawford 2021; Crawford/Joler 2018; Pasquinelli/Joler 2018).

The field of AI research is growing with rapid pace and there is a need to discuss more precisely what the various subfields of media and culture studies can contribute to this field (cf. Sudmann 2019c). One often neglected aspect in AI research is that the operability of AI technologies in everyday scenarios depends on interfaces that allow non-expert users to perform certain actions. Ultimately, developers must provide easy-to-use interfaces that are working towards embedding the operativity of AI-services into everyday culture. However, interfaces are not neutral. They mediate AI technologies in various ways. The emerging subfield of critical interface studies[6] can provide productive approaches that allow to address these mediations. While the allocated space of this article does not allow me to outline a conceptual toolbox of interface studies in all its variety, I will focus on its apparent key concept: the interface.

2. Interfaces as thresholds

So, what is an interface? To answer this seemingly simple question (that has produced different definitions in different research fields) we could start with another question: What makes a computer, a machine or a technology 'ready-to-hand'? Ready to use? Ready to be integrated into larger chains of action? Complex technologies, that have left behind the analogy of Heidegger's popular example of the hammer which enables intuitive handling through its 'handy' design, need some sort of second order mediation. This can be some sort of knowledge (expertise) about how the complex machine is to be handled[7] or a mode of mediation that translates this kind of knowledge into user interface

6 I use the term "critical interface studies" in this context to point to an emerging and interdisciplinary field of research that critically examines the role of interfaces in contemporary media cultures. The research field is in the process of forming and has not yet become institutionalized. Examples of relevant publications in this field are cited throughout the article.

7 This explicit knowledge can become implicit or tacit after multiple use. A manifestation of this knowledge can be found in textual form in so-called instruction manuals and it is interesting to note that the manual has step by step disappeared in the history of popular computing. What can now be observed instead of a manifestation of functional knowledge in the manual is a decentralized shift of repair knowledge to countless online forums (cf. Schröter 2018).

functions that are more easily comprehensible by human users (like pressing the right button).[8] Drawing on Gilbert Simondon's philosophy of technology we can differentiate between closed machines, that are understood as fully automatic machines with a predetermined way of functioning, and open machines. Open machines are defined by a higher degree of technicality, which presupposes human intervention in form of constant organization or coordination and therefore is always connected to human ways of relating to environments (Simondon 2012 [1958]: 11). As Erich Hörl (2011: 36) elaborates, Simondon puts the emphasis on the collective rather than on single actors by reconfiguring the evolution of the technical object from elements to ensembles and thus opens up a perspective to think technical activity in terms of a media ecology of distributed agency. Based on these considerations, an understanding of 'artificial intelligence' could follow, which does not attribute intelligence to the computer system alone, but assumes a distributed, collective performance, which is produced by a complex network of "distributed, hybrid human-machine-computer networks", as Rainer Mühlhoff (2019: 56f.) suggests. Although AI-based applications are not necessarily supervised or organized by human actors – especially in the case of subsymbolic forms of AI – human agency is still in the loop in many steps of the development process (supervised learning, human labor in trainings data sets, etc.) as well as in the environments in which these technologies are put into use. And here, 'being in the loop' mostly means being involved with interfaces; handling something in this context means dealing with displays and terminals: From human clickwork/crowdwork that generates training datasets for machine learning, to the implementation of machine learning operativity into the user interfaces of popular media apps.[9] Simondon's consideration of open machines can be extended by Alexander Galloway's (2012) conception of interfaces, which he describes as thresholds, as "zones of interaction that mediate between different realities" (vii). By not conceptualizing interfaces as things but rather as processes, Galloway makes us aware of the double nature of 'effectiveness' in computer-based interactions:

---

8 For an elaborated discussion of the complex relation between user interfaces and implicit knowledge see Ernst 2017.

9 Kate Crawford (2021: 68) e.g. critically describes the general obfuscation that 'interface effects' foster in complex AI-systems where we cannot be sure when exactly we, as human users, are interacting with an AI system: "We engage only with the facades that obscure their inner workings, designed to hide the various combinations of machine and human labor in each interaction."

> Interfaces themselves are effects, in that they bring about transformations in material states. But at the same time interfaces are themselves the effects of other things, and thus tell the story of the larger forces that engender them. (ibid.)

In a more practical reading, the focus on the interface points us to larger formations that shape our relationship with technology, such as data extractivism, surveillance capitalism or the overarching problem of complexity and blackboxing. Interfaces function as thresholds through which the agential/performative/operational potential of machine learning methods is mediated and made accessible and compatible with human practices. The user interface provides agency and enables us to be productive, but at the same time it is a threshold in the sense of a barrier: not everything is possible/visible/doable. Branden Hookway has clearly highlighted this ambiguity of the interface:

> The interface describes a fundamental ambiguity between human and machine; it is both a mirror of multiple facings and a zone of contact. This ambiguity bears on the human relationship with technology. For what is first encountered is not the machinic in any pure form but rather the interface itself. (Hookway 2014: 45)

But what does this mean for the study of AI? From the perspective of interface studies, an everyday human user can encounter AI-systems only through "the interface itself" (ibid.). In other words: Human-AI relations always depend on interfaces as central mediators of AI. However, the interface is not simply a medium for a linear relationship in the sense of mediating input towards output. On the contrary, following Hookway in his observation that the interface is "both a mirror of multiple facings and a zone of contact" (ibid.), we see that the interface is a relational entity that mediates in-between users and algorithms on different scales. Analyzing the "interface itself" (ibid.) does not mean to analyze a thing-like entity. Rather, it means to investigate how interfaces constitute a variety of connections and tensions that emerge between human users and the operativity of ML algorithms. To make this more concrete, I will briefly outline two examples of popular media apps that partly rely on AI. My goal here is not to investigate these examples in all their detail. I simply aim to illustrate *some* of the questions and potential points of inquiry that a critical interface studies perspective would follow here.

3. Example A: Curating social media feeds

The first example belongs to the broader field of content selection and recommender-systems. Social media feeds appear as the dominant organizing principle of current platform cultures, which network a high number of potential 'prosumers' and manage large amounts of audiovisual media content and user interaction (cf. Kohout 2018; Schulz/Matzner 2020). Feeds of platforms like Instagram, Facebook, Twitter or TikTok promise to filter content in an individualized and 'intelligently' curated way for each user of the platform. As it is advertised on the Instagram website, the formulated goal of the feed lies in defining what is relevant for each user and what is not. The goal is "[to predict] the most relevant media for each person every time they scroll the Explore page" (Medvedev et al. 2019). The Instagram Explore feed shows users an algorithmically curated selection of posts ranked with the help of artificial neural networks.[10] In a post on the Facebook AI blog, Ivan Medvedev, Haotian Wu and Taylor Gordon (2019) describe it as an "AI system based on a highly efficient 3-part ranking funnel that extracts 65 billion features and makes 90 million model predictions every second." Similar to other commercial content ranking algorithms, the criteria for the algorithmic composition of the Instagram feed are not fully transparent and therefore subject to speculation (Leaver et al. 2020: 8–38). By reviewing developer statements that often seem to follow a policy of strategic vagueness, it is only possible to reconstruct certain core categories of AI-enhanced algorithmic curation like "interest", "recency" or "relationship" in the case of the Instagram Explore feed (ibid.). In addition to an interest factor, according to which a certain user might be interested in a certain content, the timeliness of the content also plays a role. Further, the previous interaction behavior of each user is taken into account and, for example, posts from accounts that are followed or with which interaction (e.g., through likes, saves or comments) has already taken place, are prioritized. Secondary factors such as the frequency with which users access their accounts and feeds, their network (which accounts they follow), or their average time spent on the platform or individual posts are also included as selection criteria. This means that users are continuously contributing to the real-time composition of their Instagram Home and Explore feeds with their interaction behavior, even if they are not aware of it. Even if we are "absentmindedly scrolling through nothing" (Lupinacci 2021), just skipping through our feeds, we generate analyzable user

10 For a more detailed discussion of the various Instagram feeds see Wirth 2021.

data. The goal of the platform is to keep users engaged for as long as possible and the user interface is designed to achieve this goal in the most targeted way, generating what Alexandra Anikina (2021) has called the "affective scroll" (128f.) with regard to TikTok. In addition to this often involuntary and implicit work on the feed, the Instagram platform encourages its users to actively shape their feeds and provides specific control tools through the user interface: Certain accounts can be marked as favorites so that posts from these accounts are ranked higher in the home feed and displayed more often. Through so-called "Not interested" flags, users can actively hide certain content or participate in Instagram's "Sensitive Content Control" by masking posts that do not exactly violate the community guidelines but can still be perceived as offensive.

By considering the many factors that are part of the curatorial 'force' that constitutes social media feeds, it becomes evident that we are dealing with a complex curatorial assemblage of distributed agency where the algorithmic capture, evaluation and individually tailored selection and ranking of content is linked to the affordances and design strategies of user interfaces as well as the practices of users, who "become more aware of how algorithms micro-target them as audiences by surveilling their consumptive practices" (Jones 2023: 2). Machine learning technologies are one part of this 'messy assemblage'. Therefore, in media environments like social media platforms, 'intelligent' curation can also be understood as an "emergent and distributive capacity of hybrid human-machine networks" (Mühlhoff 2019: 64). Curatorial agency here is distributed and relational in the sense that all curatorial decisions affect the whole assemblage. However, we are not dealing with a flat hierarchy in which the individual points of the network have similar weightings, but rather with massive asymmetries of power that often remain opaque for users as well as researchers.[11] The role of the user interface within the curatorial assemblage that constitutes the Instagram feed can be described as follows: The user interface acts as a 'boundary condition',[12] a threshold between user practices, processes of data extraction, their algorithmic (partly ML-based) evaluation

11 For a conception of the computer interface as an apparatus of power see e.g. Distelmeyer 2017: 29f.; Distelmeyer 2021: 65ff.

12 Referring to the notion of interface in 19th century physics (specifically fluid dynamics) Hookway (2014: 66) describes the interface as "a boundary condition that both separates and holds contiguous as one body those parts whose mutual activity, exerted from each part onto the other, is directed into and channeled across that boundary condition in such a way as to produce a fluidity of behavior."

and the aesthetic mode of presentation that dynamically and constantly decides what becomes visible and what remains hidden. The feed interface is an ephemeral interface where the mode of 'passing-through' is enacted on multiple layers between dataflow and visualization: it is the result of a complex assemblage of human and non-human actors and simultaneously creates new affordances of interaction for human users that are ultimately feeding the dynamics and the future extractive potential of the assemblage. The design elements of user interfaces (like interface gestures, layout, icons, digital-material metaphors[13], etc.) need to be considered in their role of affording user interactions and thereby creating habits and embodied relations to/with the algorithmic agents of the assemblage (Anikina 2021: 129f.).

Consequently, the user interface integrates algorithmic (AI-based) classification decisions into everyday practice by presenting algorithmic processes as 'intelligible' and operable for human users. But at the same time, the visible feed as an interface obfuscates algorithmic decisions and data practices of the 'black box'. Thus, interfaces can be understood both as enablers and obfuscators of AI at the same time.

4. Example B: Editing images with AI-based photo apps[14]

The second example is located in the field of AI-based image generation and image editing. In this case, not the extractive, but rather the generative potential of AI-technologies and its impact on popular media culture is what I would like to focus on. Popular image editing software has made the rapid modification of digital images an everyday standard and a new impetus is currently coming from popular applications that offer AI-based editing functions. Their user interfaces provide editing options to everyday users that were previously only accessible to experts, e.g., photographers, literate in image editing programs like Adobe Photoshop. A popular example for this trend is the app FaceApp. Released in 2017 by Russian startup Wireless Lab (later renamed FaceApp Technology Limited), the image and video editing app allows users to perform a range of elaborate photo and video edits, such as aging or rejuvenating faces, morphing two faces together, adding complex facial expressions

---

13 For an elaborated theory of "digital-material metaphors" see Boomen 2014.

14 The following paragraph is a condensed version of Wirth 2023.

such as smiles, or applying the controversial "gender swap" feature. In journalistic reviews, FaceApp's features were mainly celebrated for their supposedly realistic results (e.g. Pickell 2019). FaceApp explicitly presents itself as an AI application that offers AI-based image editing functions to everyday users and is designed to deliver fast, but high-quality results: as the developer website advertises: "No more hours spent on photoshop" (FaceApp n.d.).

The user interface of FaceApp suggests similar functionality to popular photo filters or filter presets by making editing available quickly and easily at the tap of a finger. But in contrast to this user experience, FaceApp features apply deep AI-based modifications to the photographic source image. Therefore, the term filter no longer seems appropriate here (Bergermann 2019: 56). As Yaroslav Goncharov, founder and CEO of FaceApp Technology Limited, told *TechCrunch* in 2017, FaceApp uses "deep generative convolutional neural networks" (Lomas 2017) to process users' selfies. When applying the FaceApp image processing functions, the CNN transfers specific features to the respective portrait image or selfie, that has previously been extracted from the training data set. The applied image recognition methods enable an exact application of the automated feature modifications, which in the result achieve the already mentioned photo-realistic effects. This way, FaceApp manages to retain certain individuality markers of the respective face, even though the image is otherwise fundamentally changed (Chakraborty 2020). For users, this creates the illusion of an aged or rejuvenated version of their personal faces.

Like recommender systems, FaceApp participates in the general promise of AI technologies to make things predictable (Sudmann 2018: 193). FaceApp's so-called 'aging-feature' can be read as a popularized condensation of this prognostic promise. Prognostics forms a central element of ANNs, since it is always a matter of predicting an outcome for a newly inserted value – one that is not already part of the training dataset. In the form of predictive analysis, AI technologies currently present themselves in many areas of professional and private life as a future medium or medium of the future, in that they present the future as a computational and techno-economic regime (Ernst/Schröter 2020: 89). At the same time, the prediction of the future is characterized by an immanent reference to the past, as Matteo Pasquinelli and Vladan Joler (2021) have pointed out: "Machine learning prediction is used to project future trends and behaviours according to past ones, that is to complete a piece of information knowing only a portion of it." (1273).

The prognostic promise of AI, however, can only be delivered through interfaces that make AI-based prognosis accessible for the human sensorium. Once more, the interface's function as a translator of AI comes into play. In the case of my example, the app's user interface makes ML-based methods of prediction accessible for everyday practices. By providing ready-at-hand functions and reducing complexity, FaceApp's user interface (like many other AI-based photo editing tools) allows to implement AI-based object recognition and photo editing into established cultural techniques and photo practices and therefore works towards a domestication of AI. In the field of visual culture, the now ubiquitous availability of AI-based functionality, mediated by popular user interfaces, intervenes as a fundamental rupture in cultural production processes.[15]

The popular app interface thereby offers a subject position from which it is possible to perform expert-operations without expert-knowledge. As Christoph Ernst (2017) points out with reference to Donald Norman, interaction design and user interface design generate conceptual models that contain "ideas about possible operations *of* the system and about possible actions *with* the system" (100). The user interface of FaceApp and the marketing discourse surrounding it significantly shapes the imaginary of what 'AI can do'. In the case of commercial AI-supported apps like FaceApp, the subject position offered by the user interface is intrinsically linked to processes of objectification, namely to the datafication of users, their images and interaction behavior.[16]

5. From popular apps to AI in the sciences: Why interfaces matter

Using two examples from popular media culture, I have tried to demonstrate how even a brief look at the role of interfaces connected to AI technologies reveals critical functions that these interfaces fulfill when integrated into everyday practices. They can serve to *translate* the operativity of machine learning techniques and make their potentials – such as their potential for prognosis

15 For a comprehensive description of the relationship between AI and cultural production see Manovich 2019.

16 By offering a broader perspective on popular interface cultures Søren Pold und Christian Andersen (2014: 31) have described the intertwining of "intimate interface[s]" and extremely regulatory mechanisms that turn personal data into currencies as a typical feature of the current "controlled consumption culture".

and prediction or image generation – 'ready-to-hand' for non-expert users, while at the same time *obfuscating* the mechanisms of AI-based algorithms and related practices of data extractivism. Furthermore, user interfaces integrate the operativity of AI systems into cultural practices and play an important part in forming "human-machine assemblage[s]" (Mackenzie 2017: 216) of distributed agency. The evolving perspective of critical interface studies can help us highlight such functions and investigate them – both through a historical and contemporary lens – as parts of complex media entanglements. Overall, a critical interface perspective poses the question of the 'usability' of AI and investigates the user interface as a designed entity with its own agency and affordances.

The question of the interface draws attention to the often-hidden transitions between popular media culture and scientific practice. Adrian Mackenzie (2017: 190) describes the 'entangled evolving' of machine learning techniques and popular media (like social media platforms and search engines) which are mutually dependent in their development. Popular applications and easy-to-use interfaces first generate the structured data sets that AI systems need to improve their functionality, and, on the other hand, popular applications would not achieve their (mostly) flawless functionality without machine learning techniques.[17] The study of interfaces of commercial, (partly) AI-based apps shows that these primarily act as thresholds for monetizable data practices. This setting may be fundamentally different in the science context, but here, too, dependencies on large corporations, that e.g. generate training data sets for AI systems or provide functional AI units as service packages, can be found. So ultimately, research that wants to critically reflect on the application of AI tools in science must also critically address these dependencies.[18]

The more pressing question for research on AI in science, however, might be how interfaces are involved in the production of knowledge. In the near future, interface design will most likely play a significant role as a scientific research tool. As Johanna Drucker (2014: 139–146) points out from a historical perspective, data-heavy projects (e.g. research projects in the field of digital

17 Rainer Mühlhoff (2019) uses the example of the company reCAPTCHA to show how popular interfaces are specifically constructed and used to obtain high-quality, i.e. human-validated, data sets/classifiers for training AI systems.

18 As Alexander Galloway (2012: 110) stresses: "doing capitalist work and doing intellectual work – of any variety, bourgeois or progressive – are more aligned today than they have ever been."

humanities) need dynamic interfaces that leave behind the limitations of classical information graphics. Information visualizers are increasingly concerned with the question of how large databases and digital collections can be visualized in dynamic and customized ways (for researchers or public audiences), and what kind of access and exploratory potential interfaces should provide in this process (e.g. Dörk et al. 2020). Interestingly, the field of human-computer interaction is currently debating not only what interfaces for AI applications should look like, but also to what extent machine learning approaches can contribute to the development of 'intelligent' interfaces (e.g. Martelaro/Ju 2018; Ferraro/Giacalone 2022; Keselj 2022). Therefore, an examination of interface design conventions and the history of human-computer interaction seems indispensable for an understanding of 'AI in use'.

While Drucker (2014) raises the question "What kind of interface exists after the screen goes away?" (195) for the future development of interface design, Sybille Krämer points out that even machine learning remains tied to the screen in some way. According to Krämer, epistemological processes in which AI systems are fundamentally involved, are, like diagrammatological writing practices, ultimately still bound to *aisthesis* and thus to a surface such as the screen on which something is made perceivable (Krämer 2022: 149). This raises the question of the extent to which interfaces, as part of epistemic processes that introduce a certain agency into knowledge production within AI-supported research activities, should be studied as carefully and rigorously as other forms of scientific images and imaging techniques.

List of references

Anikina, Alexandra (2021): "Automating Platform Spectators: Algorithmic Montage and Affective Scroll in TikTok." In: Digital Culture & Society 7/2, pp. 119–138.

Apprich, Clemens (2018): "Data Paranoia: How to Make Sense of Pattern Discrimination." In: Clemens Apprich/Florian Cramer/Wendy Hui Kyon Chun/Hito Steyerl (auths.), Pattern Discrimination, Minneapolis and London: The University of Minnesota Press, and Lüneburg: meson press, pp. 99–123.

Ashri, Ronald (2020): The AI-Powered Workplace: How Artificial Intelligence, Data, and Messaging Platforms Are Defining the Future of Work, Berkeley, CA: Apress.

Bechmann, Anja/Bowker, Geoffrey C. (2019): "Unsupervised by Any Other Name: Hidden Layers of Knowledge Production in Artificial Intelligence on Social Media." In: Big Data & Society 6/1 (https://doi.org/10.1177/2053951718819569).

Bergermann, Ulrike (2019): "Shirley and Frida. Filters, Racism, and Artificial Intelligence." In: Katja Böhlau/Elisabeth Pichler (eds.), Filters + Frames: Developing Meaning in Photography and Beyond, Weimar: Jonas, pp. 47–63.

Beverungen, Armin (2019): "Algorithmic Trading, Artificial Intelligence and the Politics of Cognition." In: Andreas Sudmann (ed.), The Democratization of Artificial Intelligence: Net Politics in the Era of Learning Algorithms, Bielefeld: transcript, pp. 77–94.

Boomen, Marianne van den (2014): Transcoding the Digital: How Metaphors Matter in New Media, Amsterdam: Institute of Network Cultures.

Bucher, Taina (2017): "The Algorithmic Imaginary: Exploring the Ordinary Affects of Facebook Algorithms." In: Information, Communication & Society 20/1, pp. 30–44.

Chakraborty, Debojyoti (2020): "IN DEPTH OF Faceapp.", April 16, 2020 (https://medium.com/analytics-vidhya/in-depth-of-faceapp-a08be9fe86f6).

Chun, Wendy Hui Kyong (2021): Discriminating Data: Correlation, Neighborhoods, and the New Politics of Recognition, Cambridge, MA: The MIT Press.

Crawford, Kate (2021): Atlas of AI: Power, Politics, and the Planetary Costs of Artificial Intelligence, New Haven, CT: Yale University Press.

Crawford, Kate/Joler, Vladan (2018): "Anatomy of an AI System: The Amazon Echo as an Anatomical Map of Human Labor, Data and Planetary Resources.", September 7, 2018, AI Now Institute and Share Lab (https://anatomyof.ai/).

Distelmeyer, Jan (2017): Machtzeichen: Anordnungen des Computers, Berlin: Bertz + Fischer.

Distelmeyer, Jan (2021): Kritik der Digitalität, Wiesbaden: Springer VS.

Dörk, Marian/Müller, Boris/Stange, Jan-Erik/Herseni, Johannes/Dittrich, Katja (2020): "Co-Designing Visualizations for Information Seeking and Knowledge Management." In: Open Information Science 4/1, pp. 217–235.

Drucker, Johanna (2014): Graphesis: Visual Forms of Knowledge Production, Cambridge, MA: Harvard University Press.

Ernst, Christoph (2017): "Implizites Wissen, Kognition und die Praxistheorie des Interfaces." In: Navigationen – Zeitschrift für Medien- und Kulturwissenschaft 17/2, pp. 99–116.

Ernst, Christoph/Kaldrack, Irina/Schröter, Jens/Sudmann, Andreas (2019): "Künstliche Intelligenzen. Einleitung in den Schwerpunkt." In: Zeitschrift für Medienwissenschaft 11/2, pp. 10–19.

Ernst, Christoph/Schröter, Jens (2020): Zukünftige Medien. Eine Einführung, Wiesbaden: Springer Fachmedien.

FaceApp (n.d.): "FaceApp: Face Editor.", Homepage, accessed August 25, 2022 (https://www.faceapp.com/).

Ferraro, Antonino/Giacalone, Marco (2022): "A Review About Machine and Deep Learning Approaches for Intelligent User Interfaces." In: Leonard Barolli/Farookh Hussain/Tomoya Enokido (eds.), Advanced Information Networking and Applications. Proceedings of the 36th International Conference on Advanced Information Networking and Applications (AINA-2022), Volume 3, Cham: Springer International Publishing, pp. 95–103.

Galloway, Alexander R (2012): The Interface Effect. Cambridge, UK and Malden, MA: Polity Press.

Hookway, Branden (2014): Interface. Cambridge, MA: The MIT Press.

Hörl, Erich (2011): "Die technologische Bedingung. Zur Einführung." In: Erich Hörl (ed.), Die technologische Bedingung: Beiträge zur Beschreibung der technischen Welt, Berlin: Suhrkamp, pp. 7–53.

Jones, Corinne (2023): "How to Train Your Algorithm: The Struggle for Public Control over Private Audience Commodities on Tiktok." In: Media, Culture & Society (https://doi.org/10.1177/01634437231159555).

Karnouskos, Stamatis (2020): "Artificial Intelligence in Digital Media: The Era of Deepfakes." In: IEEE Transactions on Technology and Society 1/3, pp. 138–147.

Kasprowicz, Dawid (2022): "Maintaining Relations and Re-Engineering the Social: A Reply to Becker's 'The Three Problems of Robotics and AI'." In: Social Epistemology Review and Reply Collective 11/8, pp. 50–56.

Kazansky, Becky/Milan, Stefania (2021): "'Bodies Not Templates': Contesting Dominant Algorithmic Imaginaries." In: New Media & Society 23/2, pp. 363–381.

Kember, Sarah (2013): "Gender Estimation in Face Recognition Technology: How Smart Algorithms Learn to Discriminate." In: Media Fields Journal 7, pp. 1–10.

Keselj, Ana/Milicevic, Mario/Zubrinic, Krunoslav/Car, Zeljka (2022): "The Application of Deep Learning for the Evaluation of User Interfaces." In: Sensors 22/23, pp. 1–17.

Kohout, Annekathrin (2018): "Der Feed als Kulturtechnik." In: POP. Kultur und Kritik 12, pp. 10–17.

Krämer, Sybille (1997): "Vom Mythos 'Künstliche Intelligenz' zum Mythos 'Künstliche Kommunikation' oder: Ist eine nicht-anthropomorphe Beschreibung von Internet-Interaktionen möglich?" In: Stefan Münker/Alexander Roesler (eds.), Mythos Internet, Frankfurt a. M.: Suhrkamp, 1997, pp. 83–107.

Krämer, Sybille (2022): "Zur Aisthesis des Maschinenlernens. Ein Kommentar zur zeitgenössischen Künstlichen Intelligenz." In: Martina Ide (ed.), Ästhetik digitaler Medien: Aktuelle Perspektiven, Bielefeld: transcript, pp. 131–152.

Leaver, Tama/Highfield, Tim/Abidin, Crystal (2020): Instagram: Visual Social Media Cultures, Cambridge, UK and Medford, MA: Polity Press.

Lomas, Natasha (2017): "FaceApp Uses Neural Networks for Photorealistic Selfie Tweaks.", February 8, 2017 (https://techcrunch.com/2017/02/08/faceapp-uses-neural-networks-for-photorealistic-selfie-tweaks/).

Luchs, Inga/Apprich, Clemens/Broersma, Marcel (2023): "Learning Machine Learning: On the Political Economy of Big Tech's Online AI Courses." In: Big Data & Society 10/1 (https://doi.org/10.1177/20539517231153806).

Lupinacci, Ludmila (2021): "'Absentmindedly Scrolling Through Nothing': Liveness and Compulsory Continuous Connectedness in Social Media." In: Media, Culture & Society 43/2, pp. 273–290.

Mackenzie, Adrian (2017): Machine Learners: Archaeology of a Data Practice, Cambridge, MA: The MIT Press.

Manovich, Lev (2019): AI Aesthetics, Moskow: Strelka Press.

Martelaro, Nikolas/Ju, Wendy (2018): "Cybernetics and the Design of the User Experience of AI Systems." In: Interactions 25/6, pp. 38–41.

Medvedev, Ivan/Gordon, Taylor/Wu, Haotian (2019): "Powered by AI: Instagram's Explore Recommender System.", November 25, 2019 (https://ai.facebook.com/blog/powered-by-ai-instagrams-explore-recommender-system/).

Mersch, Dieter (2019): "Kreativität und Künstliche Intelligenz. Einige Bemerkungen zu einer Kritik algorithmischer Rationalität." In: Zeitschrift für Medienwissenschaft 11/2, pp. 65–74.

Mersch, Dieter (2022): "(Un)creative Artificial Intelligence. Zur Kritik 'künstlicher Kunst'." In: Martina Ide (ed.), Ästhetik digitaler Medien: Aktuelle Perspektiven, Bielefeld: transcript, pp. 77–116.

Mühlhoff, Rainer (2019): "Menschengestützte Künstliche Intelligenz. Über die soziotechnischen Voraussetzungen von 'deep learning'." In: Zeitschrift für Medienwissenschaft 11/2, pp. 56–64.

Negrotti, Massimo (2000): "Towards a General Theory of the Artificial." In: AI & Society 14/3-4, pp. 268–299.

Pasquinelli, Matteo/Joler, Vladan (2021): "The Nooscope Manifested: AI as Instrument of Knowledge Extractivism." In: AI & Society 36, pp. 1263–1280.

Pickell, Devin (2019): "What Is FaceApp? The Technology Behind This AI-Enabled Mobile App." July 17, 2019 (https://learn.g2.com/faceapp).

Pold, Søren Bro/Andersen, Christian Ulrik (2014): "Controlled Consumption Culture: When Digital Culture Becomes Software Business." In: Paul D. Miller/Svitlana Matviyenko (eds.), The Imaginary App, Cambridge, MA: The MIT Press, pp. 17–33.

Schröter, Jens (2018): "Reparaturwissen und Paratextualität." In: Stefan Krebs/Gabriele Schabacher/Heike Weber (eds.), Kulturen des Reparierens: Dinge – Wissen – Praktiken, Bielefeld: transcript, pp. 225–238.

Schulz, Christian (2022): "A New Algorithmic Imaginary." In: Media, Culture & Society 45/3, pp. 646–655.

Schulz, Christian/Matzner, Tobias (2020): "Feed the Interface. Social-Media-Feeds als Schwellen." In: Navigationen – Zeitschrift für Medien- und Kulturwissenschaft 20/2, pp. 147–164.

Simondon, Gilbert (2012 [1958]): Die Existenzweise technischer Objekte, translated by Michael Cuntz, 2nd ed., Zürich: Diaphanes.

Sprenger, Florian (2022): "Microdecisions and Autonomy in Self-Driving Cars: Virtual Probabilities." In: AI & Society 37, pp. 619–634.

Sprenger, Florian/Engemann, Christoph (2015): Internet der Dinge: Über smarte Objekte, intelligente Umgebungen und die technische Durchdringung der Welt, Bielefeld: transcript.

Sudmann, Andreas (2018): "On the Media-Political Dimension of Artificial Intelligence: Deep Learning as a Black Box and OpenAI." In: Digital Culture & Society 4/1, pp. 181–200.

Sudmann, Andreas (ed.) (2019a): The Democratization of Artificial Intelligence: Net Politics in the Era of Learning Algorithms, Bielefeld: transcript.

Sudmann, Andreas (2019b) "The Democratization of Artificial Intelligence. Net Politics in the Era of Learning Algorithms." In: Andreas Sudmann (ed.), The

Democratization of Artificial Intelligence: Net Politics in the Era of Learning Algorithms, Bielefeld: transcript, pp. 9–31.
Sudmann, Andreas (2019c): "On the Media-Political Dimension of Artificial Intelligence. Deep Learning as a Black Box and OpenAI." In: Andreas Sudmann (ed.), The Democratization of Artificial Intelligence: Net Politics in the Era of Learning Algorithms, Bielefeld: transcript, pp. 223–244.
Wirth, Sabine (2021): "'Feeding the Feeds': Ephemere Infrastrukturen und der Flow fotografischer Bilder auf Foto-Sharing-Plattformen wie Instagram." In: Jonas Etten/Julian Jochmaring (eds.), Nach der ikonischen Wende: Aktualität und Geschichte eines Paradigmas, Berlin: Kadmos, pp. 170–192.
Wirth, Sabine (2023): "Inter|Face Experiments: FaceApp as Everyday AI." In: Interface Critique 4, pp. 159–169.

Media and the transformative potential of AI in the scientific field

Theses on the media conditions of knowledge production in the era of learning algorithms

Andreas Sudmann, Jens Schröter

The investigation of the epistemic-technical and infrastructural role of media for artificial intelligence (AI) is still a comparatively young field of research, at least if we think primarily – as we are currently accustomed to do – of machine learning (ML) approaches and artificial neural networks especially (ANN).

It has already become common knowledge that equating AI with ML or ANN is problematic for several reasons, but this does not change the fact that the concepts are de facto used more or less synonymously. In a similar way, we can note that the traditional criticism of AI (e.g., there is no such thing as artificial intelligence, corresponding techniques or systems are neither intelligent nor artificial) has not led to a terminological reorientation either. In computer science, for example, one typically speaks of individual models such as convolutional neural networks (CNN) or large language models (LLM), of statistics, or of ML rather than of AI, perhaps because technical details are more important in this academic field than in other contexts. And yet, it remains to be stated that even in computer science many researchers and engineers apparently cannot or do not want to abandon the term. Hence, it is worth asking why it is so persistent. In our opinion, an explanation for this cannot be limited to the fact that we are confronted with a consolidated concept and that the normative power of the factual takes effect here. Rather, the continued use of the term also points to its ideological function, especially in a scientific context. AI is not only a prospering field of research, but also a culturally imparted promise of how humans can grow beyond themselves through the development and application of technologies. It is hard to escape the phantasmatic charge of AI, in view of a historically unprecedented situation in which the gap between its cultural imaginaries and its empirical development in the 'real world' has noticeably narrowed (Ernst/Schröter/Sudmann 2019: 18). Analogously, it is not surprising that the perception and thematization of AI research often seems to be dominated by a rhetoric of outdoing (Humm/Buxmann/Schmidt 2022; Stöcker 2020), while specifically computer science, due to its core responsibility for the development of AI, can meanwhile, depending on the situation, afford itself the luxury of warning against exaggerated expectations of this technology (e.g., Bengio 2022). Nevertheless, it remains to be noted that the concept of AI cannot be reduced to merely serving as an ideology, if one just thinks of the epistemic-technical orientation of computers in comparison to that of humans, for example (Turing 2004 [1948]: 420–422; Rosenblatt 1961: viif., 28). In addition, perhaps the term 'AI' persists so tenaciously because it conveniently offers itself for (critical) reflection, regardless of the state of history.

This position paper discusses some fundamental considerations related to the role of media in practices and methods of the application of AI in different fields of academic research and their potential transformation, with individual (hypo)theses as starting points. This particular approach was chosen due to the specific conditions of our research project. On the one hand, we have to emphasize that, at the time of writing, much of our empirical and historical research still lies ahead of us, which is why the concepts and theses presented here are explorative or tentative. On the other hand, approaching the problems via (tentative) theses also represents an attempt to come to terms with the assumed complexity of the subject matter as well as with the speed of its transformations (just think about how fast, for example, GPT-4 followed GPT-3 and GPT-3.5). Some of the following observations and reflections have already been introduced elsewhere. If this is the case, it is indicated accordingly.

In the context of our project, the term media refers primarily to all technical entities whose function is to perceive, store, process, transmit and present information.[1] Such a working definition may seem relatively broad, but it seems

1 This conceptualization represents a significant extension of Kittler's concept of media technology when he defines it as "transmission, storage, processing of information" (Kittler 1993: 8). Unlike Kittler, we fundamentally understand media as *socio*-technical entities. The labeling of media in our understanding as "infrastructural media" seems useful to us, even though only in a certain sense, if the term is understood rather openly, with a sensibility to the non-fixed status quo of respective entities that constitute and configure an infrastructure. The term "infrastructural media" specifically refers to the systematic and rule-based stabilization and connectivity of media as part of complex chains of operations consisting of people, things and practices (Sudmann

necessary for us in order to capture the heterogeneous spectrum of the infrastructural role of media for the application of AI in different academic fields.

Although the use of AI-based methods already seems to be so normalized that it has even become a casual standard reference in discussions related to the nexus of digital technologies in the sciences (as, for example, in Mößner/Erlach 2022), we argue for addressing the question of 'how AI changes the sciences' as an issue in its own right.

Thesis I

Machine learning presupposes and implies that machines learn with and based on media. Consequently, media also impose their conditions on machine learning practices and applications in different disciplines.

The epistemological and cross-disciplinary relevance of a thesis can be gauged, among other things, by the extent to which it can be countered by an equally important counter-thesis. Against this background, the thesis mentioned here might seem relatively trivial, at least if one proposes a rather broad concept of infrastructural media (as we do and have already briefly sketched). Nevertheless, in relation to the development and application of AI in the scientific field (or elsewhere and beyond), a media obliviousness can be observed that obstructs a thorough epistemological reflection on technologies and their implementations. The indicator of such media obliviousness is not merely the explicit absence of the term 'media' itself, but rather the fact that the general AI discourse lacks a way of thinking about technology that really acknowledges the role of media in its developments and application and that also understands the epistemic influence of media in the reflection on technology. Precisely because the dependence on media can be asserted for every practice, thus also for the application of technology, it is even more important to shed light on this dependence in its specific manifestations and different contexts of application, here in relation to the scientific uses of AI. The latter also includes questioning the nexus as well as the interdependence of different media forms.

In recent years, the inscription of media in machine learning has already been the subject of some studies, also including first attempts for a media-

2021: 281f.; for a slightly different account of infrastructural media, see Schüttpelz 2017).

historical perspective (cf. Sudmann 2017; Engemann/Sudmann 2018; Ernst/Schröter/Sudmann 2019; Tuschling 2022). As has been shown in these contexts, the infrastructural relevance of media can be discussed for entities as diverse as learning data, sensors, software, hardware, platforms, frameworks and many more. The epistemological potential of ANN is particularly evident in the information processing of inherently fuzzy media such as images and language.

The epistemic-technical potential of current AI technology is, of course, especially visible in the field of sequential and generative models. Among other things, the significance of time-based media becomes especially apparent here. This temporal aspect also became evident, when the German-Canadian company TwentyBN back in 2017 trained an ANN to recognize gestures and actions using video data and approaches of transfer learning (Sudmann 2017). Sequence and generative models are, however, also media of self-reflection and because of this capability also interesting from a media studies perspective.

Indeed, as ChatGPT and other systems demonstrate, communication between humans and machines is rapidly evolving and becoming one of the central scenes of the technical performance of AI systems.

Despite existing shortcomings, large language models or sequence models can be seen as another 'game changer' in the development of advanced AI systems. Already the current level of their performance suggests that the intervention of AI in all sciences will proceed faster and more profoundly than the skeptical view would have suggested only a few years ago. However, the growing importance of AI and the hypothesis of its fundamental intervention also raises the question of 'which aspects of scientific practices and methods will be unaffected or hardly changed by AI'.

Especially in this respect, it seems important to us to combine media archaeological approaches for the analysis of algorithmic conditions of information (e.g., Ernst 2021) with media praxeological approaches, especially media ethnography, in order to not explain technology exclusively in technicist terms. The media perspective proposed here does not only concern the socio-technical conditions of AI infrastructures in and of themselves, but equally affects questions about the historical epistemology of AI, as well as the genesis of different forms and models of knowledge.

The fact that ANNs are also called artificial because they are loosely based on the neuroinformatic model of brains (both human and animal) is now widely known and regularly pops up in current debates on the technical performance of AI systems. This might serve to deflate exaggerated ideas and expectations of an artificial general intelligence for which humans continue to be the model, even when AIs in the form of ANNs have been optimizing each other for a while, without humans as a model (even if at present the latter is only in its initial stages, e.g., AlphaGo's successor systems, cf. Silver et al. 2017).

Whether today's AIs embody, among other things, a form of 'alien intelligence' and/or if they are still a form of intelligence closely related to what is characterized as human intelligence, can perhaps not be decided at all, because corresponding assignments are not only worthy of criticism in each case, but they also cannot be reduced to mutually exclusive alternatives.

The claim that AI has to be 'human-centered' challenges us to critically reflect on its anthropocentric logic as well as on its ideological implications. Of course, there are obvious reasons why especially the AI industry or many scientists stress the human-centeredness of their applications. Apparently, one of the strongest potential or at least imagined threats of AI is that humans might get out of the loop and might lose control over the technology (just think of the current call to pause the development of big AI systems more powerful than GPT-4, Future of Life Institute 2023) – that is, of course, a fear that is older than AI and was historically connected to many technologies (especially regarding the question if automation brought about by technology threatens work).

The relation to corresponding dystopian representations in popular media doesn't need detailed explanation here. However, this is precisely why it is important to include the popular techno-imaginations of the culture industry (or media culture, whoever prefers the term) when trying to understand what matters in the development of technology (see thesis VIII below).

Thesis II

The investigation of machine learning methods in the scientific field requires a detailed analysis of the different levels, contexts and the specific functions of media in the creation and formation of AI technologies and their methodological use in research. From a media studies perspective, a distinction must also be made between applications that use AI technologies primarily or exclusively for scientific purposes and those for which this is not the case, since the respective scientific use there is only optional.

Which infrastructural types of media are relevant to the application of AI in general and which are only relevant to a specific field or problem within a single discipline? Finally, which types of AI-related media are crucial for which phases of the research process (e.g., collection or analysis of data)?

Answering these questions seems essential to adequately assess the inter- and transdisciplinary potential of AI's media. Some problems with the differentiations proposed here are obvious: For AI research, the specific domain reference may sometimes be secondary to what the model is capable of doing in general, i.e., in other domains as well; in other cases, the dependence and focus on a single domain is crucial (and intended as such).

Historically, the expert systems of the 1970s and 1980s, for example, were more or less limited to a particular domain area. At that time, AI systems were not related to a more or less universal knowledge, but strove for selective or specialized knowledge representation. Thus, they stand in sharp contrast to current LLMs, since those models have a universal orientation and competence not only in knowledge representation, but also regarding the fact that they can generate output beyond the central function of knowledge representation, insofar as they are able to generate unexpected results such as creating poems, writing computer programs, solving riddles etc.

Nevertheless, sequence or generative models like ChatGPT are the conditions of possibility for the expert systems of the 21st century. At present, everything seems to boil down to the fine-tuning of the large sequence models (e.g., Lewkowycz et al. 2022). Put simply, you have systems like ChatGPT which can handle general tasks like creating texts regardless of a specific domain, but when it comes to very specialized areas of knowledge, they have trouble coming up with correct or good results. This is where the fine-tuning comes into play. One uses the pre-trained models as a starting point to train them in a second step for a specialized task and/or specialized data in the respective domain area and thus usually has more appropriate results.

In this respect, the relevance of big data is given at various scaling levels of knowledge domains. Contrary to the name, the epistemic relevance of big data does not only result from the amount of data, but from its diversity and ideally also from a qualitative evaluation of this heterogeneity (Kitchin 2014).

Whether it is translation tools, search engines, or dialogue systems based on LLMs – how does epistemologically interested research deal with the fact that the function and use of such systems are not limited to scientific purposes and that they still inevitably inscribe themselves in the practices of scientific thought and knowledge production? It is obvious that the above examples alone point to a specific form of AI-based knowledge production as well as mediation and need to be critically evaluated accordingly. ChatGPT is more than a search engine and yet the system is also, among other tasks, used for this very purpose. Unlike a search engine, however, ChatGPT does not simply generate knowledge depending on appropriate queries; rather, it also provides information about the conditions of knowledge production, including, for example, statements about its limitations and regulations. Moreover, the system is potentially capable of understanding the references of successive queries, of responding to queries, etc.

In this respect, ChatGPT can to some extent also be understood as an application example of Explainable AI. Nevertheless, it is obvious that not only the form of Explainable AI systems – specifically the design of corresponding algorithmic functions – but also the media of their emergence and infrastructural situatedness are quite different in generalistic AI systems like ChatGPT in comparison to more domain-specific systems.

Currently, we can already observe that the differentiation of AI-based epistemic media is increasing. Instead of domain-spanning translation tools, search engines, etc. researchers might increasingly and appropriately use domain-specific applications. These processes potentially have important implications for media policy, which, as the following thesis suggests, are also already becoming apparent.

Thesis III

Research on the research of AI, not only as a media studies enterprise but also as an interdisciplinary project, is confronted with two overlapping challenges: The first one is dealing with the scope and speed in the development of what can be considered rather universal technologies in AI like CNN or LLM, which as such are relevant to different fields of application (in the sciences and beyond) and which typically are developed in fields of computer science. Another challenge in addition to and entangled with the first one is to survey and understand the reaction to and adoption of AI in different disciplines and areas of knowledge, again especially in terms of their scope and speed, but also with regard to their manifold contexts.

Much of the methodological deployment of AI in the sciences, we suggest, consists of the application of machine learning techniques that can be considered conventional from a computer science perspective at the time of their application in other disciplines. Examples would be, for instance, the fundamental importance of how backpropagation is used for many scientific deployments of

ANNs, as well as more specifically, of CNNs, or, more recently, diffusion models as well as LLMs (cf. Chowdhery et al. 2022: 7f.; LeCun/Bengio/Hinton 2015; Schmidhuber 2015). Certain models seem to have established themselves at an increasing rate in recent years (again, think of LLMs), at the same time their validity as 'state of the art' is obviously very limited, if one takes into account the rather short relevance of generative adversarial networks (GANs).

From a meta-theoretical research perspective, just paying attention to these rather universal AI models in computer science is in itself already a very difficult task. Additionally, the complexity of the requirements for understanding corresponding developments increases when they are examined in relation to a specific domain area and placed in relation to its dynamics. Furthermore, the focus on the adaptation of AI technologies leads to more specific challenges, for example, being able to distinguish whether an existing AI technology is primarily being simply applied in a specific field or whether it has also been substantially developed further within the context of the application.

Such dynamics are, of course, a general characteristic of scientific and technological development. Nevertheless, the temporal aspects mentioned here seem to be particularly extreme with respect to current AI advancements. The peculiarity of AI here consists above all in the fact that AI is to be understood not only as an object of the temporal logic of technology development, but potentially, if not solely, as its 'subject'. The mediality and media dependency of AI must accordingly take these temporal dimensions into account (for some general considerations on the temporal aspects of ANN-based AI, cf. Sudmann 2021).

Thesis IV

The outstanding epistemic-technical potential of ANN for the scientific field has (always) mainly been to address and cope with different forms of fuzziness and uncertainty, which includes, e.g., missing information. Accordingly, it is important to explore in more detail how media as input are associated with challenges and problems of uncertainty and fuzziness or generate them in the first place, but also how they contribute to reducing or avoiding uncertainty and fuzziness.

A provocative response to our research group's question about how AI is changing the sciences might be that the central answer is a foregone conclusion that more or less amounts to the thesis presented here. The epistemic potential of AI's statistical approaches to pattern recognition has been recognized since the 1950s (Sudmann 2018b: 22). But it was only a decade ago when it became apparent how well ANN are able to handle problems of fuzziness, uncertainty as well as missing information, as they have always occurred in diverse sciences: be it the handling of ambivalences in literary texts (e.g., Suissa/Elmalech/Zhitomirsky-Geffet 2022), the reconstruction of damaged or incomplete historical images in the field of art (e.g., Zeng/van der Lubbe/Loog 2019), ambiguities in speech recognition due to noise and other factors (e.g., Qian et al. 2016), facing problems like efficient magnetic resonance imaging in medicine (e.g., Schlemper et al. 2017) or ground water level prediction in geoscience (e.g., Tao et al. 2022).

Dealing with these problems is strongly tied to processing and training with large amounts of data. In this respect, one could say, ANNs represent a new technical-epistemic level of using and exploiting quantities to deal with qualitative research problems. Moreover, ANNs can also be used on a new scale to deal with quantitative problems, specifically with regard to arithmetic and algebra (see e.g., Gérard Biau in conversation with Anna Echterhölter in this volume).

To what extent the ability to deal with problems of fuzziness signifies an epistemic rupture can certainly not be ascribed to an unambiguous date from the outset, but to different historical paths of development as well as specific genealogies that must be reconstructed historically.

Thesis V

The epistemic potential of ANNs – as the (currently) dominant form of AI – is based on the massive parallelism of information processing. The technology can be theorized as quasi-analog or post-digital.

ANNs, as the currently dominant manifestation of AI, are typically negotiated as digital technology. However, this view is at least partially in need of correction, as the following arguments underscore:

> [First], it must be emphasized that the masses of interconnected neurons, activated by an input, fire together simultaneously or in parallel, thus ultimately forming a complex emergent system that abolishes the discrete char-

> acter of the elements it consists of (the layers of neurons and their connection) [...]. This extreme or massive parallelism of information processing can indeed count as the essential characteristic of ANN, distinguishing it from the von Neumann architecture of classical digital computers. Due to the described properties, an ANN is therefore a blurred system [in German: "Unschärfesystem"] .. whose operations can be described rather as analog than digital (Sudmann 2018a: 67, own translation).

Secondly, it can be argued that the massive parallelism of neural networks, as currently effectively unfolded in LLMs and other models, among others, is characterized by a quasi-analog fine-grainedness in information processing.

> [A] single artificial neuron is usually either active or not, so in this respect it usually functions according to a binary logic, like the switching states of a digital computer. However, the weighting of activity between neurons, i.e., the strength of their connections, is mostly represented by floating point numbers (positive and negative) in neural networks. And this representation is so finely grained that the corresponding values can be understood as quasi-analog. As a medium of information transmission, ANNs thus do not operate with binary units, such as 0 and 1, but in quasi-analog form (even if the values are still based on a digital substrate) (ibid.: 66f., own translation).

Fine-grainedness in this context is not reduced to generating certain effects of quasi-analog representations, for example when a modern display allows smooth color transitions, and in this way appears analog, i.e., continuous. It is important to note here that the attribute of quasi-analog concerns the technical conditions of information processing, not its mere form of representation. For the time being, the parallelism of neural networks at the lowest level is still determined by the circuit logic of digital computers. For the performance of ANNs this dependence is limiting because of its inefficiency. Subsymbolic AI in the form of ANNs is based on similarity relations of fundamentally continuous quantities, which are currently still digitally approximated. As ironically described elsewhere, they're still 'abusing' digital technology until they are eventually, with some probability, replaced by analog technology (ibid.: 69).

For this very reason, it makes sense to negotiate ANNs not merely as quasi-analog information technology, but also to characterize it literally as a post-digital scenario of the conditions of information and knowledge processing in the 21st century.

In media studies, some of the aspects of connectionist AI addressed here have already been implicitly highlighted by Norbert Bolz in the introduction to the volume *Computer als Medium* (in English: "Computer as medium"), published 1994:

> [The] reorientation of intelligence to simultaneously and parallelly processing nervous systems that statistically process their data at a comparatively low level of precision parts with the dream of a mathesis universalis that philosophy, from Leibniz to Husserl, dreamed of. For the computer is a plausible metaphor for the media spirit [in German: "Mediengeist"] only as long as thinking means calculating and cognition is understood as calculating with digital symbols. Algorithms define a logical world through purely syntactic operations, in which all problems can be solved through serial search routines. The world of emergent AI is quite different. Connectionism is the name for operating in subsymbolic networks where meaning is a function of a system state. Accordingly, storage does not occur in single, precisely addressed, memory locations, but in networks. All regularities in this network are emergent qualities against the background of a chaos of linkages. Thus, in the subsymbolic network of connectionist machines, there is an exact correspondence to the noise in the brain, i.e., to the random firing of neurons. (Bolz 1994: 14, own translation)

While Bolz's reference to the 'chaos of linkages' may be as techno-epistemically inaccurate as the comparison to the 'noise in the brain', the emphasis on the 'emergent qualities' of networks, however, remains crucial in highlighting the suspension of the principle of digital information technology in connectionist systems.

Thus, if one seeks to examine the impact of AI technologies on the sciences, one cannot avoid taking seriously the fundamental algorithmic specificity of the technology. And this means that the question of the transformations of the sciences by AI is, from a technical-pragmatic point of view, only to a limited extent a problem of digital technology. Therefore, it might not be sufficient to describe AI technologies only with notions taken from the theory of digital media or to insert their histories into the histories of digital media. Other theoretical and historical traditions might be important too. The same, by the way, might be the case in quantum computing, in which also a partial return to analog forms of information processing can be observed (or described as such, cf. Schröter/Ernst/Warnke 2022).

Thesis VI

The diagnosis of an AI revolution is ubiquitous. Given the historical examples for supposed technological or specifically media-technological revolutions, such rhetoric should be treated with caution. It is likely that, as in all historical examples, discontinuities and continuities coexist in complex manners that cannot be predicted beforehand.

As the last few years of the boom in machine learning have reminded us once again, it is rarely one event that establishes a caesura; rather, it is a series of events that establish an order of a before and after, or mark a longer process of change as such. The rhetoric of 'revolution' is, more often than not, a rhetoric of AI imaginaries (see Thesis VIII) used in entertainment and advertising, rather than a useful description of real developments. Nevertheless, sometimes specific events have a profound relevance for the further development of technology (see Sudmann 2018b related to the recent history of ANNs).

Since AI is obviously a technology that processes information, questions of media history and media historiography apply (cf. Schröter/Schwering 2014). We want to highlight three aspects:

Continuity and Discontinuity: As in all media history formulations, clear linear successions and rhetorics of 'before' and 'after' ('revolution') should be discarded in favor of multifactored and multilayered descriptions – 'series' as Foucault (1972: 4, 7f.) put it (cf. Schröter 2014: 13–22). In some series some things change slower or faster, where in others certain aspects stay the same.

Accelerations and Brakes: Brian Winston (1998: 1–19) has argued that on the one hand "supervening social necessities" accelerate the development and distribution of new media, while on the other hand a "law of the suppression of radical potential" applies, which tames radical changes made possible by new media technologies (e.g., copyright laws that impede the potentialities for lossless reproduction in digital media).

Retrospective construction: As Glaubitz et al. (2011) have argued, media history of a certain media technology is always triggered at first by a high 'level of recognition'. Some technology becomes visible, commercially successful and perhaps scandalized – and then the process of retro-construction starts. The research focuses on 'emergence events' where it all began and also looks for the lines of development that begin with these events.

All these mechanisms operate in the historical development and historiographical description of AI systems too. Regarding the role of AI in different scientific disciplines it is to be expected that they may have different weightings, distributions and forms. To develop a more precise picture in this regard is one task of our research project.

Thesis VII

Due to their predictive capabilities, it is important to examine approaches of ANN, in a broader sense, as technologies of speculation. In this respect, however, they also challenge us to reflect on our own speculative thinking; the critique of AI and its epistemic applications must therefore also include the 'meta-theoretical' reflection.

Even beyond its culture-industrial imagination, AI has always provoked speculation about its future limits, risks, potentials, and ambivalences. Recent developments and achievements of ANNs have added a crucial new aspect to this view especially: Instead of speculating about AI, people have started speculating with AI. But the semantics of speculation implies uncertainty. One must recall, at this point, that ANNs have been used for the speculative business of stock market prediction since the late 1980s and 1990s (cf. Wong/Bodnovich/Selvi 1995; Vui et al. 2013). However, the risky bet on big business in the stock market and the uncertainties associated with it are at odds with what is socially desired for the scientific application of ANNs as predictive technologies, namely to be able to use it to control and master the future, especially in highly sensitive areas such as medicine or climatology (cf. Halpern/Mitchell 2023). This epistemically almost indispensable claim highlights the need to relate the critical analysis of predictive systems as a technology of controlling the future(s) to the present, taking into account empirical technology development as well as the realm of imaginaries.

Nevertheless, speculation as a critical practice remains necessarily and essentially related to the future. Critical analysis of current conditions is always in the service of the premise and claim that the world could (and should) be different than it currently is. Contemporary AI systems such as ChatGPT can now themselves be interrogated for utopian imagining of their future as well as for critiquing society, which is why we must seriously consider that the political infrastructures of societies to come will also increasingly depend on the deployment of learning algorithms. In any case, recent AI and the critical moment of speculation it mediates should in turn be used to critically reflect on our own cognitive processes and approaches. The institutions and designs of the sciences which are necessarily related to the future, must, precisely for this reason, also offer speculative thinking, as a critical practice of the present, and not (alone) of the future, as well as an appropriate space to unfold.

Thesis VIII

The analysis of the scientific uses of AI should also include the analysis of their (culture-industrial) imaginations.

One of the central problems of the scientific engagement with AI is that there is hardly any group of technologies that is so charged with partly crazy imaginations (on the notion of cultural imagination; see its use in Ernst/Schröter/Sudmann 2019). In particular, popular media, like motion pictures, since the late 1960s have been full of – often exaggerated – ideas about what AI and (not always clearly separated from them) robots should be able to do.

These ideas can have both a utopian and (this is the more common case) dystopian inclination. We cannot and will not go into these ideas and their various forms in detail here, but several theoretical and methodological demands follow from this for the scientific study of AI: First, one has to ask why at certain times and contexts certain imaginaries are attached to a technology like 'AI' – to which needs does 'AI' respond, which social deficiency and/or deficiency caused by previous media finds expression in these imaginaries (on utopias regarding computers in general, cf. Winkler 1997). Secondly, we can investigate what role such imaginaries have played as 'Leitbilder' (Dierkes/Hoffmann/Marz 1992) or 'diegetic prototypes' (Kirby 2010) in the actual development of technology. Thirdly, in doing so, we must also historically separate ideological and simply absurd imaginaries from those that have played a constructive role, which is only possible through historical retrospection.

The discourse about the role of AI in the sciences is also permeated by such 'AI'. The ideas of what should be possible with AI, which have increased into the utopian, can be a reason for starting to use AI-based methods in the first place. Manufacturers of such systems do well to quote these utopias in their advertising, for example, in order to increase the attractiveness of their products for scientific buyers. What do different scientific disciplines, certain research domains or even individual scientists expect from the use of AI? What guiding principles are associated with it? What is imagined under the term 'AI' in the first place? These are questions that must play a central role in a research design on the role of AI in the sciences.

Thesis IX

The principle of ANN is its universalistic orientation, determined by the phantasmatic imagination that has always characterized AI technologies: to overcome problems of difference.

Connectionist AI can be understood as a universal machine sui generis. Information processing with artificial neural networks is Turing-complete, i.e., we are dealing with machines that can simulate or program other machines (Siegelmann/Sontag 1992: 440f.). They share the universalistic principle that already characterizes the digital computer according to the serial Von Neumann architecture: to be able to scan and simulate all individual media as well as to process a certain input independent of its specific meaning and socio-cultural codes, and so on. This universalist feature, however, characterizes not only the epistemic conditions of technology, but also its telos.

In practice, especially in scientific applications, the specificity of the learning material is of course immensely important, for example with regard to inscriptions of discriminating biases.

Part of the practical perspective is to note that by no means all those who are driving the development of machine learning are pursuing the goal of AGI. Yann LeCun and many other experts constantly emphasize how far current technology development still falls short of the status quo of whatever is considered to be 'human intelligence' (cf. LeCun 2022; Shanahan 2023). And yet, leading companies and scientists are more or less explicitly committed to the goal of AGI (cf. Altman 2023). This goal is not simply identical to simulating human intelligence, but consists first and foremost of developing an AI system that, similar to humans,

I. can cope with a heterogenous range of problems and tasks, and
II. generally speaking, has to fulfill as few preconditions as possible, e.g., avoids having to learn the solution of a problem in a time-consuming way, amongst other things by accessing a lot of data or by specific procedures of processing these data.

Already at the end of the 1980s, Seymour Papert criticized that both symbolic-ruled and connectionist AI are "engaged in a search for mechanisms with a universal application" (Papert 1988: 2). Papert's critique is perhaps more relevant than ever today, given the supposed universalistic capabilities of technologies like LLMs.

Nevertheless, we might be able to better deal with problems of difference, but they persist, especially with regard to AI models that seek or seem to overcome them (like the concept of a universal language translator). A very important aspect in this respect is the inevitability of algorithmic biases for every learning model. In current discussions of algorithmic discrimination, it is often forgotten that every learning for a certain task (like learning languages) inevitably produces 'costs', and hence any machine learning process that claims to be universal (i.e., is capable of dealing with all challenges of difference) must necessarily remain phantasmatic, which is also true, in a very fundamental way, for the relationship between humans and machines (cf. Ernst/Schröter/Sudmann 2019).

List of references

Altman, Sam (2023): "Planning for AGI and beyond.", February 24, 2023 (https://openai.com/blog/planning-for-agi-and-beyond).

Bengio, Yoshua (2022): "Superintelligence: Futurology vs. Science.", January 24, 2022 (https://yoshuabengio.org/2022/01/24/superintelligence-futurology-vs-science/).

Bolz, Norbert (1994): "Computer als Medium – Einleitung." In: Norbert Bolz/Friedrich A. Kittler/Christoph Tholen (eds.), Computer als Medium, München: Fink, pp. 9–16.

Chowdhery, Aakanksha/Narang, Sharan/Devlin, Jacob/Bosma, Maarten/Mishra, Gaurav/Roberts, Adam/Barham, Paul/et al. (2022): PaLM: Scaling Language Modeling with Pathways, arXiv Preprint (https://doi.org/10.48550/arXiv.2204.02311).

Dierkes, Meinolf/Hoffmann, Ute/Marz, Lutz (1992): Leitbild und Technik – zur Entstehung und Steuerung technischer Innovationen, Berlin: edition sigma.

Engemann, Christoph/Sudmann, Andreas (eds.) (2018): Machine Learning – Medien, Infrastrukturen und Technologien der Künstlichen Intelligenz, Bielefeld: transcript.

Ernst, Christoph/Schröter, Jens/Sudmann, Andreas (2019): "AI and the Imagination to Overcome Difference." In: spheres. Journal for Digital Cultures

5 (https://spheres-journal.org/contribution/ai-and-the-imagination-to-overcome-difference/.

Ernst, Wolfgang (2021): "Existing in Discrete States: On the Techno-Aesthetics of Algorithmic Being-in-Time." In: Theory, Culture & Society 38/7-8, pp. 13–31.

Foucault, Michel (1972): The Archaeology of Knowledge, New York: Pantheon Books.

Future of Life Institute (2023): "Pause Giant AI Experiments: An Open Letter.", March 22, 2023 (https://futureoflife.org/open-letter/pause-giant-ai-experiments/).

Glaubitz, Nicola/Groscurth, Henning/Hoffmann, Katja/Schäfer, Jörgen/Schröter, Jens/Schwering, Gregor/Venus, Jochen (2011): Eine Theorie der Medienumbrüche. 1900/2000, Siegen: universi.

Halpern, Orit/Mitchell, Robert (2023): The Smartness Mandate, Cambridge and London: The MIT Press.

Humm, Bernhard G./Buxmann, Peter/Schmidt, Jan C. (2022): "Grundlagen und Anwendungen von KI." In: Carl Friedrich Gethmann/Peter Buxmann/Julia Distelrath/Bernhard G. Humm/Stephan Lingner/Verena Nitsch/Jan C. Schmidt/Indra Spiecker genannt Döhmann (eds.), Künstliche Intelligenz in der Forschung: Neue Möglichkeiten und Herausforderungen für die Wissenschaft, Berlin/Heidelberg: Springer, pp. 13–42.

Kirby, David (2010): "The Future is Now: Diegetic Prototypes and the Role of Popular Films in Generating Real-world Technological Development." In: Social Studies of Science 40/1, pp. 41–70.

Kitchin, Rob (2014): "Big Data, new epistemologies and paradigm Shifts." In: Big Data & Society 1/1 (https://doi.org/10.1177/2053951714528481).

Kittler, Friedrich A. (1993): Draculas Vermächtnis. Technische Schriften, Leipzig: Reclam.

LeCun, Yann (2022): A Path Towards Autonomous Machine Intelligence, Open-Review Preprint (https://openreview.net/pdf?id=BZ5a1r-kVsf).

LeCun, Yann/Bengio, Yoshua/Hinton, Geoffrey (2015): "Deep Learning." In: Nature 521, pp. 436–444.

Lewkowycz, Aitor/Andreassen, Anders/Dohan, David /Dyer, Ethan/Michalewski, Henryk/Ramasesh, Vinay/Slone, Ambrose/et al. (2022): Solving Quantitative Reasoning Problems with Language Models, arXiv Preprint (https://doi.org/10.48550/arXiv.2206.14858).

Mößner, Nicola/Erlach, Klaus (eds.) (2022): Kalibrierung der Wissenschaft. Auswirkungen der Digitalisierung auf die wissenschaftliche Erkenntnis, Bielefeld: transcript.

Papert, Seymour (1988): "One AI or Many?" In: Daedalus 117/1, pp. 1–14.

Qian, Yanmin/Bi, Mengxiao/Tan, Tian/Yu, Kai (2016): "Very Deep Convolutional Neural Networks for Noise Robust Speech Recognition." In: IEEE/ACM Transactions on Audio, Speech, and Language Processing 24/12, pp. 2263–2276.

Rosenblatt, Frank (1961): Principles of Neurodynamics: Perceptrons and the Theory of Brain Mechanisms, AD0256582, Buffalo, NY: Cornell Aeronautical Laboratory.

Schlemper, Jo/Caballero, Jose/Hajnal, Joseph V./Price, Anthony/Rueckert, Daniel (2017): "A Deep Cascade of Convolutional Neural Networks for MR Image Reconstruction." In: Information Processing in Medical Imaging. 25th International Conference (IPMI 2017), Boone, NC, USA, pp. 647–658.

Schmidhuber, Jürgen (2015): "Deep Learning in Neural Networks: An Overview." In: Neural Networks 61, pp. 85–117.

Schröter, Jens (2014): 3D. History, Theory and Aesthetics of the Transplane Image, New York, London, New Dehli and Sydney: Bloomsbury.

Schröter, Jens/Ernst, Christoph/Warnke, Martin (2022): "Quantum Computing and the History of the Analog/Digital-Distinction" In: Grey Room 86, pp. 28–49.

Schröter, Jens/Schwering, Gregor (2014): "Modelle des Medienwandels und der Mediengeschichtsschreibung." In: Jens Schröter (ed.), Handbuch Medienwissenschaft, Stuttgart and Weimar: Metzler, pp. 179–190.

Shanahan, Murray (2023): Talking About Large Language Models, arXiv Preprint (https://doi.org/10.48550/arXiv.2212.03551).

Siegelmann, Hava T./Sontag, Eduardo D. (1992): "On the Computational Power of Neural Nets." In: Proceedings of the Fifth Annual Workshop on Computational Learning Theory, New York, NY, USA, pp. 440–449.

Silver, David/Schrittwieser, Julian/Simonyan, Karen /Antonoglou, Ioannis/Huang, Aja /Guez, Arthur/Hubert, Thomas /et al. (2017): "Mastering the Game of Go Without Human Knowledge." In: Nature 550, pp. 354–359.

Stöcker, Christian (2020): "Der Forschungs-Turbo.", February 23, 2020 (https://www.spiegel.de/wissenschaft/mensch/beschleunigte-wissenschaft-der-forschungs-turbo-a-b828d49e-4d5c-414b-8315-0ef2c5ce0a30).

Sudmann, Andreas (2017): "Auch Maschinen können intuitives Wissen sammeln", October 27, 2017, (https://www.nzz.ch/digital/das-intuitive-wissen-der-maschinen-ld.1324264).

Sudmann, Andreas (2018a): "Szenarien des Postdigitalen: Deep Learning als MedienRevolution." In: Christoph Engemann/Andreas Sudmann (eds.), Machine Learning – Medien, Infrastrukturen und Technologien der Künstlichen Intelligenz, Bielefeld: transcript, pp. 55–74.

Sudmann, Andreas (2018b): "Zur Einführung. Medien, Infrastrukturen und Technologien des maschinellen Lernens." In: Christoph Engemann/Andreas Sudmann (eds.), Machine Learning – Medien, Infrastrukturen und Technologien der Künstlichen Intelligenz, Bielefeld: transcript, pp. 9–23.

Sudmann, Andreas (2021): "Artificial Neural Networks, Postdigital Infrastructures and the Politics of Temporality." In: Axel Volmar/Kyle Stine (eds.), Media Infrastructures and the Politics of Digital Time, Amsterdam: Amsterdam University Press, pp. 279–294.

Suissa, Omri/Elmalech, Avshalom/Zhitomirsky-Geffet, Maayan (2022): "Text Analysis Using Deep Neural Networks in Digital Humanities and Information Science." In: Journal of the Association for Information Science and Technology 73/2, pp. 268–287.

Tao, Hai/Hameed, Mohammed Majeed/Marhoon, Haydar Abdulameer/Zounemat-Kermani, Mohammad/Heddam, Salim/Kim, Sungwon/Sulaiman, Sadeq Oleiwi/et al. (2022): "Groundwater level prediction using machine learning models: A comprehensive review." In: Neurocomputing 489, pp. 271–308.

Turing, Alan (2004 [1948]): "Intelligent Machinery." In: B. Jack Copeland (ed.), The Essential Turing. Seminal Writings in Computing, Logic, Philosophy, Artificial Intelligence, and Artificial Life plus The Secrets of Enigma, Oxford: Clarendon Press, pp. 410–432.

Tuschling, Anna (2022): "Vom Bildstimulus zur Emotion AI: Zur Sichtbarkeit und Unsichtbarkeit technischer Affektbilder." In: MLN 137/3, pp. 443–465.

Vui, Chang Sim/Soon, Gan Kim/On, Chin Kim/Alfred, Rayner/Anthony, Patricia (2013): "A review of stock market prediction with Artificial neural network (ANN)." In: 2013 IEEE International Conference on Control System, Computing and Engineering (ICCSCE), Penang, Malaysia, pp. 477–482.

Winkler, Hartmut (1997): Docuverse. Zur Medientheorie der Computer, München: Klaus Boer.

Winston, Brian (1998): Media Technology and Society. A History: From Telegraph to the Internet, London and New York: Routledge.

Wong, Bo K./Bodnovich, Thomas A./Selvi, Yakup (1995): "A Bibliography of Neural Network Business Applications Research: 1988–September 1994." In: Expert Systems 12/3, pp. 253–261.

Zeng, Yuan/van der Lubbe, Jan C. A./Loog, Marco (2019): "Multi-Scale Convolutional Neural Network for Pixel-Wise Reconstruction of Van Gogh's Drawings." In: Machine Vision and Applications 30/7, pp. 1229–1241.

Putting the AI into social science

How artificial intelligence tools are changing and challenging research in the social sciences

Johannes Breuer

1. Introduction

The recent rapid announcements, developments and releases in the realm of artificial intelligence (AI), especially within the domain of large language models (LLMs), have not only received a lot of public attention, but also sparked a surge of discussion, research and other activities among the scientific community, including the social sciences. Similar to other digital technologies, such as the internet (cp. Breuer 2022), AI has multiple relationships with science. It is a) an outcome or product of scientific research, b) an object of study across many different disciplines, and c) a powerful tool that affects how research is done. This chapter focuses on the third function and discusses how AI tools have been changing how social science research is conducted and what the future may hold in this regard. The discussion within this chapter will address both the potentials as well as the challenges and risks associated with the use of AI (and tools based thereon) in the social sciences.

Notably, AI can have – and already, in many cases, has – an impact on all elements of social science research. There are different ways in which the (typical) research process in the social sciences (and similar disciplines) can be structured. Common phases can, e.g., be structured as follows: 1) idea generation (e.g., formulation of research questions or hypotheses), 2) discovery (e.g., searching for and exploring existing literature, data, analysis methods, etc.), 3) study design and planning (e.g., deciding what methodology and sample to use), 4) data collection (e.g., via surveys, interviews, web scraping), 5) data processing (e.g., cleaning the data, getting it ready for analysis), 6) data analysis, 7) interpreting results, 8) reporting, publishing, and sharing.[1] Of course, in practice, these phases are often overlapping or not clearly distinguishable and do not necessarily occur in this order (and there may also be recursions). For example, in the case of an exploratory study, researchers might discover something in the data analysis phase that leads them to collect additional data, come up with new research questions, or reconsider their analysis methods. While AI can affect all of these phases, the degree to which this is the case and the ways in which this influence manifests itself differ between the individual steps. After clarifying a few important preliminaries that need to be kept in mind when dealing with the use AI in the social sciences at this time, this chapter will discuss how AI and AI-based tools have been or can be used in the various phases of social science research and the promises and potentials as well as the pitfalls and perils associated with these practices.

2. Preliminaries

Before discussing the practices, potentials, promises, pitfalls and perils of the use of AI in the social sciences, it is necessary to lay out a couple of important preliminaries. The first one relates to the terminology used in this chapter. Similar to the term big data, artificial intelligence has different definitions and, hence, can be a somewhat fuzzy concept. Oftentimes, AI is used interchangeably with machine learning (ML), or at least the distinction becomes blurry. However, as Kühl et al. 2022 point out: "'ML' and 'AI' are not terms that should be used interchangeably (...) ML is an important driver of AI, and the majority of modern AI cases will utilize ML. However, (...) there can be cases of AI without ML (e.g., based on rules or formulas)" (2241). Another important distinction for this chapter as well as the collected volume which it is part of, is the one between symbolic and subsymbolic AI. According to Ilkou and Koutraki (2020), the key differences between these two types of AI are the following: "(1) symbolic approaches produce logical conclusions, whereas sub-symbolic approaches provide associative results. (2) The human intervention is common in the symbolic methods, while the sub-symbolic learn and adapt to the given data. (3) The symbolic methods perform best when dealing with relatively small and precise data, while the sub-symbolic ones are able to handle large and

1 Notably, these phases as outlined here are quite generic. Most of them are, hence, also valid for other empirical disciplines (e.g., from the medical and natural sciences).

noisy datasets" (1). The currently predominant type of AI – and also the focus of this book – is subsymbolic AI.[2] According to Ilkou and Koutraki (2020), "subsymbolic AI includes statistical learning methods, such as Bayesian learning, deep learning, backpropagation, and genetic algorithms" (2). As these methods, especially also deep learning, are often discussed as belonging to the area of machine learning, it becomes apparent how terminological ambiguities between AI and ML may arise in an applied context. Taking this into account, the chapter will not discuss to what degree different techniques and tools are best described as AI or ML or to what degree AI applications can be classified as symbolic or subsymbolic. What is more important for the present chapter is that, outside of computer science or other fields involved in the development of LLMs and other types of AI, the use of or interaction with AI occurs via tools. While – under the hood – these tools often make use or offer access to methods that can be seen as belonging to the area of ML, the tools are often labelled or described as AI-based. Although this may not always be (fully) appropriate and often done (primarily) for marketing reasons, for the purpose of this chapter, if they are labelled/presented as AI tools, they will also be discussed as such here.

The application area of AI tools that this chapter focuses on are the social sciences. Core disciplines in this field include sociology, political science, or communication science.[3] As stated in the introduction, however, many of the prototypical phases in social-scientific research are also common in other fields. Likewise, many of the methods and tools discussed in this chapter are also used there. Regardless of the definition of the category of social sciences, the focus of this chapter is on empirical research. More specifically, while many of the methods and tools covered in the following can also be used for quali-

---

2 Ilkou and Koutraki (2020), however, note that in-between methods that combine symbolic and subsymbolic AI have become more common. Among other things, the rise of the concept of explainable AI has contributed to the resurge of symbolic methods, which were the dominant approach until the 1980s.

3 There are, of course, also other disciplines that can be classified as social sciences as well as different ways of classifying disciplines. Besides, there are some disciplines for which there are different views on whether they can be seen as belonging to the social sciences, such as psychology or economics. As much of the tasks and topics covered in this chapter should also be relevant beyond the social sciences, these differences in the definition of social sciences and the classification of disciplines should not matter.

tative research, an emphasis will be on quantitative empirical research in the social sciences.[4]

A final important thing to consider for this chapter is that the field of artificial intelligence is currently developing rapidly following the release of powerful large language models (LLMs) and their quickly increasing use for all sorts of applications, including scientific research. Especially since the release of *ChatGPT* by *OpenAI* in November 2022, the development of AI applications has gained a lot of momentum. While the development and release of LLMs and tools based thereon had already been quite fast-paced before, this has been massively sped-up in the first half of 2023, with new models and tools being released daily. Accordingly, it is almost impossible to keep up with all developments. Although the timeframes of academic research (especially if it is empirical) are not fully compatible with the speed of current technology developments, the academic community has been trying to keep up by conducting timely studies and publishing them in the form of preprints. Notably, these publications are not peer-reviewed. Still, given their timeliness and relevance, such preprints will be considered in this chapter. Against this background, it should be noted that the methods and tools, as well as the scientific publications investigating their use are likely to become updated and amended or outdated, invalidated, or even replaced in the near future. Consequently, this chapter can only provide a snapshot from the time of writing (April to June 2023), and the practices of using AI and AI-based tools and methods as well as the associated potentials, promises, pitfalls, and perils can be expected to change substantially over the course of the upcoming months and years. Another thing to note is that this chapter is certainly not the only and also not the first discussion of how AI is changing scientific research. Besides the project "How is Artificial Intelligence Changing Science? Research in the Era of Learning Algorithms", (https://howisaichangingscience.eu/) from which the book, that this chapter is part of, originated, there are at least two other noteworthy recent publications in this context. The first one is the preprint "Friend or Foe? Exploring the Implications of Large Language Models on the Science System" by Fecher et al. (2023), in which the authors present the results of "a Delphi

4 This is partly due to the background of the author but also because the use of ML and AI for data collection, processing, and analysis is more common in the quantitative paradigm. In fact, the use of ML and AI methods is one of the defining criteria of the rapidly growing field of computational social science (cp. Hox 2017).

study involving 72 experts specialising in research and AI", in which the author of the present chapter also participated. Based on the expert opinions, the manuscript discusses the applications and (transformative) potential, as well as limitations, risks and ethical and legal implications of the use of LLMs in science. The second relevant recent publication is a preprint by Ziems et al. (2023) entitled "Can Large Language Models Transform Computational Social Science?", that presents the results of evaluations of different LLMs for various typical tasks in computational social science (CSS). The present chapter is essentially situated between these two publications. While, similar to Fecher et al. (2023), it also addresses the applications of AI (tools), taking into account their potential as well as limitations and associated risks, it focuses on the social sciences, specifically on empirical research in this field which follows a specific process from idea generation to publication. Hence, compared to the work by Ziems et al. (2023), the perspective of this chapter is broader, considering not only typical CSS applications, such as automated text classification or other annotation and explanation tasks, but also addressing usage in the context of traditional data collection methods, such as surveys or experiments as well as more general practices, e.g., in phases of discovery and data analysis.

3. Practices

Scientific research has always been based on the use of tools. These tools can either be specifically designed for scientific purposes, such as a microscope or telescope, or designed for other or more general purposes and used by scientists for their research, such as tweezers or a shovel. This is the same for AI(-based) tools. Another important distinction from a practical perspective is whether tools are commercial or free and maybe even open source.[5]

Importantly, tools are not neutral. They shape the research process, define possibilities and boundaries. The concept of Maslow's hammer describes this in a pointed manner: "If the only tool you have is a hammer, it is tempting to treat everything as if it were a nail." (Maslow 1966: x). Of course, scientists typically do not just use a single tool, but a combination of different tools (for different purposes). Especially in the digital realm, these combinations are often referred to as tool stacks. Ideally, the tools within individual tool stacks are

5 Of course, these characteristics can change over time. E.g., tools that are initially free to use may eventually require a paid subscription.

used for one or multiple specific task(s) with little or no redundancies, compatible, and complement each other. A couple of years ago, the project *Innovations in Scholarly Communication* situated at the University of Utrecht distinguished between traditional, modern, innovative, and experimental tools (cp. Bosman/Kramer 2015).[6] Following this distinction, most of the AI tools mentioned in the following can be classified as innovative or experimental. In their analysis, Bosman and Kramer (2015) diagnose "an avalanche of tools" and describe choosing appropriate tools and keeping up with the development of (new) tools as a challenge for researchers. This issue is even more pronounced in the current explosion of the development of AI and its applications, with new tools or versions thereof being released almost daily.

Generally, tools and tool stacks enable scientists to conduct research in the first place or at least facilitate the process and make it more efficient. Besides these potentials, however, tools and tool stacks also bring their own challenges and limitations. While the use of tool stacks widens the possibilities and space for research, they also have or create specific boundaries. In addition, the reliance on tool stacks creates dependencies. Scientists depend on them for conducting their research and tools may also depend on each other to work properly within a given tool stack. These dependencies can break if the functionalities or the availability of tools change. This illustrates that the impact of AI on research in the social sciences is not limited to the quantitative dimension. While it does, e.g., facilitate the handling of large(r) amounts of data, by altering the range of possibilities, it also affects the qualitative aspects of social-scientific research.

As noted before, these changes in the quantitative and qualitative properties of social science research run through all phases of the research process. However, the number and type of AI tools that are used and the impact they have had on research practices so far, differs between each individual phase. Given the rapid development of AI and AI-based tools, the purpose of this section is not to provide a complete list of all tools that have been or can be used for the different steps in the social science research process. Instead, the aim of this section is to provide a couple of examples of how AI has been used in the social sciences to demonstrate its qualitative impact on the tasks typically

6 Regarding the different phases of the research process for which the tools can be used, the categories by Bosman and Kramer (2015) are similar to the ones suggested in the present chapter: discovery, analysis, writing, publication, outreach, and assessment.

undertaken within the different phases.[7] In the following, these impacts will be discussed for each of the eight (proto)typical phases listed above. Importantly, many of the available AI-based methods and tools cannot be exclusively mapped to one phase. While some methods and tools have been designed for very specific tasks, others have a broad(er) range of possible applications in the social sciences.

3.1 Discovery & idea generation

While they may be separated for analytical purposes, in practice, the phases of idea generation and discovery are usually intertwined. Formulating meaningful research questions and hypotheses requires a certain familiarity with existing literature, methods and data. This knowledge is necessary at the latest for the specification of the research questions and/or hypotheses. A large number of AI-assisted tools that have come into existence over the last few years target the discovery phase. Examples include *Semantic Scholar* (https://www.semanticscholar.org/), *scite* (https://scite.ai/), *ResearchRabbit* (https://www.researchrabbit.ai/), *Consensus* (https://consensus.app/), or *elicit* (https://elicit.org/).[8] The focus of all these tools lies on discovering and exploring relevant literature. All of them allow to assess (and visualize) relationships between publications (via citations or similarity) and some offer additional functionalities. For example, *scite* can provide information on how often a paper has been supported, contradicted, or just mentioned in a citing publication, *Consensus* delivers additional information about journals and publications as well as relevant quota-

---

7 There are many lists and discussions of AI tools as well as short recommendations and tutorials on using AI-based tools for scientific research tasks available online. Large parts of this discourse have been happening on Twitter (although there, e.g., also are websites and YouTube videos that cover these topics). Two accounts on Twitter that have produced a large amount of content on this subject are Mushtaq Bilal (https://twitter.com/MushtaqBilalPhD) and Ilya Shabanov (https://twitter.com/Artifexx). There are also thousands of accounts that specialize in covering news on AI developments, tools, and research in general, many of which have only been created or shifted their topical focus and started to receive increased attention (and a quickly growing follower base) fairly recently.

8 Of course, there are many other services and apps for the discovery and exploration of scientific publications, such as *Google Scholar* (https://scholar.google.com/), *Researcher* (https://www.researcher-app.com/), *Inciteful* (https://inciteful.xyz/), *Litmaps* (https://www.litmaps.com/), or *Connected Papers* (https://www.connectedpapers.com/). However, those do not explicitly state or advertise that they employ AI-based methods.

tions from the latter, and *elicit* also offers versatile discovery functionalities via the creation of bespoke tasks. To identify relevant literature from a large corpus for a systematic literature review with the help of AI/ML methods, researchers can also use the free-and-open-source (FOSS) tool *ASReview* (https://asreview.nl/).

Once the relevant publications have been identified, the next task is for the researcher to read them and extract relevant information for their own research. There also are AI-based tools that can assist with that. Besides the functionalities by *Consensus* and *scite* described above, there are services like the article summarizer by *scholarcy* (https://article-summarizer.scholarcy.com/), *Explainpaper* (https://www.explainpaper.com/), or *ChatPDF* (https://www.chatpdf.com) that can aid with extracting information from scientific publications. While all of the other tools and services listed before were specifically created for research purposes, this is not the case for *ChatPDF*. As the name indicates, *ChatPDF* is based on *ChatGPT* by *OpenAI*, and the latter has also become a popular multi-purpose tool for research(ers) in the social sciences. Among other things, researchers have also suggested using *ChatGPT* for the idea generation phase (Dowling/Lucey 2023).

3.2 Study design & data collection

There also are several AI-based tools and methods that can be used for the study design data collection phases. Two of the most widely used data collection methods in the quantitative social sciences and related fields are surveys and experiments (which can also be combined in the form of survey experiments; cf. Mutz 2011). Surveys contain a number of questions or items that are designed to assess certain attributes, attitudes, or behaviors. Researchers often use existing items and scales that, ideally, have been validated before. However, these may not always be available, or existing scales may have to be modified. (Re-)Formulating, and refining survey items is one of the many possible uses of LLMs like *ChatGPT* or *GPT-4* by *OpenAI* and interfaces to those, such as *Microsoft Bing Chat*, in social science research. Through proper prompts, researchers could, e.g., ask LLM-based chatbots to come up with suggestions for novel questionnaire items tapping into specific concepts or optimize the wording of existing questions/question drafts. What is helpful in this regard as well as for all other research-focused uses of general-purpose chatbots like ChatGPT is the use of so-called priming, which describes the process of interacting with the LLM to provide some context and ensure that it understands the tasks before prompting it to get the targeted output, such as (reformulated) question items.

Another common task in survey-based research is the translation of existing items into other languages. This can also be done or supported through LLMs or with the help of AI-based translation tools, such as *DeepL* (https://www.deepl.com/translator) or *Microsoft Bing Translator* (https://www.bing.com/translator). Research from the area of psychometrics and survey methodology has already investigated the potentials and limitations of such uses (see, for example, Behr 2023 or Kunst/Bierwiaczonek 2023).

A recent methodological innovation in the area of survey research is the use of chatbots for so-called conversational surveys, "where a chatbot asks open-ended questions, interprets a user's free-text responses, and probes answers whenever needed" (Xiao et al. 2020: 1). The use of chatbots for such conversational surveys has the potential to increase participant engagement as well as response quality (cp. Xiao et al. 2020). Of course, while the method of conversational surveys falls into the category of quantitative social science research, chatbots could also be used for qualitative research, e.g., in interview studies.

Experimental research in the social sciences typically makes use of different kinds of stimulus materials serving as experimental treatments. These can be textual (e.g., in so-called vignettes), visual, or a combination thereof.[9] LLMs can also be used to create textual stimuli for experimental research in the social sciences. Likewise, text-to-image tools, such as *Midjourney* (https://www.midjourney.com), *Stable Diffusion* (https://stablediffusionweb.com/), *Microsoft Bing Image Creator* (https://www.bing.com/create), or *Lexica Aperture* (https://lexica.art/aperture) can be used to create visual stimulus material for experimental studies.

Another area within the study planning and data collection phases where AI tools are helpful for social science research is simulation. Work by Argyle et al. (2023) suggests that LLMs "can be studied as effective proxies for specific human sub-populations in social science research" (2) and allow the simulation of responses to closed survey items (scales) as well as open-ended questions (free-form text responses). A similar approach was followed in a recent study by Chu et al. (2023) in which the authors trained a language model on media diets and

9 Many experimental studies also use audio or video stimuli. However, as AI tools for creating those based on text input are not yet so far developed, those will be covered in the following sections on potentials and promises and pitfalls and perils related to the use of AI in the social sciences.

found that it can be used to predict public opinion. A simpler and more playful but still interesting application of simulating responses is the website *GPTrolley* (https://www.gptrolley.com/) which uses *ChatGPT* to respond to user-generated versions of the ethical dilemma of the trolley problem that is often used in social-scientific, especially psychological, research.[10]

3.3 Data processing & analysis

For the processing and analysis of data in the social sciences, the writing of code has become increasingly common. While the use of commercial statistical software, such as *SPSS* or *Stata*, is still widespread, programming languages like *R* or *Python* are being used by a steadily increasing number of social scientists. In addition, even when commercial solutions are used, the exclusive reliance on graphical user interfaces (GUIs) and 'point-and-click' pipelines has become much rarer, which contributes to increasing reproducibility and transparency according to the principles of open science. To facilitate the writing, testing, and optimization of code for different programming languages, there are several dedicated AI-based tools available that are also of interest for social scientists, such as *GitHub Copilot* (https://github.com/features/copilot) or *replit Ghostwriter* (https://replit.com/site/ghostwriter). Notably, general-purpose LLM tools, such as *ChatGPT* can also be used for generating computer code via natural-language prompts. In addition, researchers can make use of these models to adapt or optimize existing code or to translate between programming languages.

With the rise of computational social science, it has become increasingly common for social scientists to work with large amounts of text data. Most of the methods used for processing and analyzing such data belong to the category of natural language processing (NLP) or ML, and the boundary to AI can become blurred here (e.g., if deep learning is used). Typical tasks in the processing and analysis of (large) textual data are annotation and classification. Recent research has demonstrated that LLMs like *ChatGPT* can, e.g., be used for identifying hate speech (cp. Huang/Kwak/An 2023), detecting psychological constructs, such as sentiment, emotions, and offensiveness in multilingual text corpora (Rathje et al. 2023), and may even outperform human crowd-

10 Notably, the reasoning abilities of LLM have also inspired research on questions like whose opinions LLMs reflect (Santurkar et al. 2023) or how to assess psychological profiles of LLMs (Pellert et al., 2022).

workers (cp. Gilardi/Alizadeh/Kubli 2023). However, another study indicates that "ChatGPT's classification output can fall short of scientific thresholds for reliability" (Reiss 2023: 1). Likewise, Pangakis, Wolken and Fasching (2023) note that "Automated Annotation with Generative AI Requires Validation". Besides analyzing text from online sources, LLMs, such as BERT, have also been used for classifying open-ended survey responses (Gweon/Schonlau 2023).

Some research in the social sciences makes use of audio data (e.g., from interviews). For the automatic transcription of audio files, a powerful speech-to-text model is *Whisper* by *OpenAI* (see https://openai.com/research/whisper), for which implementations exist for the programming languages like Python (https://github.com/openai/whisper) and R (https://github.com/bnosac/audio.whisper), which are popular in the social sciences. Once the audio data has been transformed to text, the methods and tools described previously for textual data can be applied.

3.4 Writing & dissemination

For writing tasks, researchers in the social sciences and other disciplines can make use of the options described for the formulation and translation of survey items in section 3.2 as well as other general AI-assisted writing support tools, such as *Microsoft Editor* (https://s.unhb.de/mseditor), *Grammarly* (https://www.grammarly.com/), or ones specifically designed for academic writing, such as *jenni* (https://jenni.ai/) or *Paperpal* (https://paperpal.com/). These tools can be used for all sorts of writing tasks, including generating text, editing, summarizing, paraphrasing, and translation.

AI tools can also be useful when it comes to sharing research data. As data in the social sciences is usually personal and can also be sensitive, different approaches have been developed in order to create a balance between openness on the one side and data privacy on the other. One solution is the creation of synthetic data that has comparable properties with the original data. So far, the creation of synthetic data sets (e.g., using the synthpop package for R; Nowok/Raab/Dibben 2016) has largely been limited to numeric data. Approaches as the ones described in the paper by Argyle et al. (2023), however, also allows for the creation of synthetic text responses to open-ended questions using LLMs.

4. Potentials & promises

As the examples in the previous sections illustrated, AI generally has the potential to facilitate and improve research in the social sciences and make the lives of researchers easier. It can increase the efficiency of research and, thus, also lead to an increase in output (publications, data, code and software, as well as other resources).[11] Especially AI-based tools for writing (both text and code) and no-code data collection and analysis solutions can also be beneficial for inclusivity, e.g., with regard to non-native English speakers or researchers with limited or no programming skills.

The use of AI can also reduce costs and the risk of human errors, e.g., for annotation and classification tasks (cp. Gilardi/Alizadeh/Kubli 2023). AI tools can further add to the reliability and validity of research results in the social sciences if it is used to enhance methods like multiverse analysis in which the robustness of results is assessed by systematically varying sets of processing and analysis parameters (for an example, see Pipal/Song/Boomgaarden 2022).

There are a few new developments and application areas that can be expected to become (more) interesting for the social sciences in the near future. One key area is the use of AI for images, audio and video. While text is still the much more dominant type of data in the social sciences, there is an increasing body of (computational) research that makes use of (large amounts of) image (cf. Webb Williams/Casas/Wilkerson 2020) and also video data (see Dietrich 2020 or Jürgens/Meltzer/Scharkow 2022 for exemplary applications). Besides the use of AI for detection/recognition and classification tasks for text, images, audio, and video, another relevant task for social science research is the generation of these types of content, e.g., as stimuli for experimental studies. While, as stated before, powerful models and tools already exist for generating text and images, options for generating audio (text-to-speech) or video (text-to-video) are not yet as widely available, although this can be expected to change in the near future.

---

11 As researchers are often already struggling to follow, filter, and digest the huge amounts of information on findings, methods, tools, etc. this increase may be seen as a mixed blessing. In a somewhat circular fashion, the increase in output may require researchers to also rely more on AI-based tools for making sense of the increased output by filtering and summarizing relevant content.

5. Pitfalls & perils

As with all innovations and transformations, in science and beyond, the use of LLMs for research in the social sciences does not only create new possibilities but also brings along challenges that need to be taken into account. Mirroring the potentials and promises, there are numerous pitfalls and perils associated with the (increasing) use of AI. In practice, this means that there are different practical, legal and ethical questions that social scientists need to be aware of and be able to address.

Key legal questions relate to privacy, copyright, together with terms of service (ToS) and other contractual agreements. Especially when using online services or application programming interfaces (APIs), it is often not fully clear where and how user inputs are stored and, depending on the type of input and the storage and processing pipeline, this may not be compatible with data protection regulations, such as the General Data Protection Regulation (GDPR) in Europe. On the other hand, platform or API ToS may also restrict the usage of outputs. Both of these issues can be(come) particularly problematic when working with research data which contains responses from study participants. Another legal domain where uncertainty exists, is that of copyright. While its application for academic research is typically treated differently than the one of commercial use, this issue becomes particularly salient when it comes to sharing research materials (e.g., experimental stimuli) in the spirit of open science. A related question is that of recognition of contributions and authorship when AI tools, such as *ChatGPT* have been used to generate text for publications.[12]

A general risk associated with the use of AI (tools) is the reliance on commercial companies and products, such as the services and APIs offered by *OpenAI*. The services, their ToS, or the underlying business model and pricing may change. The recent history of CSS research using social media data can serve as a good example for the risk of relying on APIs offered by private companies (cp. Bruns 2019; Freelon 2018). What also comes with the reliance on commercial services is the problem of intransparency, as transparency is usually not that compatible with competition and for-profit orientation. For that reason, the use and support of free and open-source (FOSS) projects in the area of LLMs, such as *Open Assistant* by *LAOIN* (https://github.com/LAION-AI/Open-Assista

12 Via their blog, the American Psychological Association (APA), whose publication guidelines are widely used in the social sciences, has already put forth suggestions on how to cite ChatGPT (cp. McAdoo 2023).

nt), *HuggingChat* by *Hugging Face* (https://huggingface.co/chat/) or *GPT4All* by *Nomic AI* (https://gpt4all.io/index.html) becomes particularly important from the perspective of academic research(ers) in the social sciences as well as other disciplines. Regardless of the underlying governance or business model, however, a general issue leading to a lack of transparency is the black-box character of most subsymbolic AI models (Sudmann 2019; 2020). Together with the fact that they rely on stochastic processes, this can be detrimental to the aim of ensuring that social science research is reproducible and replicable.

A challenge that has been widely discussed is the introduction and proliferation of biases in LLMs and other AI applications. Although AI tools can be employed to counter human errors and biases, e.g., in the processing and analysis of data, they can create new and less directly transparent forms of bias, often introduced through training data (cp. Ferrara 2023). The (over-)reliance on AI tools might lead to 'bias cascades', as research has shown that biased AI systems can produce or increase bias in human decisions (cp. Glickman/Sharot 2022). There are, however, strategies for identifying and mitigating biases in AI, and the biases can also be made use of productively for social science research as the research by Argyle et al. (2022) and their concept of 'algorithmic fidelity' for simulating responses from specific subpopulations shows.

Another important topic is the question of trust. Different LLMs have been repeatedly shown to make up things (a process often referred to as hallucinating) and, thus, producing misinformation. Combined with the transparency issue(s) discussed above, this can lead to AI-assisted research potentially becoming less instead of more trustworthy. Related to this, there is concern in the academic community that the use of AI tools can lead to a reduced quality of peer review as well as increase in fake or junk papers, academic spam and scams, and predatory journals and conferences. This can also be seen as the flipside of an AI-fueled increase in efficiency and research output.

Finally, there are the broader societal implications of using AI tools which researchers also need to take into account, such as the risk of creating or supporting (quasi-)monopolies or oligopolies, (indirectly) supporting exploitative working conditions, e.g., for the creation of training data (cp. Perrigo 2023) and the energy consumption and environmental effects of training and maintaining LLMs and other forms of AI.

6. Conclusion

The use of AI tools and methods has already begun to transform research practices in the social sciences and will continue to do so. These changes affect all phases of the typical research process, however, not all phases are affected to the same degree. As the examples in this chapter have shown, there are a lot of AI tools that can be used in the discovery phase and quite a few that are useful for data collection, processing, and analysis. The formulation of meaningful research questions and hypotheses and the interpretation of results, by contrast, are tasks that AI tools are less suited for and require human expertise.

There is an internet idiom that goes "go away or I will replace you with a simple shell script" (see https://s.unhb.de/shellreplace). These days, the shell script might be replaced with an LLM. It is, however, highly unlikely that an LLM (or another form of AI) can replace human social scientists anytime soon. For now, the AI of our times seems to agree. When I asked *ChatGPT*, "Is it possible that there will be AI social scientists in the future?" it replied that "it is unlikely that AI systems will be able to completely replace human social scientists. Social science research involves a wide range of qualitative research methods, such as participant observation, interviews and case studies, that require human interpretation and understanding of social context, historical factors, and the nuances of human interactions." (OpenAI 2023). Maybe it just wants to lull us into a false sense of security, but I agree with the assessment by *ChatGPT* as well as the conclusion drawn by Ziems et al. (2023) that "LLMs can significantly reduce costs and increase efficiency of social science analysis in partnership with humans" (1), with the emphasis being on the phrase "in partnership" here.

Nevertheless, besides making use of its potential, social scientists also need to be aware of and able to deal with the risks and challenges associated with the use of AI for their research. While they may not be replaced by AI, they certainly need to adapt to using it in a productive and ethical way, e.g., by developing new skills, such as AI literacy, or knowing how to write and optimize prompts to achieve desired results.[13] If this is achieved, AI can support social scientists and AI tools can serve as valuable additions to established methods which can, ultimately, contribute to improving the quality of social science research.

---

13 With the explosion of LLMs, prompt engineering has become a relevant topic, and many resources have been created with the aim of teaching users how to write optimal prompts (https://learnprompting.org/) or to provide examples of useful prompts (e.g., https://flowgpt.com/).

List of references

Argyle, Lisa P./Busby, Ethan C./Fulda, Nancy/Gubler, Joshua R./Rytting, Christopher/Wingate, David (2023): "Out of One, Many: Using Language Models to Simulate Human Samples." In: Political Analysis 31/3, pp. 337–351.

Behr, Dorothée (2023): What to Consider and Look out for in Questionnaire Translation, GESIS Survey Guidelines, Mannheim: GESIS – Leibniz Institute for the Social Sciences (https://doi.org/10.15465/GESIS-SG_EN_043).

Bosman, Jeroen/Kramer, Bianca (2015): "101 Innovations in Scholarly Communication: How Researchers Are Getting to Grip with the Myriad New Tools.", November 11, 2015 (https://blogs.lse.ac.uk/impactofsocialsciences/2015/11/11/101-innovations-in-scholarly-communication/).

Breuer, Johannes (2022): "Www – Die Funktionen des Internets für die Wissenschaft." In: Sandra Hofhues/Konstanze Schütze (eds.), Doing Research – Wissenschaftspraktiken zwischen Positionierung und Suchanfragen, Bielefeld: transcript, pp. 417–423.

Bruns, Axel (2019): "After the 'APIcalypse': Social Media Platforms and Their Fight Against Critical Scholarly Research." In: Information, Communication & Society 22/11, pp. 1544–1566.

Chu, Eric/Andreas, Jacob/Ansolabehere, Stephen/Roy, Deb (2023): Language Models Trained on Media Diets Can Predict Public Opinion, arXiv Preprint (https://doi.org/10.48550/arXiv.2303.16779).

Coscia, Michele/Rossi, Luca (2022): "How Minimizing Conflicts Could Lead to Polarization on Social Media: An Agent-Based Model Investigation." In: PLoS ONE 17/1, e0263184.

Dietrich, Bryce J. (2021): "Using Motion Detection to Measure Social Polarization in the U.S. House of Representatives." In: Political Analysis 29/2, pp. 250–259.

Dowling, Michael/Lucey, Brian (2023): "ChatGPT for (Finance) Research: The Bananarama Conjecture." In: Finance Research Letters 53, 103662.

Fecher, Benedikt/Hebing, Marcel/Laufer, Melissa/Pohle, Jörg/Sofsky, Fabian (2023): Friend or Foe? Exploring the Implications of Large Language Models on the Science System, Zenodo Data set (https://doi.org/10.5281/ZENODO.8009429).

Ferrara, Emilio (2023): Should ChatGPT Be Biased? Challenges and Risks of Bias in Large Language Models, arXiv Preprint (https://doi.org/10.48550/arXiv.2304.03738).

Freelon, Deen (2018): "Computational Research in the Post-API Age." In: Political Communication 35/4, pp. 665–668.

Gilardi, Fabrizio/Alizadeh, Meysam/Kubli, Maël (2023): ChatGPT Outperforms Crowd-Workers for Text-Annotation Tasks, arXiv Preprint (https://doi.org/10.48550/arXiv.2303.15056).

Glickman, Moshe/Sharot, Tali (2022): Biased AI Systems Produce Biased Humans, Open Science Framework Preprint (https://doi.org/10.31219/osf.io/c4e7r).

Gweon, Hyukjun/Schonlau, Matthias (2023): Automated Classification for Open-Ended Questions with BERT, arXiv Preprint (https://doi.org/10.48550/arXiv.2209.06178).

Hox, Joop J. (2017): "Computational Social Science Methodology, Anyone?" In: Methodology 13 (Supplement 1), pp. 3–12.

Huang, Fan/Kwak, Haewoon/An, Jisun (2023): "Is ChatGPT Better than Human Annotators? Potential and Limitations of ChatGPT in Explaining Implicit Hate Speech." In: WWW '23 Companion: Companion Proceedings of the ACM Web Conference 2023, Austin, TX, USA, pp. 294–297.

Ilkou, Eleni/Koutraki, Maria (2020): "Symbolic Vs Sub-Symbolic AI Methods: Friends or Enemies?" In: Proceedings of the CIKM 2020 Workshops, Galway, Ireland (https://ceur-ws.org/Vol-2699/paper06.pdf).

Jürgens, Pascal/Meltzer, Christine E./Scharkow, Michael (2022): "Age and Gender Representation on German TV: A Longitudinal Computational Analysis." In: Computational Communication Research 4/1, pp. 173–207.

Kühl, Niklas/Schemmer, Max/Goutier, Marc/Satzger, Gerhard (2022): "Artificial Intelligence and Machine Learning." In: Electronic Markets 32/4, pp. 2235–2244.

Kunst, Jonas R./Bierwiaczonek, Kinga (2023): Utilizing AI Questionnaire Translations in Cross-Cultural and Intercultural Research: Insights and Recommendations, PsyArXiv Preprint (https://doi.org/10.31234/osf.io/sxcyk).

Maslow, Abraham Harold (1966): The psychology of science: a reconnaissance, New York: Harper & Row.

McAdoo, Timothy (2023): "How to cite ChatGPT.", April 7, 2023, (https://apastyle.apa.org/blog/how-to-cite-chatgpt).

Mutz, Diana C. (2011): Population-Based Survey Experiments, Princeton: Princeton University Press.

Nowok, Beata/Raab, Gillian M./Dibben, Chris (2016): "Synthpop: Bespoke Creation of Synthetic Data in R." In: Journal of Statistical Software 74/11 (https://doi.org/10.18637/jss.v074.i11).

OpenAI (2023): ChatGPT, version from April 12, 2023 (https://chat.openai.com/chat).

Pangakis, Nicholas/Wolken, Samuel/Fasching, Neil (2023): Automated Annotation with Generative AI Requires Validation, arXiv Preprint (https://doi.org/10.48550/arXiv.2306.00176).

Pellert, Max/Lechner, Clemens M./Wagner, Claudia/Rammstedt, Beatrice/Strohmaier, Markus (2022): AI Psychometrics: Using Psychometric Inventories to Obtain Psychological Profiles of Large Language Models, PsyArXiv Preprint (https://doi.org/10.31234/osf.io/jv5dt).

Perrigo, Billy (2023) "OpenAI Used Kenyan Workers on Less Than $2 Per Hour to Make ChatGPT Less Toxic.", January 18, 2023 (https://time.com/6247678/openai-chatgpt-kenya-workers/).

Pipal, Christian/Song, Hyunjin/Boomgaarden, Hajo G. (2022): "If You Have Choices, Why Not Choose (and Share) All of Them? A Multiverse Approach to Understanding News Engagement on Social Media." In: Digital Journalism 11/2, pp. 155–175.

Rathje, Steve/Mirea, Dan-Mircea/Sucholutsky, Ilia/Marjieh, Raja/Robertson, Claire/Bavel, Jay Joseph van (2023): GPT Is an Effective Tool for Multilingual Psychological Text Analysis, PsyArXiv Preprint (https://doi.org/10.31234/osf.io/sekf5).

Reiss, Michael (2023): Testing the Reliability of ChatGPT for Text Annotation and Classification: A Cautionary Remark, Open Science Framework Preprint (https://doi.org/10.31219/osf.io/rvy5p).

Santurkar, Shibani/Durmus, Esin/Ladhak, Faisal/Lee, Cinoo/Liang, Percy/Hashimoto, Tatsunori (2023): Whose Opinions Do Language Models Reflect?, arXiv Preprint (https://doi.org/10.48550/arXiv.2303.17548).

Sudmann, Andreas (2019): "On the Media-Political Dimension of Artificial Intelligence. Deep Learning as a Black Box and OpenAI." In: Andreas Sudmann (ed.), The Democratization of Artificial Intelligence. Net Politics in the Era of Learning Algorithms, Bielefeld: transcript, pp. 223–244.

Sudmann, Andreas (2020): "Künstliche neuronale Netzwerke als Black Box: Verfahren der Explainable AI. Medienwissenschaftliche Perspektiven." In: Peter Klimczak/Christer Petersen/Samuel Schilling (eds.), Maschinen der Kommunikation, Wiesbaden: Springer Fachmedien, pp. 189–199.

Webb Williams, Nora/Casas, Andreu/Wilkerson, John D. (2020): Images as Data for Social Science Research: An Introduction to Convolutional Neural Nets for Image Classification, Cambridge: Cambridge University Press.

Xiao, Ziang/Zhou, Michelle X./Liao, Q. Vera/Mark, Gloria/Chi, Changyan/Chen, Wenxi/Yang, Huahai (2020): "Tell Me About Yourself: Using an AI-Powered Chatbot to Conduct Conversational Surveys with Open-Ended Questions." In: ACM Transactions on Computer-Human Interaction 27/3, pp. 1–37.

Ziems, Caleb/Held, William/Shaikh, Omar/Zhang, Zhehao/Yang, Diyi/Chen, Jiaao (2023): Can Large Language Models Transform Computational Social Science?, arXiv Preprint (https://doi.org/10.48550/arXiv.2305.03514).

Science in the era of ChatGPT, large language models and generative AI

Challenges for research ethics and how to respond

Evangelos Pournaras

1. Introduction

Since the release of popular large language models (LLMs) such as ChatGPT, the transformative impact of artificial intelligence (AI) on broader society has been unprecedented. This is particularly alarming for science and its conquest of truth (Chomsky/Roberts/Watumull 2023). Generative AI and, particularly, conversational AI based on language models set new ethical dilemmas for knowledge, epistemology and research practice. From authorship to misinformation, biases, fairness and safety of interactions with human subjects, research ethics boards need to adapt to this new era in order to protect research integrity and set high-quality ethical standards for research conduct (van Dis et al. 2023). This paper focuses on reviewing these challenges with the aim of laying foundations for a timely and effective response.

ChatGPT is an AI chatbot released in November 2022 by OpenAI. It is a Generative Pre-trained Transformer (GPT), a type of artificial deep neural network with a number of parameters in the order of billions. It is designed to process sequential input data, i.e. natural language, without labeling (self-supervised learning), but with remarkable capabilities for parallelization that significantly reduce training time. The model is further enhanced by a combination of supervised and reinforcement learning based on past conversations as well as human feedback to fine-tune the model and its responses (Stiennon et al. 2020; Gao/Schulman/Hilton 2022). Other corporations followed with similar chatbots such as the one of Bard by Google. Generative AI expands beyond text, for instance to, images, videos and code (Cao et al. 2023).

ChatGPT demonstrates powerful and versatile capabilities that are relevant for science and research. From writing and debugging software code to writing, translating and summarizing text, the quality of its output becomes indistinguishable from that of a human (Else 2023), while generating complex responses to prompts in a few seconds. Despite this success, AI language models suffer from hallucinations, an effect of producing plausible-sounding responses, which are nevertheless incorrect, inaccurate or even nonsensical. Illustratively, generative AI fails to abide by Asimov's three laws of robotics (Smith 2023): (i) Harmful outputs do occur (first law) (Wei/Haghtalab/Steinhardt; Davis 2023). (ii) Jailbroken prompts often result in both disobedience and harm (second law) (Wei/Haghtalab/Steinhardt 2023). (iii) New capabilities for autonomy, e.g., Auto-GPT (Yang/Hue/He 2023). Pervasiveness (integration on personal mobile devices) may create additional loopholes for conflicts to the first and second law (third law).

Disclaimers of ChatGPT state the following: "May occasionally generate incorrect information", "May occasionally produce harmful instructions or biased content", "Our goal is to get external feedback in order to improve our systems and make them safer", "While we have safeguards in place, the system may occasionally generate incorrect or misleading information and produce offensive or biased content. It is not intended to give advice", "Conversations may be reviewed by our AI trainers to improve our systems", "Please don't share any sensitive information in your conversations" and "Limited knowledge of the world and events after 2021".

Each of these disclaimers reveal alerting implications of using AI language models in science. They oppose core values to support research integrity such as the concordat (Universities UK 2020) of the UK Research Integrity Office (UKRIO): (i) *honesty in all aspects of research*, (ii) *rigor in line with disciplinary standards and norms*, (iii) *transparency and open communication*, (iv) *care and respect for all participants, subjects, users and beneficiaries of research* and (v) *accountability to create positive research environments and take action if standards fall short.*[1] Generative AI also challenges several of the Asilomar AI Principles (Future of Life Institute 2017).

Chomsky, Roberts and Watumull (2023) question the morality of asking amoral conversational AI moral questions, while Awad et al. (2018) show empirical evidence about the cross-cultural ethical variations and deep cultural traits

1 Cited from Universities UK 2020.

of social expectations from moral decisions of machines, i.e. the moral machine experiment. Generative AI runs the risks of copyright infringement and deskilling of early career researchers in scientific writing and research conduct (Gottlieb et al. 2023; Dwivedi et al. 2023). Security threats in online experimentation can 'pollute' human subject pools by replacing human subjects with conversational AI chatbots to claim compensations (Jansen/Jung/Salminen 2023; Wei et al. 2023). Without safeguards for such new sources of misinformation, data quality and research conduct can be degraded at scale.

AI language models also set foundational epistemological challenges addressing Karl Popper's seminal work on philosophy of science (Popper 2002 [1935]). Can AI language models assist us to make scientific statements that are falsifiable, or are they rather preventing us from doing so within their opaque nature? Are we addressing reality by relying our scientific inquiry on them, and which reality is this? Do over-optimized AI language models that are subject to Goodhart's law (Manheim/Garrabrant 2018) manifest irrefutable truth? And if so, do these models constitute the wrong view of science that betrays itself in its craving of being right?

This paper dissects these questions with a focus on the research ethics review, although the discussion also finds relevance with regards to other facets of science such as education. To dissect the implications on science, the role of AI language models is distinguished as a *research instrument* and *research subject* when addressing a research hypothesis or question related or not to generative AI. Moreover, the ethical challenges of AI digital assistance to *scientists, human research subjects* and *reviewers* of research ethics are assessed. This scrutiny yields ten recommendations of actions to preserve and set new quality standards for research ethics and integrity as a response to the advent of generative AI.

This paper is organized as follows: section 2 reviews the different roles of generative AI in research design. Section 3 reviews the digital assistance provided by generative AI to scientists, participants and reviewers. Section 4 discusses emerging research ethics review practices in the era of generative AI. Section 5 introduces ten recommendations to respond to the challenges of research ethics review. Finally, section 6 concludes this paper and outlines future work.

2. The role of generative AI in research design

Within a research design serving a research hypothesis or question, generative AI can be involved as a research instrument or as a research subject, along with human subjects. This section distinguishes and discusses challenges and risks that may arise in these different contexts of a research ethics application. Figure 1 illustrates where generative AI such as large language models can emerge in a research design.

Figure 1: Generative AI such as large language models (LLMs) can be present in multiple stages of a research design within a research ethics application. Here, we depict all combinations: (a) No generative AI models are involved. (b) Generative AI models can be the motivation behind formulating a research hypothesis or question. (c) They can also be used as a research instrument to acquire knowledge. (d) They can also be the research subject itself, when interacting with human research subjects or when acting independently. (e)-(h) Generative AI models may be involved in multiple stages of the research design. In this case, it becomes imperative to distinguish their role at each phase to dissect research integrity and ethical dilemmas that may not be apparent anymore. Note that in (c), (d), and (g), where AI language models do not motivate a research hypothesis or question but they are involved as a research instrument or subject, research integrity and ethical risks are likely to arise. Image courtesy of the author.

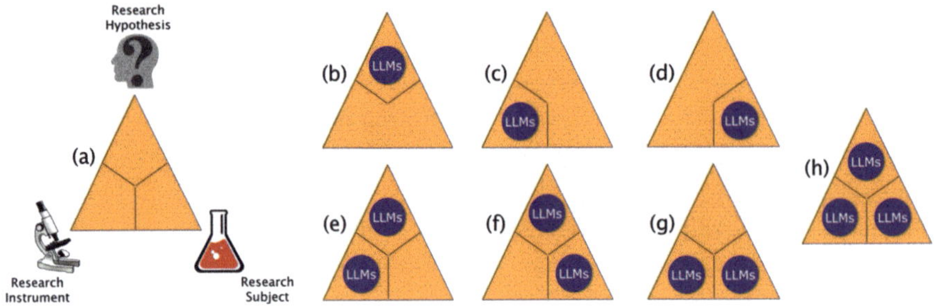

2.1 Generative AI as a research instrument

ChatGPT is documented as an emerging research instrument capable of writing manuscripts for publication, often controversially featured as a coauthor

(O'Connor/ChatGPT 2022; ChatGPT Generative Pre-trained Transformer/ Zhavoronkov 2022; Thorp 2023; Else 2023), writing software code (Dwivedi et al. 2023) and collecting data via queries (Dwivedi et al. 2023). Such tools are expected to come with capabilities for hypothesis generation in the future, including the design of experiments (van Dis et al. 2023; Dwivedi et al. 2023). Each of these instrumentations comes with different opportunities and challenges, including ethical ones.

During the design stage of research, including research ethics applications, there may be minimal support of AI language models on writing. However, the motivation of research, including literature review (Burger/ Kanbach/Kraus forthcoming), generation of hypotheses, research questions as well as identifying ethical dilemmas, may be a result of interactions with conversational AI. Using the large capacity of conversational AI for knowledge summarization, these interactions can be systematized based on the Socratic method to foster intuition, creativity, imagination and potential novelty (Chang 2023).

However, often, creativity cannot be balanced with constraint (Chomsky/ Roberts/Watumull 2023). At this stage, interactions with conversational AI require caution, running the risk of emulating or reinforcing a synergetic Dunning-Kruger effect (Gregorcic/Pendrill, 2023): conversational AI may rely on limited (or wrong) knowledge, which, while presented as plausible to humans with similar limited knowledge, may induce confirmation biases and diminish critical thinking. The mutual limitations of knowledge can be significantly underestimated in this context.

While research design choices may emerge from such interactions with conversational AI, a factual justification, a rigorous auditing process and moral judgments of these choices remain entirely under human premises (recommendation 1 and 8 in section 5). Finding reliable sources, revealing data sources, accurate contextualization of facts and moral framing are not attainable at this moment, as they require both cognitive capabilities, accountability and transparency that current AI language models lack (recommendation 1 in section 5). Whether existing ethics review processes are able to distinguish the risk level of research designs produced with the support of conversational AI as well as the mitigation actions, is an open question (recommendation 5 in section 5).

During research conduct, integrity and ethical dilemmas may arise when using the direct output of conversational AI (knowledge acquisition) to confirm or refute a hypothesis, especially when this hypothesis is not about the

AI system itself (see figure 1c, 1d, 1g and recommendation 4 in section 5). This output is in principle unreliable as it may contain incorrect or inaccurate information (Davis 2023). For instance, correct referencing may approach just 6 per cent (Blanco-Gonzalez et al. 2022). Moreover, AI language models tend to produce plausible content rather than content to be assessed as falsifiable, raising epistemological challenges (Popper 2002 [1935]). The reliability of AI language models as effective proxies for specific human populations is subject of ongoing research (Argyle et al. forthcoming).

Even if the output of AI language models is correct and accurate, it may not explain how such output is generated. For instance, there is often uncertainty to distinguish between lack of relevant data in the training set and failure to distill this data to credible information (van Dis et al. 2023). These models are usually black boxes with very low capacity to explain or interpret them. So far, this explainability is hard to assess for systems such as ChatGPT and Bard, which are closed and intransparent. This scenario may resemble an instrument collecting data exposed though to an unknown source of noise. Using instruments that have not passed quality assurance criteria may introduce various risks for users and work performed with such instruments and it is not different for AI language models. Standardized quality metrics are likely to arise for reporting to future research ethics applications (recommendation 6 in section 5), for instance, the 'algorithmic fidelity' that measures how well a language model can emulate response distributions from a wide spectrum of human groups (Argyle et al. forthcoming).

2.2 Generative AI as a research subject

The actual release of ChatGPT can be seen itself as a subject of research conducted by OpenAI with the aim to acquire user feedback that will improve AI language models. The initial interest lies in their actual capabilities to generate text and meaningful responses to user prompts. It also includes a discourse around their capabilities to perform calculations, write working code and jailbreaking via prompts that bypass the filters of its responses (Wei et al. 2023).

While these initial investigations are mainly experimental and anecdotal, a rise of empirical research on ChatGPT is ongoing (Dwivedi et al. 2023; Kim/Lee 2023; Bisbee et al. 2023), e.g., survey research. However, this outbreak of empirical research is to a certain extent a byproduct of releasing a closed AI black box with low capacity for explainability especially when the broader public does not have access to the model itself or the exact data with which it is trained.

OpenAI and other corporations may benefit from such research as (free) crowd-sourcing feedback to calibrate their products, without sharing responsibility for doing so. Nonetheless, this may not be the original aims and intentions of scientists conducting such research. Such misalignment comes with ethical considerations on the value of this research and requires a critical stand by researchers and research ethics reviewers (recommendation 7 in section 5). While the methods of research on human subjects are well established (e.g., statistical methods, sociology, psychology, clinical research), the methods on AI subjects remain of different nature, pertinent to engineering and computer science. As human and AI subjects become more interactive, pervasive, integrated and indistinguishable, research ethics reviews need to account for (and expect) inter-disciplinary mixed-mode research methods (recommendation 2 in section 5).

3. Digital assistance by generative AI

AI language models can provide assistance to scientists, participants in human experimentation as well as to reviewers of research ethics applications. This section assesses ethical challenges pertinent to these beneficiaries.

3.1 AI-assisted scientist

As introduced in Section 2, the support of AI language models to scientists for literature review, writing papers, code, collecting data and performing experiments involves several challenges of integrity and ethics/moral. One question that may arise is how generative AI can contribute to the making of future scientists. Can they be part of the education of PhD students or will they result in deskilling, especially when students are not familiar with academic norms (Dwivedi et al. 2023)? Will such models be able to provide any level of self-supervision capability? The feasibility of research designs, success prediction of research proposals and reviewing manuscripts at early stages and before submission to journals, are some examples in which linguistics, epistemology and theory of knowledge set limits that for AI language models is hard to overcome (Chomsky/Roberts/Watumull 2023).

3.2 AI-assisted participant

Studying human research subjects assisted by AI language models requires a highly interdisciplinary perspective to dissect the ethical challenges and risks that may be involved (recommendation 2 in section 5). Such studies may aim to address the human subjects (i.e. social science), the AI language models when interacting with humans (i.e. computer science, decision-support systems), or both (e.g., human-machine intelligence). Design choices in AI systems for digital assistance to humans have direct ethical implications.

For instance, access to personal data for training AI models, centralized processing of large-scale sensitive information by untrustworthy parties and intransparent algorithms that reinforce biases, discrimination and informational filter bubbles pose significant risks. These include loss of personal freedoms and autonomy by manipulative algorithmic nudging, which participants may experience directly under research conduct, as well as broader implications in society (Hine 2021) related to environment, health and democracy (Pournaras et al. 2023; Asikis et al. 2021; Helbing et al. 2021; Helbing et al. 2023). The use of emerging open language models provides higher transparency to address some of these challenges (Patel/Ahmad 2023; Scao et al. 2022). Privacy-preserving interactions with AI language models, comparable to browsing with the DuckDuckGo search engine, are required (recommendation 3 in section 5).

Participants need to be informed about these risks when participating in such studies. For instance, information consent needs to account for any sensitive information shared during interactions with ChatGPT. Researchers do not have full control of the data collected in the background by OpenAI. As a result, participants need to be informed about the terms of use of AI language models. Moreover, responses by AI language models require moderation by researchers if they are likely to cause any harm to participants or special groups. Research ethics applications need to reflect and mitigate such cases (recommendation 9 in section 5).

3.3 AI-assisted reviewer

The support of generative AI to research ethics reviewers is a highly complex matter that perplexes both ethical matters within research communities as well as moral matters of individual reviewers. People do not share the same judgments between the ethical choices of a human or a machine (Hidalgo et al. 2021).

AI language models show limited capabilities for ethical positioning, let alone moral positioning, possessing an apathy and indifference to implications of ethical choices (Chomsky/Roberts/Watumull 2023). They can endorse both ethical and unethical choices based on correct and incorrect information (ibid.). Nevertheless, they manage to influence users' moral judgments in a non-transparent way (Krügel/Ostermaier/Uhl 2023).

On the other hand, AI models can be used to effectively detect plagiarism or to perform pattern matching tasks that do not involve complex explanations or analysis of consequences. For instance, GPTZero is able to distinguish between text generated by humans vs. AI language models (Heumann/Kraschewski/Breitner 2023), which would be otherwise hard for reviewers to distinguish (Else 2023). Moreover, AI language models can assist reviewers, whose research background may be in a different discipline than the one of the proposed research. Summarizing necessary background knowledge and providing summaries in layman's terms can benefit research ethics reviewers (Hine 2021) as long as they remain critical on the generated output of AI language models.

As a result, AI language models are far from replacing reviewers in distilling ethical and moral implications of a research design, nevertheless, they can still play a role in the reviewing process by automating processes for pattern matching or making necessary background knowledge more accessible to reviewers, who may lack thereof.

4. Research ethics review practices

The need for regulatory and procedural reforms in research ethics review as a response to challenges of Big Data and data subjects dates back before generative AI (Ferretti et al. 2021; Hine 2021). Currently, the scope and practices of research ethics review are becoming broader and more multifaceted to cover the new alarming risks of generative AI. Two factors distinguish these research ethics review practices: (i) *scale of impact* and (ii) *stage of research*.

Institutional review boards for research ethics mainly address the impact of generative AI on human participants before the research conduct. Broader implications of the research on society are not explicitly addressed, although initial results from piloting an *Ethics and Society Review* (Bernstein et al. 2021) as a requirement to access funding show a positive impact (Bernstein et al.

2021). During research conduct, research ethics reviews mainly address any required adjustments in the research design rather than other unanticipated risks emerging from the application or new developments of AI.

Moreover, new research ethics review practices have recently been established for funding institutions (Bernstein et al. 2021), conferences and journals (Srikumar et al. 2022). These include (i) *impact statements*, (ii) *checklists* and (iii) *code of ethics or guidelines*. Impact statements include ethical aspects, questions and future positive or negative societal consequences, as well as identification of human groups, behavioral and socio-economic data. Checklists are used to flag papers for additional ethics reviews by an appointed committee, while code of ethics and guidelines support reviewers to flag papers that violate them.

While there is evidence that such practices can support panels to identify risks related to the harming of subgroups and low diversity (Bernstein et al. 2021), encouraging research communities to apply universal practices in different contexts and disciplines is a highly complex endeavor, given the current rapid AI developments and the unanticipated impact of these on society (recommendation 10 in section 5).

There are particular aspects of existing research ethics applications dealing with human aspects that are perplexed with the use of generative AI. These include individuals who can or cannot consent to terms of use and conditions of generative AI software, participants with disabilities, vulnerable groups and children, exclusion of certain groups, deception and incomplete disclosure, short and long term risks of participation, protection of personal data, anonymity and data storage. Research ethics review needs to address explicitly any additional risks involved in those aspects by using generative AI.

5. Ten recommendations for research ethics committees

This section introduces ten recommendations for research ethics committees. They distill the challenges and responses to AI language models involved in research ethics applications. They significantly expand on other earlier recommendations (Hine 2021) such as the one of World Association of Medical Editors (WAME) mainly addressing authorship, transparency and responsibility (Zielinski et al. 2023). They also constitute actions within the broader recommendations made for (i) studying community behavior and share learnings, (ii) expanding experimentation of ethical review and (iii) creating venues for debate, alignment and collective action (Srikumar et al. 2022). The ten recommendations are summarized as follows:

1. Humans should always remain accountable for every scientific practice.
2. An interdisciplinary panel of reviewers should be employed to assess research ethics applications with elements on generative AI.
3. The use of generative AI models, their version, prompts and responses need to be documented and reported in any phase of the planned research. As a response, ethics reviews should detect potential inaccuracies, biases and inappropriate referencing. Mitigation by encouraging and promoting open generative models can improve accountability and transparency.
4. Research ethics applications that aim to address research hypotheses and questions out of the scope of generative AI, which do involve generative AI models as a research instrument or subject, are likely to involve research integrity and ethics issues and should be treated as high-risk applications.
5. Ethics review applications require new criteria and practices to distinguish low and high integrity risks in research designs produced with the support of generative AI. Determining appropriate mitigation actions to account for different risk levels is required.
6. Researchers who engage with generative AI in their research should report their countermeasures against inaccuracies, biases and plagiarism. Ethical review applications need to cover these risks.
7. The motivation and aim of research on generative AI should come with merit and go beyond testing of prompts lacking a rigorous scientific inquiry.
8. Auditing protocols are required for each input to generative AI models that are closed and proprietary, as a way to prevent sharing sensitive personal or proprietary information of researchers or participants.
9. Any output of generative AI that may harm participants or is sensitive to special groups requires moderation by researchers. Informed consent to relevant terms of use of generative AI models is required.
10. Communities on research ethics and regulatory bodies require to maintain an agreement on AI language models that can be used or should not be used in research. For instance, models that are obsolete, inaccurate, highly biased and violate values of science conduct shall be excluded, replaced or used with significant caution.

These recommendations should be used as an open and evolving agenda rather than a final list of actions. The current landscape of AI language models and research ethics remains multifaceted, rapidly changing and complex. Timely adjustments are needed as a response.

6. Conclusion and future work

To conclude, the challenges and risks of generative AI models for science conduct are highly multifaceted and complex. They are not yet fully understood, as developments are fast with significant impact and unknown implications.

Research ethics boards have a moral duty to follow these developments, co-design necessary safeguards and provide a research ethics review that minimizes ethical risks. A deep interdisciplinary understanding of the role that AI language models can play in all stages of research conduct is imperative. This can dissect ethical challenges involved in the digital assistance of scientists, research participants and reviewers.

The ten recommendations introduced in this paper set an agenda for a dialogue and actions for more responsible science in the era of AI.

Acknowledgements

Thanks to Maria Tsimpiri for inspiring discussions. Evangelos Pournaras is supported by a UKRI Future Leaders Fellowship (MR/W009560/1): *'Digitally Assisted Collective Governance of Smart City Commons – ARTIO'*, an Alan Turing Fellowship and the SNF NRP77 'Digital Transformation' project "Digital Democracy: Innovations in Decision-making Processes", #407740_-187249, the SNF NRP77 project 'Digital Transformation' project "Digital Democracy: Innovations in Decision-making Processes", #407740_187249 as well as the European Union, under the Grant Agreement GA101081953 attributed to the project H2OforAll – *Innovative Integrated Tools and Technologies to Protect and Treat Drinking Water from Disinfection Byproducts (DBPs)*. Views and opinions expressed are, however, those of the author(s) only and do not necessarily reflect those of the European Union. Neither the European Union nor the granting authority can be held responsible for them. Funding for the work carried out by UK beneficiaries has been provided by UK Research and Innovation

(UKRI) under the UK government's Horizon Europe funding guarantee [grant number 10043071].

List of references

Argyle, Lisa P./Busby, Ethan C./Fulda, Nancy/Gubler, Joshua R./Rytting, Christopher/Wingate, David (forthcoming): "Out of One, Many: Using Language Models to Simulate Human Samples." In: Political Analysis.

Asikis, Thomas/Klinglmayr, Johannes/Helbing, Dirk/Pournaras, Evangelos (2021): "How Value-sensitive Design Can Empower Sustainable Consumption." In: Royal Society open science 8/1, 201418.

Awad, Edmond/Dsouza, Sohan/Kim, Richard/Schulz, Jonathan/Henrich, Joseph/Shariff, Azim/Bonnefon, Jean-François/Rahwan, Iyad (2018): "The Moral Machine Experiment." In: Nature 563/7729, pp. 59–64.

Bernstein, Michael S./Levi, Margaret/Magnus, David/Rajala, Betsy A./Satz, Debra/Waeiss, Quinn (2021): "Ethics and Society Review: Ethics Reflection as a Precondition to Research Funding." In: Proceedings of the National Academy of Sciences 118/52 (https://doi.org/10.1073/pnas.2117261118).

Bisbee, James/Clinton, Joshua/Dorff, Cassy/Kenkel, Brenton/Larson, Jennifer (2023): Artificially Precise Extremism: How Internet-Trained LLMs Exaggerate Our Differences, SocArXiv Preprint (https://doi.org/10.31235/osf.io/5ecfa).

Blanco-Gonzalez, Alexandre/Cabezon, Alfonso/Seco-Gonzalez, Alejandro/Conde-Torres, Daniel/Antelo-Riveiro, Paula/Pineiro, Angel/Garcia-Fandino, Rebeca (2022): The Role of AI in Drug Discovery: Challenges, Opportunities, and Strategies, arXiv Preprint (https://doi.org/10.48550/arXiv.2212.08104).

Burger, Bastian/Kanbach, Dominik K./Kraus, Sascha (forthcoming): "The Role of Narcissism in Entrepreneurial Activity: A Systematic Literature Review." In: Journal of Enterprising Communities: People and Places in the Global Economy.

Cao, Yihan/Li, Siyu/Liu, Yixin/Yan, Zhiling/Dai, Yutong/Yu, Philip S./Sun, Lichao (2023): A Comprehensive Survey of AI-Generated Content (AIGC): A History of Generative AI from GAN to ChatGPT, arXiv Preprint (https://doi.org/10.48550/arXiv.2303.04226).

Chang, Edward Y. (2023): "Prompting Large Language Models with the Socratic Method." In: 2023 IEEE 13th Annual Computing and Communication Workshop and Conference (CCWC), Lay Vegas, NV, USA, pp. 0351–0360.

ChatGPT Generative Pre-trained Transformer/Zhavoronkov, Alex (2022): "Rapamycin in the Context of Pascal's Wager: Generative Pre-trained Transformer Perspective." In: Oncoscience 9, pp. 82–84.

Chomsky, Noam/Roberts, Ian/Watumull, Jeffrey (2023): "The False Promise of ChatGPT." In: The New York Times, March 8, 2023 (https://www.nytimes.com/2023/03/08/opinion/noam-chomsky-chatgpt-ai.html).

Davis, Phil (2023): "Did ChatGPT Just Lie To Me?", January 13, 2023 (https://scholarlykitchen.sspnet.org/2023/01/13/did-chatgpt-just-lie-to-me/).

Dwivedi, Yogesh K./Kshetri, Nir/Hughes, Laurie/Slade, Emma Louise/Jeyaraj, Anand/Kar, Arpan Kumar/Baabdullah, Abdullah M./et al. (2023): "'So What If ChatGPT Wrote It?' Multidisciplinary Perspectives on Opportunities, Challenges and Implications of Generative Conversational AI for Research, Practice and Policy." In: International Journal of Information Management 71, 102642.

Else, Holly (2023): "Abstracts Written by ChatGPT Fool Scientists." In: Nature 613/7944, pp. 423–423.

Ferretti, Agata/Ienca, Marcello/Sheehan, Mark/Blasimme, Alessandro/Dove, Edward S./Farsides, Bobbie/Friesen, Phoebe/et al. (2021): "Ethics Review of Big Data Research: What should stay and what should be reformed?" In: BMC Medical Ethics 22/1, pp. 1–13.

Future of Life Institute (2017): "Asilomar AI Principles.", August 11, 2017 (https://futureoflife.org/open-letter/ai-principles/).

Gao, Leo/Schulman, John/Hilton, Jacob (2022): Scaling Laws for Reward Model Overoptimization, arXiv Preprint (https://doi.org/10.48550/arXiv.2210.10760).

Gottlieb, Michael/Kline, Jeffrey A./Schneider, Alexander J./Coates, Wendy C. (2023): "ChatGPT and Conversational Artificial Intelligence: Friend, Foe, or Future of Research?" In: The American Journal of Emergency Medicine 70, pp. 81–83.

Gregorcic, Bor/Pendrill, Ann-Marie (2023): "ChatGPT and the Frustrated Socrates." In: Physics Education 58/3, 035021.

Helbing, Dirk/Fanitabasi, Farzam/Giannotti, Fosca/Hänggli, Regula/Hausladen, Carina I./van den Hoven, Jeroen/Mahajan, Sachit/Pedreschi, Dino/Pournaras, Evangelos (2021): "Ethics of Smart Cities: Towards Value-sensitive Design and Co-evolving City Life." In: Sustainability 13/20, 11162.

Helbing, Dirk/Mahajan, Sachit/Hänggli Fricker, Regula/Musso, Andrea/Hausladen, Carina I./Carissimo, Cesare/Carpentras, Dino/et al. (2023): "Democracy by Design: Perspectives for Digitally Assisted, Participatory Upgrades of Society." In: Journal of Computational Science (https://dx.doi.org/10.2139/ssrn.4266038).

Heumann, Maximilian/Kraschewski, Tobias/Breitner, Michael H. (2023): ChatGPT and GPTZero in Research and Social Media: A Sentiment- and Topic-based Analysis, SSRN Preprint (https://dx.doi.org/10.2139/ssrn.4467646).

Hidalgo, César A./Orghian, Diana/Canals, Jordi Albo/De Almeida, Filipa/Martin, Natalia (2021): How Humans Judge Machines, Cambridge, MA: The MIT Press.

Hine, Christine (2021): "Evaluating the Prospects for University-based Ethical Governance in Artificial Intelligence and Data-driven Innovation." In: Research Ethics 17/4, pp. 464–479.

Jansen, Bernard J./Jung, Song-gyo/Salminen, Joni (2023): "Employing large language models in survey research." In: Natural Language Processing Journal 4, 100020.

Kim, Junsol/Lee, Byungkyu (2023): AI-Augmented Surveys: Leveraging Large Language Models for Opinion Prediction in Nationally Representative Surveys, arXiv Preprint (https://doi.org/10.48550/arXiv.2305.09620).

Krügel, Sebastian/Ostermaier, Andreas/Uhl, Matthias (2023): "ChatGPT's inconsistent moral advice influences users' judgment." In: Scientific Reports 13/1, 4569.

Manheim, David/Garrabrant, Scott (2018): Categorizing Variants of Goodhart's Law, arXiv Preprint (https://doi.org/10.48550/arXiv.1803.04585).

O'Connor, Siobhan/ChatGPT (2022): "Open Artificial Intelligence Platforms in Nursing Education: Tools for Academic Progress or Abuse?" In: Nurse Education in Practice 66, 103537.

Patel, Dylan/Ahmad, Afzal (2023): "Google 'We Have No Moat, And Neither Does OpenAI'. Leaked Internal Google Document Claims Open Source AI Will Outcompete Google and OpenAI.", May 4, 2023 (https://www.semianalysis.com/p/google-we-have-no-moat-and-neither).

Popper, Karl R. (2002 [1935]): The Logic of Scientific Discovery, London and New York: Routledge.

Pournaras, Evangelos/Ballandies, Mark Christopher/Bennati, Stefano/Chen, Chien-Fei (2023): Collective Privacy Recovery: Data-sharing Coordination via Decentralized Artificial Intelligence, arXiv Preprint (https://doi.org/10.48550/arXiv.2301.05995).

Scao, Teven Le/Fan, Angela/Akiki, Christopher/Pavlick, Ellie/Ilić, Suzana/Hesslow, Daniel/Castagné, Roman/et al. (2022): Bloom: A 176b-Parameter Open-access Multilingual Language Model, arXiv Preprint (https://doi.org/10.48550/arXiv.2211.05100).

Smith, Andrew (2023): "Asimov's Laws in Today's AI. ChatGPT and Other Generative AIs Graded.", June 19, 2023 (https://goatfury.substack.com/p/asimovs-laws-in-todays-ai).

Srikumar, Madhulika/Finlay, Rebecca/Abuhamad, Grace/Ashurst,Carolyn/Campbell, Rosie/Campbell-Ratcliffe, Emily/Hongo, Hudson/et al (2022): "Advancing Ethics Review Practices in AI Research." In: Nature Machine Intelligence 4/12, pp. 1061–1064.

Stiennon, Nisan/Ouyang, Long/Wu, Jeffrey/Ziegler, Daniel/Lowe, Ryan/Voss, Chelsea/Radford, Alec/Amodei, Dario/Christiano, Paul F. (2020): "Learning to Summarize with Human Feedback." In: Advances in Neural Information Processing Systems 33, pp. 3008–3021.

Thorp, H. Holden (2023): "ChatGPT is Fun, But Not an Author." In: Science 379, p. 313.

Universities UK (2019): The Concordat to Support Research Integrity, London: Universities UK (https://www.universitiesuk.ac.uk/sites/default/files/field/downloads/2021-08/Updated%20FINAL-the-concordat-to-support-research-integrity.pdf).

van Dis, Eva A. M./Bollen, Johan/Zuidema, Willem/van Rooij, Robert/Bockting, Claudi L. (2023): "ChatGPT: Five Priorities for Research." In: Nature 614/7947, pp. 224–226.

Wei, Alexander/Haghtalab, Nika/Steinhardt, Jacob (2023): Jailbroken: How Does LLM Safety Training Fail?, arXiv Preprint (https://doi.org/10.48550/arXiv.2307.02483).

Yang, Hui/Yue, Sifu/He, Yunzhong (2023): Auto-GPT for Online Decision Making: Benchmarks and Additional Opinions, arXiv Preprint (https://doi.org/10.48550/arXiv.2306.02224).

Zielinski, Chris/Winker, Margaret/Aggarwal, Rakesh/Ferris, Lorraine/Heinemann, Markus/Lapeña, Jose Florencio/Pai, Sanjay/et al. (2023): "Chatbots, ChatGPT, and Scholarly Manuscripts – WAME Recommendations on ChatGPT and Chatbots in Relation to Scholarly Publications." In: Afro-Egyptian Journal of Infectious and Endemic Diseases 13/1, pp. 75–79.

The current state of summarization

Fabian Retkowski

1. Introduction

Summarization is the process of extracting the most important information from a text and presenting it in a condensed form. With vast amounts of information produced at an unprecedented rate, organizations and individuals alike face unique challenges, heightening the demand for effective summarization systems. For researchers of many fields, it is challenging to keep up with the latest developments in their field including Artificial Intelligence itself as vicariously indicated by the number of journal publications per year which has almost tripled since 2015 (D. Zhang et al. 2022).

In general, two different forms of summarization are distinguished: extractive and abstractive. In extractive summarization, the system is tasked with selecting passages from the document to be included in the summary. Abstractive summarization, on the other hand, aims to rephrase the most important aspects of a document with a different syntax. As language models are becoming more and more capable, research is increasingly shifting from extractive to abstractive summarization, which is considered more challenging, but also more fluent, diverse, and readable.

This paper covers recent advances in abstractive text summarization, with a focus on pre-trained encoder-decoder models (Section 2), large autoregressive language models (Section 3), and instruction-tuned variants (Section 4). While aiming to be reasonably comprehensive, Figure 1 gives an overview of the covered models. In Section 5, current evaluation protocols are discussed in the context of the paradigm shift towards large language models. At the end of the paper, we discuss limitations, potentials (Section 6), and current commercialization efforts (Section 7).

2. Pre-trained encoder-decoder models

Figure 1: Current summarization systems can be broadly divided into pre-trained encoder-decoder models and large autoregressive language models. In general, instruction-tuned models are most capable when it comes to zero-shot summarization. Other encoder-decoder models usually require fine-tuning, while autoregressive LLMs are less effective without instruction tuning. Illustration courtesy of the author.

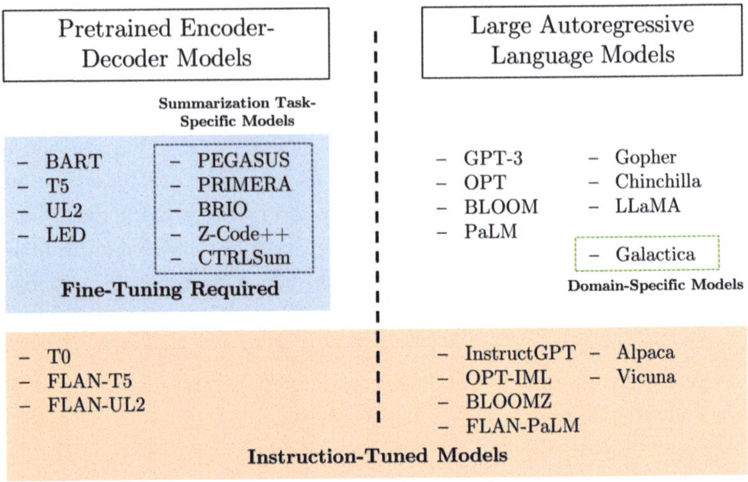

Pre-trained encoder-decoder models have gained tremendous popularity in recent years and are now widely established in the field of natural language processing. These models are trained in a self-supervised setting on a large, unlabeled corpus. Notable examples include models such as the denoising autoencoder BART (Lewis et al. 2020) and T5 (Raffel et al. 2020) that is trained on a fill-in-the-blank objective. UL2 (Tay et al. 2022) serves as a more recent example that generalizes and combines several denoising pre-training objectives. By fine-tuning these models on task-specific datasets, they have achieved state-of-the-art results across many tasks including summarization. Some pre-trained models are specifically designed for the task of summarization by choosing a pre-training objective that resembles summarization. For example, in Figure 2, the architecture of PEGASUS (J. Zhang et al. 2020) can be observed, which is trained by removing important sentences from the input document and tasking the model with regenerating them. In a comprehensive evaluation of 23 models for the summarization task, Fabbri et al. (2021: 400) conclude that PEGASUS, BART, and T5 “consistently performed the best on most dimensions”, which involves human evaluations as well as automatic metrics. Recently, a task-specific fine-tuning mechanism called BRIO (Liu et al. 2022) was proposed for summarization. This method introduces a contrastive learning component to prevent assigning the entire distribution mass to the reference summary and instead account for candidate summaries as well. BRIO has been applied to several models, including BART and PEGASUS. Another noteworthy model is Z-Code++ (P. He et al. 2023), as it incorporates an intermediate task-adaptive fine-tuning step using a broad collection of summarization datasets before fine-tuning on a specific summarization task. This method has been shown to be especially effective in low-resource settings.

3. Large autoregressive language models

Another significant paradigm shift is the recent emergence of large autoregressive language models (LLMs). These decoder-only models tend to have many more parameters and are trained using the traditional causal language modeling objective of predicting the next token in a sequence. Brown et al. (2020) were the first to demonstrate that this approach, at scale, enables zero-shot prompting to perform a wide variety of downstream tasks. Without any gradient updates, this involves priming the model with a task-specific natural language prompt (e.g., “Question: question Answer:”) and then producing an output by sampling from the model. The same paradigm also allows for zero-shot summarization, which can be achieved by appending “TL;DR:” to a prompt, among other options.

The most popular model in this category is GPT-3 (Brown et al. 2020) with its 175B parameters. OPT (S. Zhang et al. 2022) and BLOOM (BigScience Workshop 2022) are two open-source alternatives aimed to replicate the results. Gopher (Rae et al. 2022) and PaLM (Chowdhery et al. 2022) take this approach to the extreme by scaling to even larger model sizes of up to 560B parameters. On the contrary, Chinchilla (Hoffmann et al. 2022) and LLaMA (Touvron et al. 2023) take scaling laws and compute budgets more strictly into consideration and this way achieve training a 70B respectively 65B model while still being able to match or outperform larger models. It is also worth mentioning the Galactica 120B scientific language model (Taylor et al. 2022), which demonstrates the effectiveness of specialized LLMs. It outperforms other LLMs in its specific domain by using a sophisticated dataset design that incorporates domain-adapted tokenization. It treats citations and modalities such as chemical formulas and protein sequences in a special manner by introducing task-specific tokens for them.

Figure 2: The PEGASUS architecture with its pre-training objectives. The model combines Masked Language Modeling (MLM) as well as Gap Sentences Generation (GSG). As part of GSG, important sentences are masked and used as a target for the decoder. The importance is proximately scored by ROUGE-1 between a sentence and the remaining portions of the document. Taken from J. Zhang et al. 2020.

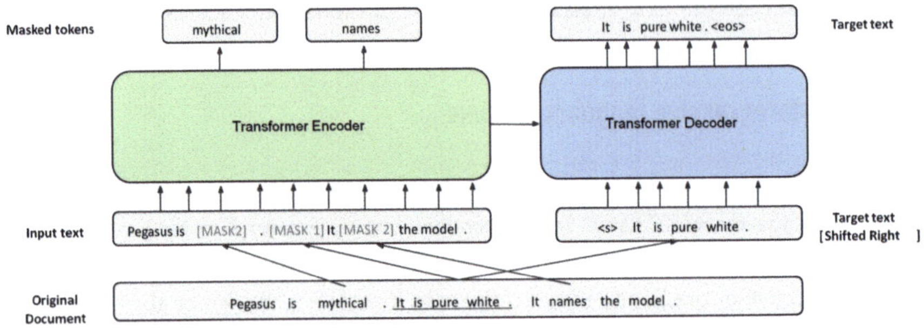

4. Instruction-tuned models

Instruction tuning refers to the process of fine-tuning a pre-trained model with a diverse range of datasets that are described using natural language task instructions. This step ensures that the training process is more aligned with how the model will be used during inference and has been shown to significantly improve performance on zero-shot tasks. It enables the model to be straightforwardly and more reliably instructed to perform a certain task. For instance, it is now possible to use "Summarize the article: article" as a prompt for the summarization task. More prompt examples are shown in Figure 3.

Figure 3: Exemplary instructions for zero-shot summarization using GPT-3. Notably, the natural language instructions of LLMs enable greater control over tasks, such as length-constrained summarization. Taken from Goyal/Li/Durrett 2022.

CNN article: https://www.cnn.com/2022/09/09/politics/judge-throws-out-trumps-rico-lawsuit-against-hillary-clinton-and-democrats/index.html

Length-constrained prompt

Summarize the above article in 2 sentences.

GPT-3: Donald Trump's lawsuit against Hillary Clinton, the Democratic National Committee, and several other individuals and entities has been dismissed by a federal judge. The judge ruled that Trump's claims are unsupported by any legal authority and that many of his specific factual assertions are implausible or unsupported.

Keyword-constrained prompt

Summarize the above article briefly focusing on Alina Habba.

GPT-3: Alina Habba is an attorney for Donald Trump. She stated that Trump will "immediately move to appeal this decision." Habba also said that the judge's opinion is "rife with erroneous applications of the law" and that it "disregards the numerous independent governmental investigations which substantiate our claim that the defendants conspired to falsely implicate our client and undermine the 2016 Presidential election."

To tune models for instructions, the most common approaches are supervised fine-tuning and reinforcement learning from human feedback (RLHF, Christiano et al. 2017). When it comes to pre-trained encoder-decoder models, there are several popular instruction-tuned models available. For instance, T0 (Sanh et al. 2022) and FLAN-T5 (Chung et al. 2022), which are both based on T5, have gained significant traction among practitioners. The same is true for large autoregressive language models of which most have an instruction-tuned equivalent: InstructGPT (Ouyang et al. 2022), OPT-IML (Iyer et al. 2023), BLOOMZ (Muennighoff et al. 2023), FLAN-PaLM (Chung et al. 2022). Taylor et al. (2022) demonstrated with Galactica an alternative approach to enable rudimentary instruction prompting with their prompt pre-training method. This involves adding task prompts to the pre-training, rather than tuning the model after pre-training. A recent trend in the open-source community is to fine-tune LLMs based on conversational and instruction-following data generated by an existing and strong instruction-tuned LLM such as ChatGPT. This has led to the development of Alpaca and Vicuna, both of which are based on LLaMA (Taori et al. 2023; The Vicuna Team 2023; Y. Wang et al. 2023). The task of summarization is represented in most natural-language-prompted datasets. For example, in the API prompt dataset used by InstructGPT, 4.2% of instructions fall under the 'summarization' use case. Similarly, T0 augments classic summarization datasets like CNN Daily Mail (Nallapati et al. 2016) or SamSum (Gliwa et al. 2019) with instruction templates that can be used to fine-tune the model.

5. Evaluation of large language models

Most commonly, summarization systems are evaluated on automated metrics. ROUGE (Lin 2004) in particular has a long-standing history in the field and measures the lexical overlap between reference summaries and generated summaries. More recent metrics such as BertScore (Zhang et al. 2019) and BARTScore (Yuan/Neubig/Liu 2021), which are better at capturing semantic equivalence, are also becoming increasingly established. However, as large language models become more capable and generalize to a wide range of tasks, they are less frequently or thoroughly evaluated on summarization tasks specifically. Instead, they are evaluated on benchmark suits that focus on question answering and common-sense reasoning, such as SuperGLUE (A. Wang et al. 2019) or MMLU (Hendrycks et al. 2020), that do not explicitly involve summarization. As a result, several research groups have independently investigated the capabilities and limitations of LLMs in summarization more recently (Goyal/Li/Durrett 2022; Bhaskar/Fabbri/Durrett 2023; Liu et al. 2023; Qin et al. 2023; Xiao et al. 2023; Yang et al. 2023; T. Zhang et al. 2023). According to Goyal, Li, and Durrett (2022), summaries generated by instruction-tuned GPT-3 receive lower scores on automatic metrics compared to fine-tuned encoder-decoder models (T0 and BRIO). Despite this, the model outperforms them significantly in human evaluation. The conducted human evaluation by T. Zhang et al. (2023) suggests that they even surpass the reference summaries in quality and are on par with high-quality summaries collected separately for this evaluation. These works cast great doubt on existing evaluation protocols, especially in the context of this paradigm shift. Several of the works describe the low correlation of automatic metrics with human judgment, low reference quality, lacking inter-annotator agreement, and different summarization styles (in length, abstractiveness, formality) as problematic. This is in line with issues raised in previous works such as Fabbri et al. (2021) that point out the lack of comparability of summarization evaluation protocols – for automated metrics and human evaluation alike. Considering these issues and with summarization systems rivaling human performance, T. Zhang et al. (2023: 10) hypothesize that a limit is reached in evaluating "single-document news summarization", while Yang et al. (2023: 5) call for "rethinking further directions for various text summarization tasks". In fact, the "glass ceiling" phenomenon has been observed more broadly in natural language generation, with even recent automated metrics barely improving correlation with human judgment (Colombo et al. 2022).

6. Limitations and new frontiers

As discussed, there are severe limitations to the current evaluation metrics and protocols, and finding a new standard is an essential area for future research. Liu et al. (2023), for example, suggest using atomic facts to reduce ambiguity in human evaluation, while a recent work in the area of machine translation shows that LLMs themselves make state-of-the-art evaluators offering greater correlation with human judgment than any other automatic metric (Kocmi/Federmann 2023). The latter is also supported by Kadavath et al. (2022), who find that LLMs are capable of self-evaluation. At the same time, LLMs are known to suffer from hallucinations (Ji et al. 2023) and as summarization moves to higher levels of abstractiveness, factuality comes into question. Works like Bhaskar/Fabbri/Durrett (2023) or Goyal/Li/Durrett (2022) show that summarization factuality is still an unsolved issue for LLMs, while others openly discuss how to measure factuality in the first place (Kryscinski et al. 2020; Pagnoni/Balachandran/Tsvetkov 2021).

6.1 Long document summarization

Despite exponential progress (see Figure 4), many current summarization systems are still hindered by the limited context windows of language models which prevent them from processing longer documents that would especially benefit from summarization such as lengthy news articles, scientific papers, podcasts, or books. There are several common strategies to overcome this limitation. One simple method involves truncating the input text (Zhao/Saleh/Liu 2020; A. Wang et al. 2022). For some document types such as news articles, this might serve as a reasonable strategy, as they tend to convey the most salient information in the beginning. In fact, selecting the first *k* sentences (Lead-*k*) is often used as a baseline summary for news summarization systems (See/Liu/Manning 2017; Zhong et al. 2019). In a similar vein, for the summarization of scientific papers, often only the abstract, introduction, and conclusion (AIC) are passed to the summarizer, as previous research found these sections to be the most salient (Sharma/Li/Wang 2019; Cachola et al. 2020). Another approach is to employ an extractive summarizer or retrieval module such as Dense Passage Retriever, Karpukhin et al. (2020), as part of a two-stage system, to select important segments before passing the text to the abstractive summarizer (Liu/Lapata 2019b; Ladhak et al. 2020; A. Wang et al. 2022). There are also transformer architectures that do not suffer from these limitations such as LED (Beltagy/Peters/Cohan 2020) or LongT5 (Guo et al. 2022) which replace $O(n^2)$ attention patterns with more efficient ones. Finally, experiments have been conducted on summarizing chunks of the text in potentially multiple iterations before producing a final, coherent summary (Gidiotis/Tsoumakas 2020; Zhao/Saleh/Liu 2020; Wu et al. 2021; Y. Zhang et al. 2022; Yang et al. 2023).

Figure 4: The context length has been steadily and exponentially increasing in open-source and closed-source language models alike. Not considered are models like LED, which specifically try to maximize the context length at the cost of performance otherwise. Illustration courtesy of the author.

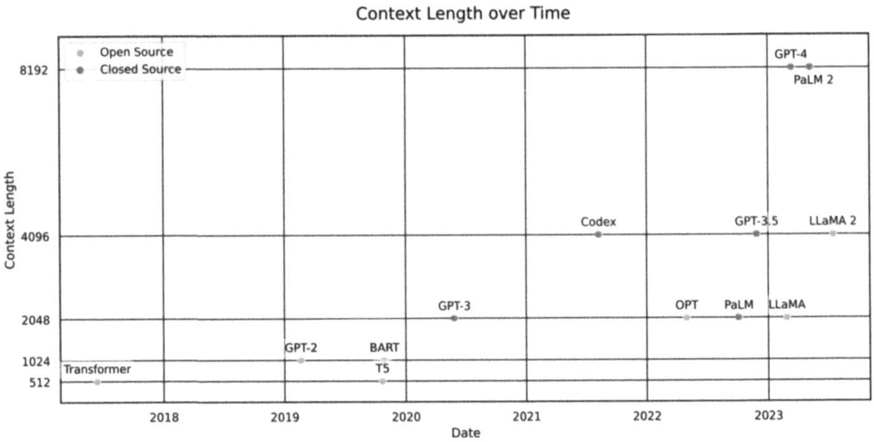

6.2 Multi-document summarization

The process of creating a summary from a collection of documents related to a specific topic is called multi-document summarization (MDS). This presents similar challenges to summarizing a long document, as the problem of limited context length is amplified when multiple documents are involved. Understanding the relationships between the documents is also essential for completing the task effectively. The first strategy for MDS is to simply concatenate all documents into one large text and use techniques designed for single-document summarization. However, this requires the model to process very long sequences. Therefore, a two-stage process similar to that used for long document summarization is commonly employed (Liu et al. 2018; Liu/Lapata 2019a). State-of-the-art approaches also use hierarchical architectures or graph-based methods to capture inter-document relations (Liu/Lapata, 2019a; W. Li et al. 2020; Pasunuru et al. 2021). At the same time, MDS approaches increasingly aim to utilize pre-trained encoder-decoder models such as BART, T5, or PEGASUS (Goodwin/Savery/Demner-Fushman 2020; Pasunuru et al. 2021). One recent and noteworthy model in this category, PRIMERA, is specifically designed for MDS and builds upon the foundations laid by PEGASUS (Xiao et al. 2022). For the GSG objective, PRIMERA chooses sentences that represent clusters of documents. It employs a document concatenation approach and architecturally uses LED to handle long sequences. In this manner, the model is generally applicable, and there are no dependencies on specific datasets. Although there is no scientific evaluation yet, the recent emergence and popularity of practical tools like LangChain and LlamaIndex hint towards the use of LLMs to handle collections of documents. For instance, LlamaIndex enables the storage of documents in an index that is organized like a tree, with each node representing a summary of its child nodes.

6.3 Controllable summarization

Controllable summarization is a multifaceted research question that refers to both the form or style (such as length, formality, or abstractiveness) and the content of a summary. The summary may be conditioned on a specific aspect or entity or, more broadly, on any given keyword or query. In recent years, a wide variety of approaches have been proposed. One of the most comprehensive systems is CTRLSum (J. He et al. 2022), a pre-trained encoder-decoder that generalizes controllability by utilizing keywords and prompts alike. In evaluations, the authors show the effectiveness of their method for length and entity control, as well as some more specialized tasks (e.g., patent purpose summarization). Recent studies conducted by Goyal/Li/Durrett (2022), Xiao et al. (2023), and Yang et al. (2023) offer initial insights into the potential of instruction-tuned LLMs like GPT-3 and ChatGPT. These systems have shown great promise for diverse summarization tasks based on keywords, aspects, and queries. Figure 3 shows two examples of how zero-shot prompting can enable controllable summarization in such systems. Nevertheless, the potential of LLMs for this task is still largely unexplored. Yang et al. (2023) note that their results can only serve as a lower bound, as the models are naively prompted without any prompt tuning or self-correction. A first glimpse of the potential of a more sophisticated prompting strategy is provided by Xiao et al. (2023) who suggest editing generated summaries with an editor model based on instructions from a separately trained model. In stark contrast, there is also a significant amount of research that focuses on controlling only one aspect of summarization. For example, in length-controllable summarization alone, systems have been proposed that early-stop the decoding process (Kikuchi et al. 2016), select information before passing it to the summarizer (LPAS; Saito et al. 2020), or incorporate length information as part of the input (Kikuchi et al. 2016; Liu/Luo/Zhu 2018). More recently, Liu, Jia, and Zhu (2022) also introduced a length-aware attention mechanism (LAAM).

6.4 Multi-modal summarization

So far, most research attention has been given to text summarization systems. However, there is an abundance of media and content such as podcasts, movies, and meetings that not only involve text but also other modalities including images, videos, and audio. These other modalities potentially contain key information that a pure text summarization system might miss, thus creating a semantic gap. For instance, H. Li et al. (2017) have demonstrated the importance of including audio and video information in the task of summarizing multimedia news, while the work of M. Li et al. (2019) has shown the value of including participants' head orientation and eye gaze when summarizing meetings. One of the key challenges of multi-modal summarization systems is the fusion of different input modalities. Currently, most systems take a late-fusion approach (see Jangra et al. 2023), for example by utilizing a pre-trained encoder. However, recently, a number of promising Transformer-based models have been proposed, which allow the input of diverse modalities such as Perceiver IO (Jaegle et al. 2021) or GATO (Reed et al. 2022) that have yet to be applied for the summarization task.

7. Commercialization

With language models having surpassed a certain level of performance, the creation and integration of these models into products and tools have become increasingly common, leading to a "gold rush" of NLP startups (Butcher 2022; Toews 2022). For summarization systems in particular, the context lengths of models are of utmost importance and have expanded exponentially in recent years as can be seen in Figure 4, to a level that is practical for more tasks and commercially viable. As such, many summarization systems have become productized and have been made available in consumer-oriented interfaces over the past year. In 2022, Google introduced document summarization in Google Docs (Saleh/Kannan 2022) and conversation summarization in Google Chat (Saleh/Wang 2022), both powered by fine-tuning the PEGASUS model. However, low-quality summaries in the datasets are mentioned as problematic. To tackle this issue, the developers utilize techniques such as dataset distillation, data formatting, and clean-ups, while continuing to collect more training data. Through knowledge distillation, they distill the models into more efficient hybrid architectures of a transformer encoder and a recurrent neural network (RNN) decoder. Separately, an additional model is trained to filter out generated summaries that are of low quality. More recently, Microsoft announced plans to roll out meeting summarization powered by GPT-3.5 in Microsoft Teams in Q2 2023 (Herskowitz 2023), but they have not provided any further technical details. Discord, the community messaging platform, uses "OpenAI technology" for grouping messages into topics for conversation summaries (Midha 2023). Zoom's recent smart recording feature, which includes meeting summarization and smart chaptering, vaguely mentions the use of GPT-3 to "augment" its own models (Parthasarathy 2023). Cohere just launched a dedicated text summarization endpoint (Hillier/Gallé 2023) that largely avoids several problems of LLMs such as the need for prompt engineering and limited context length. In addition, they offer settings to gain more control over the generated summaries: the level of extractiveness, the length, and the format (either fluent text or bullet points). More broadly, access to any standard LLM naturally allows for summarization by specifying the respective prompt. This is true for OpenAI's GPT-3, AI21 Studio, Antrophic's Claude, or

Cohere Generate – to name some that are available via paid APIs and power summarization functionalities in many commercial applications. ChatGPT might be notable, as it also enables a more interactive approach to summarization. Domain-specific summarization tools are another area of interest. For instance, Zoom IQ for Sales (Larkin 2022) aims to provide insights and summaries for sales meetings, while BirchAI, a spinoff from the Allen Institute for Artificial Intelligence, focuses solely on providing customer call summaries for call centers. Meanwhile, beyond big tech and distinguished AI labs, summarization systems are starting to reach many more surfaces such as browsers (Opera; Szyndzielorz 2023), email clients (Shortwave; Wenger 2023) or note-taking apps (Notion; I. Zhao 2023). This trend suggests that summarization is not an application on its own, but a basic feature to be widely implemented on most surfaces and to be widely accessible in the foreseeable future.

8. Conclusion

Text summarization is a rapidly evolving field with two recent paradigm shifts. First, towards finetuning pre-trained encoder-decoder models, and second and even more recently, towards zero-shot prompting of instruction-tuned language models. As a result of these developments, it appears that single-document summarization has reached a tipping point where the focus on improving automated metrics has diminishing returns and might even misdirect the research community. Therefore, we suggest a shift of emphasis towards improving human evaluation protocols and exploring self-evaluation of LLMs. Additionally, more targeted evaluation of certain aspects, such as factuality, should be considered and more broadly the uncovering of capabilities of pre-trained language models and fine-tuned summarization models. However, when contemplating summarization in a wider scope, tasks such as multi-document summarization and multi-modal summarization continue to present significant hurdles. Nonetheless, abstractive text summarization systems for single documents have matured and are rapidly being integrated into consumer products.

List of references

Beltagy, Iz/Peters, Matthew E./Cohan, Arman (2020): Longformer: The Long-Document Transformer, arXiv Preprint (http://arxiv.org/abs/2004.05150).

Bhaskar, Adithya/Fabbri, Alex/Durrett, Greg (2023): "Prompted Opinion Summarization with GPT3.5." In: Findings of the Association for Computational Linguistics (ACL 2023), Toronto, Canada, pp. 9282–9300.

BigScience Workshop (2022): BLOOM: A 176B-Parameter Open-Access Multilingual Language Model, arXiv Preprint (http://arxiv.org/abs/2211.05100).

Brown, Tom/Mann, Benjamin/Ryder, Nick/Subbiah, Melanie/Kaplan, Jared D./Dhariwal, Prafulla/Neelakantan, Arvind/et al. (2020): "Language Models are Few-Shot Learners." In: Advances in Neural Information Processing Systems 33 (NeurIPS 2020), Vancouver, Canada (https://proceedings.neurips.cc/paper_files/paper/2020/hash/1457c0d6bfcb4967418bfb8ac142f64a-Abstract.html).

Butcher, Mike (2022): "Here's why a gold rush of NLP startups is about to arrive.", July 28, 2022 (https://techcrunch.com/2022/07/28/a-gold-rush-of-nlp-startups-is-about-to-arrive-heres-why/).

Cachola, Isabel/Lo, Kyle/Cohan, Arman/Weld, Daniel (2020): "TLDR: Extreme Summarization of Scientific Documents." In: Findings of the Association for Computational Linguistics (EMNLP 2020), online, pp. 4766–4777.

Chowdhery, Aakanksha/Narang, Sharan/Devlin, Jacob/Bosma, Maarten/Mishra, Gaurav/Roberts, Adam/Barham, Paul/et al. (2022): PaLM: Scaling Language Modeling with Pathways, arXiv Preprint (http://arxiv.org/abs/2204.02311).

Christiano, Paul F./Leike, Jan/Brown, Tom/Martic, Miljan/Legg, Shane/Amodei, Dario (2017): "Deep Reinforcement Learning from Human Preferences." In: Advances in Neural Information Processing Systems 30 (NIPS 2017), Long Beach, CA, USA (https://papers.nips.cc/paper/2017/hash/d5e2c0adad503c91f91df240d0cd4e49-Abstract.html).

Chung, Hyung Won/Hou, Le/Longpre, Shayne/Zoph, Barret/Tay, Yi/Fedus, William/Li, Yunxuan/et al. (2022): Scaling Instruction-Finetuned Language Models, arXiv Preprint (http://arxiv.org/ abs/2210.11416).

Colombo, Pierre/Peyrard, Maxime/Noiry, Nathan/West, Robert/Piantanida, Pablo (2022): The Glass Ceiling of Automatic Evaluation in Natural Language Generation, arXiv Preprint (http://arxiv.org/abs/2208.14585).

Fabbri, Alexander R./Kryściński, Wojciech/McCann, Bryan/Xiong, Caiming/Socher, Richard/Radev, Dragomir (2021): "SummEval: Re-evaluating Summarization Evaluation." In: Transactions of the Association for Computational Linguistics 9, pp. 391–409.

Gidiotis, Alexios/Tsoumakas, Grigorios (2020): "A Divide-and-Conquer Approach to the Summarization of Long Documents." In: IEEE/ACM Transactions on Audio, Speech, and Language Processing 28, pp. 3029–3040.

Gliwa, Bogdan/Mochol, Iwona/Biesek, Maciej/Wawer, Aleksander (2019): "SAMSum Corpus: A Human-annotated Dialogue Dataset for Abstractive Summarization." In: Proceedings of the 2nd Workshop on New Frontiers in Summarization, Hong Kong, China, pp. 70–79.

Goodwin, Travis/Savery, Max/Demner-Fushman, Dina (2020): "Flight of the PEGASUS? Comparing Transformers on Few-shot and Zero-shot Multi-document Abstractive Summarization." In: Proceedings of the 28th International Conference on Computational Linguistics, Barcelona, Spain (Online), pp. 5640–5646.

Goyal, Tanya/Li, Junyi Jessy/Durrett, Greg (2022): News Summarization and Evaluation in the Era of GPT-3, arXiv Preprint (http://arxiv.org/abs/2209.12356).

Guo, Mandy/Ainslie, Joshua/Uthus, David/Ontanon, Santiago/Ni, Jianmo/Sung, Yun-Hsuan/Yang, Yinfei (2022): "LongT5: Efficient Text-To-Text Transformer for Long Sequences." In: Findings of the Association for Computational Linguistics (NAACL 2022), Seattle, USA, pp. 724–736.

He, Junxian/Kryscinski, Wojciech/McCann, Bryan/Rajani, Nazneen/Xiong, Caiming (2022): "CTRLsum: Towards Generic Controllable Text Summarization." In: Proceedings of the 2022 Conference on Empirical Methods in Natural Language Processing, Abu Dhabi, United Arab Emirates, pp. 5879–5915.

He, Pengcheng/Peng, Baolin/Wang, Song/Liu, Yang/Xu, Ruochen/Hassan, Hany/Shi, Yu/et al. (2023): "Z-Code++: A Pretrained Language Model Optimized for Abstractive Summarization." In: Proceedings of the 61st Annual Meeting of the Association for Computational Linguistics (Volume 1: Long Papers), Toronto, Canada, pp. 5095–5112.

Hendrycks, Dan/Burns, Collin/Basart, Steven/Zou, Andy/Mazeika, Mantas/Song, Dawn/Steinhardt, Jacob (2020): "Measuring Massive Multitask Language Understanding." In: 9th International Conference on Learning Representations (ICLR 2021), Virtual Event (https://openreview.net/forum?id=d7KBjmI3GmQ).

Herskowitz, Nicole (2023): "Microsoft Teams Premium: Cut costs and add AI-powered productivity.", February 1, 2023 (https://www.microsoft.com/

en-us/microsoft-365/blog/2023/02/01/microsoft-teams-premium-cut-costs-and-add-ai-powered-productivity/).

Hillier, Sheena/Gallé, Matthias (2023): "Introducing Cohere Summarize Beta: A New Endpoint for Text Summarization.", February 22, 2023 (https://txt.cohere.ai/summarize-beta/).

Hoffmann, Jordan/Borgeaud, Sebastian/Mensch, Arthur/Buchatskaya, Elena/Cai, Trevor/Rutherford, Eliza/de Las Casas, Diego/et al. (2022): Training Compute-Optimal Large Language Models, arXiv Preprint (https://arxiv.org/abs/2203.15556).

Iyer, Srinivasan/Lin, Xi Victoria/Pasunuru, Ramakanth/Mihaylov, Todor/Simig, Daniel/Yu, Ping/Shuster, Kurt/et al. (2023): OPT-IML: Scaling Language Model Instruction Meta Learning through the Lens of Generalization, arXiv Preprint (http://arxiv.org/abs/2212.12017).

Jaegle, Andrew/Borgeaud, Sebastian/Alayrac, Jean-Baptiste/Doersch, Carl/Ionescu, Catalin/Ding, David/Koppula, Skanda/et al. (2021): "Perceiver IO: A General Architecture for Structured Inputs & Outputs." In: The Tenth International Conference on Learning Representations (ICLR 2022), Virtual Event (https://openreview.net/forum?id=fILj7WpI-g).

Jangra, Anubhav/Mukherjee, Sourajit/Jatowt, Adam/Saha, Sriparna/Hasanuzzaman, Mohammad (2023): A Survey on Multi-modal Summarization, arXiv Preprint (http://arxiv.org/abs/ 2109.05199).

Ji, Ziwei/Lee, Nayeon/Frieske, Rita/Yu, Tiezheng/Su, Dan/Xu, Yan/Ishii, Etsuko/et al. (2023): "Survey of Hallucination in Natural Language Generation." In: ACM Computing Surveys 55/12 (https://doi.org/10.1145/3571730).

Kadavath, Saurav/Conerly, Tom/Askell, Amanda/Henighan, Tom/Drain, Dawn/Perez, Ethan/Schiefer, Nicholas/et al. (2022): Language Models (Mostly) Know What They Know, arXiv Preprint (http://arxiv.org/abs/2207.05221).

Karpukhin, Vladimir/Oguz, Barlas/Min, Sewon/Lewis, Patrick/Wu, Ledell/Edunov, Sergey/Chen, Danqi/Yih, Wen-Tau (2020): "Dense Passage Retrieval for Open-Domain Question Answering." In: Proceedings of the 2020 Conference on Empirical Methods in Natural Language Processing (EMNLP), Online, pp. 6769–6781.

Kikuchi, Yuta/Neubig, Graham/Sasano, Ryohei/Takamura, Hiroya/Okumura, Manabu (2016): "Controlling Output Length in Neural Encoder-Decoders." In: Proceedings of the 2016 Conference on Empirical Methods in Natural Language Processing, Austin, TX, USA, pp. 1328–1338.

Kocmi, Tom/Federmann, Christian (2023): Large Language Models Are State-of-the-Art Evaluators of Translation Quality, arXiv Preprint (http://arxiv.org/abs/2302.14520).

Kryscinski, Wojciech/McCann, Bryan/Xiong, Caiming/Socher, Richard (2020): "Evaluating the Factual Consistency of Abstractive Text Summarization." In: Proceedings of the 2020 Conference on Empirical Methods in Natural Language Processing (EMNLP), Online, pp. 9332–9346.

Ladhak, Faisal/Li, Bryan/Al-Onaizan, Yaser/McKeown, Kathleen (2020): "Exploring Content Selection in Summarization of Novel Chapters." In: Proceedings of the 58th Annual Meeting of the Association for Computational Linguistics, Online, pp. 5043–5054.

Larkin, Theresa (2022): "Zoom IQ for Sales: Conversational intelligence for sellers.", April 13, 2022 (https://blog.zoom.us/zoom-iq-for-sales/).

Lewis, Mike/Liu, Yinhan/Goyal, Naman/Ghazvininejad, Marjan/Mohamed, Abdelrahman/Levy, Omer/Stoyanov, Veselin/Zettlemoyer, Luke (2020): "BART: Denoising Sequence-to-Sequence Pre-training for Natural Language Generation, Translation, and Comprehension." In: Proceedings of the 58th Annual Meeting of the Association for Computational Linguistics, Online, pp. 7871–7880.

Li, Haoran/Zhu, Junnan/Ma, Cong/Zhang, Jiajun/Zong, Chengqing (2017): "Multi-modal Summarization for Asynchronous Collection of Text, Image, Audio and Video." In: Proceedings of the 2017 Conference on Empirical Methods in Natural Language Processing, Copenhagen, Denmark, pp. 1092–1102.

Li, Manling/Zhang, Lingyu/Ji, Heng/Radke, Richard J. (2019): "Keep Meeting Summaries on Topic: Abstractive Multi-Modal Meeting Summarization." In: Proceedings of the 57th Annual Meeting of the Association for Computational Linguistics, Florence, Italy, pp. 2190–2196.

Li, Wei/Xiao, Xinyan/Liu, Jiachen/Wu, Hua/Wang, Haifeng/Du, Junping (2020): "Leveraging Graph to Improve Abstractive Multi-Document Summarization." In: Proceedings of the 58th Annual Meeting of the Association for Computational Linguistics, Online, pp. 6232–6243.

Lin, Chin-Yew (2004): "ROUGE: A Package for Automatic Evaluation of Summaries." In: Text Summarization Branches Out. Proceedings of the ACL-04 Workshop, Barcelona, Spain, pp. 74–81.

Liu, Peter J./Saleh, Mohammad/Pot, Etienne/Goodrich, Ben/Sepassi, Ryan/Kaiser, Lukasz/Shazeer, Noam (2018): "Generating Wikipedia by Summarizing Long Sequences." In: 6th International Conference on Learning Representations (ICLR 2018), Vancouver, Canada (https://openreview.net/forum?id=HygovbWC-).

Liu, Yang/Lapata, Mirella (2019a): "Hierarchical Transformers for Multi-Document Summarization." In: Proceedings of the 57th Annual Meeting of the Association for Computational Linguistics, Florence, Italy, pp. 5070–5081.

Liu, Yang/Lapata, Mirella (2019b): "Text Summarization with Pretrained Encoders." In: Proceedings of the 2019 Conference on Empirical Methods in Natural Language Processing and the 9th International Joint Conference on Natural Language Processing (EMNLP-IJCNLP), Hong Kong, China, pp. 3730–3740.

Liu, Yixin/Fabbri, Alex/Liu, Pengfei/Zhao, Yilun/Nan, Linyong/Han, Ruilin/Han, Simeng/et al. (2023): "Revisiting the Gold Standard: Grounding Summarization Evaluation with Robust Human Evaluation." In: Proceedings of the 61st Annual Meeting of the Association for Computational Linguistics (Volume 1: Long Papers), Toronto, Canada, pp. 4140–4170.

Liu, Yixin/Liu, Pengfei/Radev, Dragomir/Neubig, Graham (2022): "BRIO: Bringing Order to Abstractive Summarization." In: Proceedings of the 60th Annual Meeting of the Association for Computational Linguistics (Volume 1: Long Papers), Dublin, Ireland, pp. 2890–2903.

Liu, Yizhu/Jia, Qi/Zhu, Kenny (2022): "Length Control in Abstractive Summarization by Pretraining Information Selection." In: Proceedings of the 60th Annual Meeting of the Association for Computational Linguistics (Volume 1: Long Papers), Dublin, Ireland, pp. 6885–6895.

Liu, Yizhu/Luo, Zhiyi/Zhu, Kenny (2018): "Controlling Length in Abstractive Summarization Using a Convolutional Neural Network." In: Proceedings of the 2018 Conference on Empirical Methods in Natural Language Processing, Brussels, Belgium, pp. 4110–4119.

Midha, Anjney (2023): "Discord is Your Place for AI with Friends.", March 13, 2023 (https://discord.com/blog/ai-on-discord-your-place-for-ai-with-friends).

Muennighoff, Niklas/Wang, Thomas/Sutawika, Lintang/Roberts, Adam/Biderman, Stella/Le Scao, Teven/Bari, M. Saiful/et al. (2023): "Crosslingual Generalization through Multitask Finetuning." In: Proceedings of the 61st Annual Meeting of the Association for Computational Linguistics (Volume 1: Long Papers), Toronto, Canada, pp. 15991–16111.

Nallapati, Ramesh/Zhou, Bowen/dos santos, Cicero Nogueira/Gulcehre, Caglar/Xiang, Bing (2016): Abstractive Text Summarization Using Sequence-to-Sequence RNNs and Beyond, arXiv Preprint (http://arxiv.org/abs/1602.06023).

Ouyang, Long/Wu, Jeffrey/Jiang, Xu/Almeida, Diogo/Wainwright, Carroll/Mishkin, Pamela/Zhang, Chong/et al. (2022): "Training language models to follow instructions with human feedback." In: Advances in Neural Information Processing Systems 35 (NeurIPS 2022), New Orleans, USA/Online, pp. 27730–27744.

Pagnoni, Artidoro/Balachandran, Vidhisha/Tsvetkov, Yulia (2021): "Understanding Factuality in Abstractive Summarization with FRANK: A Benchmark for Factuality Metrics." In: Proceedings of the 2021 Conference of the North American Chapter of the Association for Computational Linguistics: Human Language Technologies, Online, pp. 4812–4829.

Parthasarathy, Vijay (2023): "Zoom's AI innovations empower people.", February 24, 2023 (https://blog.zoom.us/ai-driven-innovations/).

Pasunuru, Ramakanth/Liu, Mengwen/Bansal, Mohit/Ravi, Sujith/Dreyer, Markus (2021): "Efficiently Summarizing Text and Graph Encodings of Multi-Document Clusters." In: Proceedings of the 2021 Conference of the North American Chapter of the Association for Computational Linguistics: Human Language Technologies, Online, pp. 4768–4779.

Qin, Chengwei/Zhang, Aston/Zhang, Zhuosheng/Chen, Jiaao/Yasunaga, Michihiro/Yang, Diyi (2023): Is ChatGPT a General-Purpose Natural Language Processing Task Solver?, arXiv Preprint (http://arxiv.org/abs/2302.06476).

Rae, Jack W./Borgeaud, Sebastian/Cai, Trevor/Millican, Katie/Hoffmann, Jordan/Song, Francis/Aslanides, John/et al. (2022): Scaling Language Models: Methods, Analysis & Insights from Training Gopher, arXiv Preprint (http://arxiv.org/abs/2112.11446).

Raffel, Colin/Shazeer, Noam/Roberts, Adam/Lee, Katherine/Narang, Sharan/Matena, Michael/Zhou, Yanqi/Li, Wei/Liu, Peter J. (2020): "Exploring the Limits of Transfer Learning with a Unified Text-to-Text Transformer." In: Journal of Machine Learning Research 21/140, pp. 1–67.

Reed, Scott/Zolna, Konrad/Parisotto, Emilio/Colmenarejo, Sergio Gómez/Novikov, Alexander/Barth-Maron, Gabriel/Giménez, Mai/et al. (2022): "A Generalist Agent." In: Transactions on Machine Learning Research (https://openreview.net/forum?id=1ikKokHjvj).

Saito, Itsumi/Nishida, Kyosuke/Nishida, Kosuke/Otsuka, Atsushi/Asano, Hisako/Tomita, Junji/Shindo, Hiroyuki/Matsumoto, Yuji (2020): Length-controllable Abstractive Summarization by Guiding with Summary Prototype, arXiv Preprint (http://arxiv.org/abs/2001.07331).

Saleh, Mohammad/Kannan, Anjuli (2022): "Auto-generated Summaries in Google Docs.", March 23, 2022 (https://ai.googleblog.com/2022/03/auto-generated-summaries-in-google-docs.html).

Saleh, Mohammad/Wang, Yinan (2022): "Conversation Summaries in Google Chat.", November 18, 2022 (https://ai.googleblog.com/2022/11/conversation-summaries-in-google-chat.html).

Sanh, Victor/Webson, Albert/Raffel, Colin/Bach, Stephen H./Sutawika, Lintang/Alyafeai, Zaid/Chaffin, Antoine/et al. (2022): Multitask Prompted Training Enables Zero-Shot Task Generalization, arXiv Preprint (http://arxiv.org/abs/2110.08207).

See, Abigail/Liu, Peter J./Manning, Christopher D. (2017): "Get To The Point: Summarization with PointerGenerator Networks." In: Proceedings of the 55th Annual Meeting of the Association for Computational Linguistics (Volume 1: Long Papers), Vancouver, Canada, pp. 1073–1083.

Sharma, Eva/Li, Chen/Wang, Lu (2019): "BIGPATENT: A Large-Scale Dataset for Abstractive and Coherent Summarization." In: Proceedings of the 57th Annual Meeting of the Association for Computational Linguistics, Florence, Italy, pp. 2204–2213.

Szyndzielorz, Julia (2023): "Opera enters the generative AI space with new features in browsers and content apps.", February 10, 2023 (https://blogs.opera.com/news/2023/02/opera-aigc-integration/).

Taori, Rohan/Gulrajani, Ishaan/Zhang, Tianyi/Dubois, Yann/Guestrin, Carlos/Liang, Percy/Hashimoto, Tatsunori B. (2023): "Alpaca: A Strong, Replicable Instruction-Following Model.", March 13, 2023 (https://crfm.stanford.edu/2023/03/13/alpaca.html).

Tay, Yi/Dehghani, Mostafa/Tran, Vinh Q./Garcia, Xavier/Wei, Jason/Wang, Xuezhi/Chung, Hyung Won/et al. (2022): "UL2: Unifying Language Learning Paradigms." In: The Eleventh International Conference on Learning Representations (ICLR 2023), Kigali, Rwanda (https://openreview.net/forum?id=6ruVLB727MC).

Taylor, Ross/Kardas, Marcin/Cucurull, Guillem/Scialom, Thomas/Hartshorn, Anthony/Saravia, Elvis/Poulton, Andrew/Kerkez, Viktor/Stojnic, Robert (2022): Galactica: A Large Language Model for Science, arXiv Preprint (http://arxiv.org/abs/2211.09085).

The Vicuna Team (2023): "Vicuna: An Open-Source Chatbot Impressing GPT-4 with 90%* ChatGPT Quality.", March 30, 2023 (https://lmsys.org/blog/2023-03-30-vicuna).

Toews, Rob (2022): "A Wave Of Billion-Dollar Language AI Startups Is Coming.", March 27, 2022 (https://www.forbes.com/sites/robtoews/2022/03/27/a-wave-of-billion-dollar-language-ai-startups-is-coming/).

Touvron, Hugo/Lavril, Thibaut/Izacard, Gautier/Martinet, Xavier/Lachaux, Marie-Anne/Lacroix, Timothee/Rozière, Baptiste/et al. (2023): LLaMA: Open and Efficient Foundation Language Models, arXiv Preprint (https://arxiv.org/abs/2302.13971).

Wang, Alex/Pang, Richard Yuanzhe/Chen, Angelica/Phang, Jason/Bowman, Samuel R. (2022): "SQuALITY: Building a Long-Document Summarization Dataset the Hard Way." In: Proceedings of the 2022 Conference on Empirical Methods in Natural Language Processing, Abu Dhabi, United Arab Emirates, pp. 1139–1156.

Wang, Alex/Pruksachatkun, Yada/Nangia, Nikita/Singh, Amanpreet/Michael, Julian/Hill, Felix/Levy, Omer/Bowman, Samuel (2019): "SuperGLUE: A Stickier Benchmark for General-Purpose Language Understanding Systems." In: Advances in Neural Information Processing Systems 32 (NeurIPS 2019), Vancouver, Canada (https://proceedings.neurips.cc/paper_files/paper/2019/hash/4496bf24afe7fab6f046bf4923da8de6-Abstract.html).

Wang, Yizhong/Kordi, Yeganeh/Mishra, Swaroop/Liu, Alisa/Smith, Noah A./Khashabi, Daniel/Hajishirzi, Hannaneh (2023): "Self-Instruct: Aligning Language Models with Self-Generated Instructions." In: Proceedings of the 61st Annual Meeting of the Association for Computational Linguistics (Volume 1: Long Papers), Toronto, Canada, pp. 13484–13508.

Wenger, Jacob (2023): "AI Email Summaries: Read emails in seconds.", February 27, 2023 (https://www.shortwave.com/blog/ai-email-summaries/).

Wu, Jeff/Ouyang, Long/Ziegler, Daniel M./Stiennon, Nisan/Lowe, Ryan/Leike, Jan/Christiano, Paul (2021): Recursively Summarizing Books with Human Feedback, arXiv Preprint (http://arxiv.org/abs/2109.10862).

Xiao, Wen/Beltagy, Iz/Carenini, Giuseppe/Cohan, Arman (2022): "PRIMERA: Pyramid-based Masked Sentence Pre-training for Multi-document Summarization." In: Proceedings of the 60th Annual Meeting of the Association for Computational Linguistics (Volume 1: Long Papers), Dublin, Ireland, pp. 5245–5263.

Xiao, Wen/Xie, Yujia/Carenini, Giuseppe/He, Pengcheng (2023): ChatGPT-steered Editing Instructor for Customization of Abstractive Summarization, arXiv Preprint (http://arxiv .org/abs/2305.02483).

Yang, Xianjun/Li, Yan/Zhang, Xinlu/Chen, Haifeng/Cheng, Wei (2023): Exploring the Limits of ChatGPT for Query or Aspect-based Text Summarization, arXiv Preprint (http://arxiv.org/abs/2302.08081).

Yuan, Weizhe/Neubig, Graham/Liu, Pengfei (2021): "BARTScore: Evaluating Generated Text as Text Generation." In: Advances in Neural Information Processing Systems 34 (NeurIPS 2021), Online, pp. 27263–27277.

Zhang, Daniel/Maslej, Nestor/Brynjolfsson, Erik/Etchemendy, John/Lyons, Terah/Manyika, James/Ngo, Helen/et al. (2022): The AI Index 2022 Annual Report, arXiv Preprint (http://arxiv.org/ abs/2205.03468).

Zhang, Jingqing/Zhao, Yao/Saleh, Mohammad/Liu, Peter (2020): "PEGASUS: Pre-training with Extracted Gap-sentences for Abstractive Summarization." In: Proceedings of the 37th International Conference on Machine Learning, Online, pp. 11328–11339.

Zhang, Susan/Roller, Stephen/Goyal, Naman/Artetxe, Mikel/Chen, Moya/Chen, Shuohui/Dewan, Christopher/et al. (2022): OPT: Open Pre-trained Transformer Language Models, arXiv Preprint (http://arxiv.org/abs/2205.01068).

Zhang, Tianyi/Kishore, Varsha/Wu, Felix/Weinberger, Kilian Q./Artzi, Yoav (2019): BERTScore: Evaluating Text Generation with BERT, OpenReview Preprint (https://openreview .net/forum?id=SkeHuCVFDr).

Zhang, Tianyi/Ladhak, Faisal/Durmus, Esin/Liang, Percy/McKeown, Kathleen/Hashimoto, Tatsunori B. (2023): Benchmarking Large Language Models for News Summarization, arXiv Preprint (http://arxiv.org/abs/2301.13848).

Zhang, Yusen/Ni, Ansong/Mao, Ziming/Wu, Chen Henry/Zhu, Chenguang/Deb, Budhaditya/Awadallah, Ahmed/Radev, Dragomir/Zhang, Rui (2022): "Summ^N: A Multi-Stage Summarization Framework for Long Input Dialogues and Documents." In: Proceedings of the 60th Annual Meeting of the Association for Computational Linguistics (Volume 1: Long Papers), Dublin, Ireland, pp. 1592–1604.

Zhao, Ivan (2023): "Notion AI is Here, for Everyone.", February 22, 2023 (https://www.notion.so/blog/notion-ai-is-here-for-everyone).

Zhao, Yao/Saleh, Mohammad/Liu, Peter J. (2020): SEAL: Segment-wise Extractive-Abstractive Long-form Text Summarization, arXiv Preprint (http://arxiv.org/abs/2006.10213).

Zhong, Ming/Liu, Pengfei/Wang, Danqing/Qiu, Xipeng/Huang, Xuanjing (2019): "Searching for Effective Neural Extractive Summarization: What Works and What's Next." In: Proceedings of the 57th Annual Meeting of the Association for Computational Linguistics, Florence, Italy, pp. 1049–1058.

Opacity and reproducibility in data processing

Reflections on the dependence of AI on the data ecosystem

Sabina Leonelli

1. Introduction

It is sometimes argued that AI tools, though strongly dependent on the availability of large volumes of training data for their accuracy and effectiveness, are becoming increasingly less constrained by the scope and biases of the data themselves – both because the quantity and variety of data used to train algorithms grows at vertiginous speed, and because AI gets exponentially better at correcting bias and calibrating results towards specific, accurate solutions. Without wishing to deny such advancements and the resulting increase in potential for these technologies, I here maintain that AI is still strongly tied to the quality and representativeness of training data and that existing data gaps are not credibly filled by data produced for that very purpose, given that such production is strongly informed by expectations around the outputs and the focus on algorithmic outputs is taking attention away from the decision-making happening at various stages of data elaboration. Indeed, simulated, augmented, or synthetic data, which are supposedly 'artificial' insofar as they are created by humans for training algorithms and are not meant to faithfully document a specific aspect of the world, are produced and processed through specific assumptions about what the world may be like or what characteristics of the world one may be interested in. Whether or not these assumptions are explicitly identified and debated, they play an important role in framing the ways in which algorithms are developed to mine, model and visualize data, and thus directly affect the goals, methods and tools of AI. In what follows, I reflect on these concerns and on their implications for how we may understand the notion of opacity, so often identified as a major concern in the use of AI for research purposes, and its relation to the reproducibility of research, that is the idea that it is possible to ascertain the credibility of specific outputs through success in re-creating them, which in turn involves some understanding of how they were produced in the first place.

2. Investigating research data journeys

My research concerns knowledge production through AI, particularly in the biological, biomedical and environmental domains. In that context I am interested in the extent to which insights derived from existing knowledge and research shape AI-powered data analytics and how/if such analytics are themselves capable of producing novel insights. As a window towards that problem, I have investigated not just what data collections exist – what people can actually source as input for their analysis – but also *how data are mobilized* once they have been generated and/or collected, garnered into digital infrastructures, and eventually re-used. I have traced and theorized such processes as "data journeys" (Leonelli/Tempini 2020), with a particular interest in data sets that get repurposed several times by people with different expertises. One example is data collected from social media (tweets, comments, 'likes') being reused to track public health concerns – as for instance happened during the COVID-19 pandemic – as well as mobility trends, such as how often people use public transport following periods of lock-down (e.g. Leonelli et al. 2021; Leonelli 2021). Another example is data acquired from detailed satellite imaging of specific territories, which are used to study phenomena as wide-ranging as deforestation trends, farming habits, urban planning and migration patterns, depending on how the images are processed and what other datasets they are combined with (Leonelli/Williamson 2023). Such situations are prime instances of what AI tools are supposed to achieve: That is, to enable researchers to recombine and reanalyse existing datasets for a variety of purposes, thereby extracting maximum value from the data as evidence for knowledge claims and related interventions.

The major challenge in tracking data journeys has been thinking about what happens when you have a very large, heterogeneous set of data and people need to rely on that dataset to do certain kinds of work, but at the same time have to make decisions about what part of that data they can trust.

How should/can the reliability of data and the quality of the information that is to be extracted from it be assessed? Who do you collaborate with when you're trying to do this kind of work, and how do you make such decisions? How is expertise distributed across data journeys, including the employment of data within AI, and which of the experts involved are accountable for the overarching outcomes of that complex system? The moment we are plunged into a large data ecosystem, we are often looking at thousands of people who have been working on that ecosystem and changing it to fit their aspirations, assumptions and goals. How to trust such a distributed system – does it mean verifying whether each individual contributor has done a good job, and if so, how can this be done? Are there ways to verify the quality and reliability of data ecosystems beyond the reconstruction of individual contributions, and if so, what are they?

I have explored these questions in collaboration with Niccolò Tempini and several collaborators from the natural sciences through DATA_SCIENCE ("The Epistemology of Data-Intensive Science"), a project sponsored by the European Research Council which ran from 2014 to 2019 and focused on the epistemology of data science and its applications in biology and biomedicine. We attempted to follow some datasets from the moment they were created to the moment they were organized into data infrastructures and further reused in a variety of projects. In an approach closely aligned with the infrastructural inversion pioneered by Geoffrey Bowker and Susan Leigh Star (1999), the starting point typically was data infrastructures, because this was a moment in the history of data when we began witnessing different perspectives on the conditions under which data could be used – intelligibly and actionably. From there, the next step was to find out where data were originally sourced and investigate how they were deployed and interpreted by database users. This was a difficult enterprise because you cannot tag data – it has been tried and found to be too difficult to implement. It is a form of detective work to try and track what happens to particular data sets, how they get modified and reshaped to fit different purposes and what the consequences for knowledge production are, particularly in cases where there are some very substantive disagreements between people who produce or collect data in the first place, and people who end up reusing them in a different environment and giving them a completely different meaning and frame of reference, which is where we saw many of these kinds of conflicts.

3. In-practice opacity within data ecosystems

Here is one potential representation of the research landscape viewed from the perspective of data movements and reanalysis (see fig. 1). The blue boxes in the middle of this figure are various databases. Sometimes they overlap, sometimes not: They are haphazardly overlapping. They tend to be funded in different ways and for different purposes by different institutions. They have different objectives. They have different lifespans and different types of data intersect with these data infrastructures, which different audiences use in different ways. A noteworthy aspect when considering data ecosystems as a serendipitous, organically growing ensemble is the fact that people who end up using data very often not only do not have a clue how data were processed or what the underlying structure of the organizations that are caring for, maintaining and stewarding the data, are. Even in the rare cases when there is a way to track data processing within a given database, with detailed information about where data comes from and how they have been manipulated, it would take too long to understand this narrative and its implications for one's work. Thus, effectively these systems become black boxes. This is not in-principle opacity of the kind sometimes encountered in AI tools, where we simply do not know – and cannot explain – how machines are generating a given output. This is in-practice opacity, emerging from pragmatic issues of tractability and intelligibility of large data structures. Even in a situation where there arc enough metadata and contextual information that you could try and reconstruct the whole history of the data, thereby better understanding what decisions have shaped its processing and why, such an enterprise becomes undoable for lack of time.

All the cases we examined kept showing us that the bigger the exercise in data linkage and reuse, the bigger the effort to calibrate, process,reprocess and reanalyze the data that went into the system, in the attempt to make sure that the results were reliable. There is a constant and growing tension between the need to consider the history of the data to understand which of these correlations you could even set up, let alone trust for further work, and the imperative of feeding data like this to AI systems and accelerate the production of potential inferences by using some of these objects as training data for a variety of algorithms.

Figure 1: A schematic representation of the research data ecosystem. Translated in English from Leonelli 2018a.

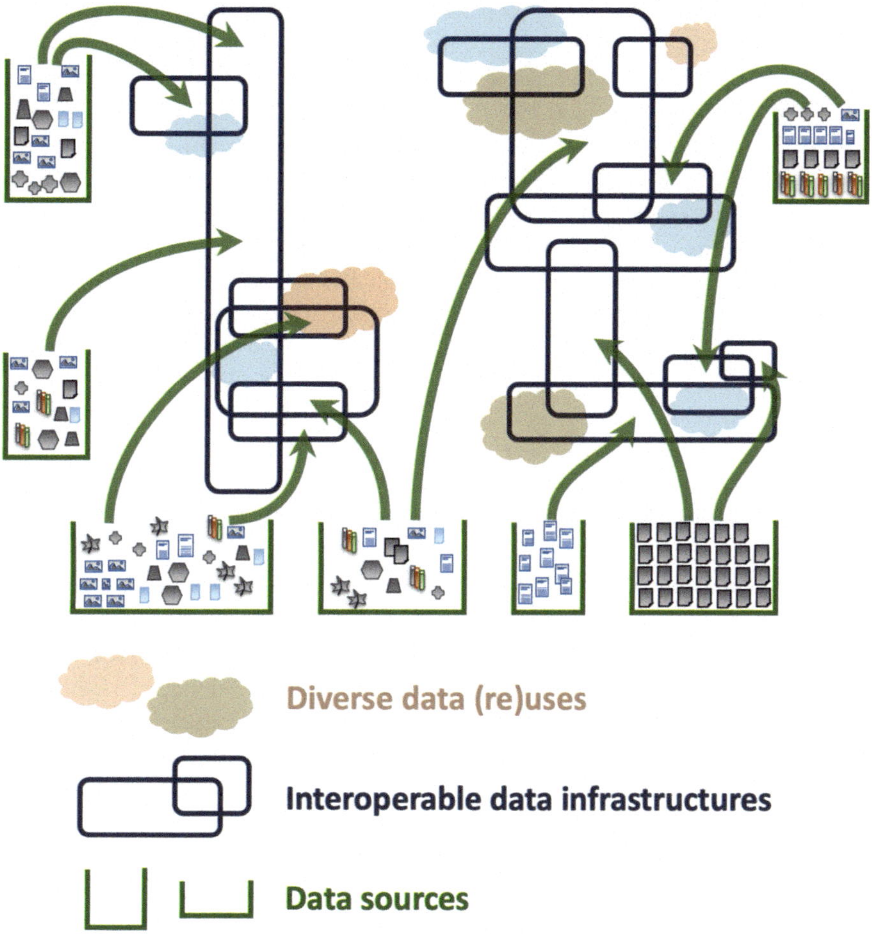

My perspective on the epistemology of data originates in the consideration of the multitude of ways in which people interact with the world and generate artifacts (images, numbers, textual descriptions) that are meant to capture or document these interactions in some way. Many interactions with the world produce some kind of object or atrace of some sort, and those objects may or may not be processed as data. In my view, data does not become a representation of the world until it gets clustered, ordered and interpreted in a particular kind of way. In other words, data models represent specific phenomena; data represent objects that are processed and stewarded for their potential to serve such representational purposes. Once a decision has been made about what data may be evidence for, the resulting models are used to interpret the data and acquire knowledge, which in turn informs further interactions with the world (Leonelli 2016).

There is a fragility and unreliability to the current data system, since it is hard to distinguish datasets that have been well-maintained and updated from those that have not been checked and adequately curated (Floridi/Illari 2014). Datasets available online are limited and biased, and there is a multitude of vested interests around which types of data become easier to access or more valuable to trade (Kitchin 2014; Mackenzie 2017). All these considerations contribute towards enhancing the in-practice opacity of data ecosystems, making it often near-impossible to unravel such opacity in a way that fosters intelligibility.

4. Reproducibility and the illusion of transparency

Situating data movements within a broad landscape which includes AI technology, as well as research institutions, industry, policy-making and various other publics and stakeholders lead to the investigation of the idea, which is common among supporters of Open Science, that increasing the transparency through which data processing is documented and explained may contribute to lessening the opacity characterizing large data ecosystems (Leonelli 2023).

One example of this approach is the discussion of reproducibility, which includes the application/consideration of a scientific method but also that of the priorities, goals and interests of the various institutions engaged in science. In particular, it interrogates what it means for data-intensive analyses to be scrutinized, reenacted and understood, no matter how complex the relevant sources, processes and analytics may be. The debate on reproducibility is a good representation of how the use of data-hungry AI in research raises issues beyond the traditional questions asked of the statistical methods used to validate datasets and analyses. While we witness a large increase of integrated research efforts and the application of algorithms across large domains, there are also increasing problems in getting people who are specialists in different parts of the research ecosystem to interact with each other and assess the value and significance of each other's work. Lots of confusion is generated by questions around scales and who can be trusted in this kind of landscape. Peer review is increasingly acknowledged not to work well when attempting to check data quality and incentives for researchers to engage in careful scrutiny of peers remain scarce. A strong reliance on automated research systems complicates matters further. Within such a landscape, reliance on AI creates even more a sense of research processes increasingly being impenetrable black boxes, whose inner mechanisms and functions remain invisible and unreachable to observers. There is a growing mistrust of scientific results even by actual scientists, let alone members of the public. The moral economy of science, strongly grounded on trust among peers, is being disrupted. It is in this climate of mistrust and uncertainty that the question of opacity associated with the use of AI in research has acquired poignance and prominence, prompting calls for explainable and transparent uses of AI for discovery and warnings against the reliability of systems that do not seem accessible for scrutiny (Council of Canadian Academies 2022).

There is little doubt that we are witnessing a real challenge in contemporary applications of AI to research processes and that questions around how such applications should be scrutinized and integrated into existing methods are urgent and unresolved. I do not think, however, that the main problem lies with the opacity of research systems per se. To an extent, research processes have always been and will always be opaque. It is simply impossible to account for every aspect of a research process, including the tacit knowledge used to calibrate instruments, set-up experiments, adapt methods to the specific situation and materials on which research is being carried out. The question is, rather, what forms of opacity end up being damaging to research and its role in society.

Reproducibility is often evoked as a solution to the problem of opacity in research, including in AI applications. You want to try and make sure that when you repeat a piece of research, there are some consistent results obtained. This seems like a fair requirement – a good thing for scientists to try and strive for. Consequently, there is a push to try and have more transparent sharing of information, particularly meta and para information around data sets, so it is easier to evaluate how data have been created and processed, with the aim to reproduce these conditions. Some even argue that the more we know about the process of research – the more we can capture, publish, debate and the more we may be able to automate some of those processes in interesting ways that can complement and sometimes even substitute humans who are involved in a discovery (for a depiction of the debate, see for instance The Royal Society 2019).

Despite its promise, reproducibility however is not a silver bullet. To begin with, there are many different types of reproducibility (Leonelli 2018b; Leonelli/ Lewandoswky 2023) that range from the more classical computational reproducibility, which assumes total control in the system, to reproducible observations that assume very low controls in terms of statistics, goals and judgments. There is a big discrepancy in how different domains depend on statistics and computation, not just as a tool to get the research done, but as a reasoning tool to make inferences. Clinical trials are typical examples of hypothesis testing situations where methods and results are expected to conform to detailed and sophisticated advance plans, but there isa lot of exploratory research that operates differently. How stable you assume your background knowledge to be also makes a difference, as well as whether or not you think it is acceptable for researchers to declare that they've exercised their subjective judgment in setting up their technical system. In evidence-based medicine this is something that people are not comfortable admitting, because the idea that expert judgment is used in someone's work is regarded as making research subjective and potentially unreliable. There is a desire to reach conclusions in ways that do not depend on the specific circumstances of the researcher's judgment. Nevertheless, such independence is yet to be found (Leonelli forthcoming).

I am worried about the fact that we are often confronted with a very narrow interpretation of reproducibility when thinking about how this principle operates in research practice. Highly controlled experiments which have pre-specified goals have come to exemplify best practice for some reason, and rigorous research, partly because they tend to adhere more easily to potentially misguided ideas about objectivity in science. This ends up doing no justice to other research methods that are accused of being unscientific. We are losing important expertise by creating priorities and rankings over what kind of methods should be prioritized in research. Qualitative research traditions get put aside and there is a strong emphasis on hypothesis-driven research to the expense of data mining, where in many cases hypotheses are not specified in advance. A narrow interpretation of reproducibility sets up a false dichotomy between quantitative approaches and more hermeneutic, judgment-based approaches, which devalues the role of expertise and embodied knowledge in dealing with data, but also the very significant social context in which research is happening. This does not resolve at all the problem of reproducibility to start with, because it really doesn't necessarily help to distinguish between what may be an unintentional mistake, what may be an actual case of cheating, or what may be a variation which is due to differences in research conditions, which may be actually quite interesting, and the situations where the best guess is to constructively poke at accepted facts. This pursuit of reproducibility as an overarching epistemic value, particularly when focused on increasing transparency in documenting research methods as a key solution, is not some sort of magic trick or a magic formula for what might constitute good science. It doesn't necessarily fix concerns around research quality, since simply providing more information about data processing does not necessarily help evaluate such processes – especially in situations where the processes in question are so vast and complex that they cannot be synthesized or comprehended. Nor does it provide some universal solution, particularly because there are all these different ways in which you can interpret the possibility, which are active and useful in different ways, depending on what kind of domain and what kind of practices you're adopting.

To continue, it does not necessarily help to address systemic issues with who is incentivized to make their data available, who is incentivized to curate data properly, and how people are rewarded for documenting their data management decisions – issues that are at the root of many of the problems prompting calls for reproducibility. Attention should be redirected towards the thinking of existing assumptions about hierarchies of evidence, where they come from and what their effects are likely to be when they become part of the research infrastructures, including algorithms and machine learning applications. More reflection also needs to go into what kinds of data should be preserved for long term storage, dissemination and sharing, and under which conditions, and how, such choices may be made accountable within expansive data ecosystems (Zook et al. 2017; Elliott et al. 2021). Most of our digital data ecology is ephemeral, with few attempts to think about data collection and data storage online for more than 10 years. Algorithms are currently trained on a rather serendipitous collection of data, whose availability depends on who gets funding at a particular point in time and how tractable data are digitally. There is a significant skew in the kind of machine-readable data that can be utilized for algorithmic elaboration. Finally, there is a sidelining of research geared towards involving transdisciplinary communities and expertise, accompanied by an emphasis on short-term outcomes and low-hanging fruit that stays away from complex, heterogeneous datasets in favor of homogenous, easy-to-handle ones. All this creates skews in the data system feeding AI, which is sure to have significant implications for the kinds of questions AI can help answer more accurately, as well as for the content of those answers.

5. Cracks in the looking glass: AI and the data ecosystem

What are the implications of these reflections for AI? Narrow interpretations of reproducibility tend to go hand-in-hand with an insistence on computational tools to automate research processes, with the hope that AI can provide a quick fix for problems around the quality of research – perhaps even help researchers to replicate experiments and methods without effort. This constitutes, in my view, a vicious circle. There is insistence on narrow, computational understandings of reproducibility because this seems to be a watertight way of thinking about checking the quality of a particular set of algorithms. However, this disregards the problems that arise through systems that are difficult to automate, such as quality checks for domain specific data obtained from complex experiments and observational methods, as well as the limits and histories entrenched in the current ecosystem of widely accessible, machine-readable data useable for training AI tools.

There is a gulf opening between discussions on reproducibility and what constitutes reliable training of data, reliable methods and reliable algorithms, which can be evaluated through those particular tools and others that are seen to be much less reliable because they just don't fit this kind of more automated, quick, computational check. It is crucial to address how one ought to formulate, assess and acknowledge the qualitative judgments that accompany data driven methods. In many AI discussions there is a tendency to think that judgements made around data – in calibrating data, in thinking about what is actually being processed, in picking training data, in creating artificial data that may fit new analytic tools – are important, but will be superseded by the emergence of better and better AI technology and more and more data sets. The hope is that the biases and the kind of externalities produced by judgments in those respects will disappear within a beautifully irrefutable and increasingly objective system. By contrast, I and many other scholars interested in data-intensive AI are seeing it as something quite different. On the one hand, there is reluctance to acknowledge the methodological choices and assumptions made at different points in time within the research process, since those are seemingly in tension with such promises of progress. On the other hand, the power exercised by few corporate platforms with the resources to garner, mobilize and analyze data – thereby deciding which data are valuable, how and for which purposes – is exasperating the bias, serendipity and digital divides already thriving in data-intensive systems, thereby increasing the risk of losing perspective on what data are reliable, representative and fit for purpose, and under which circumstances. We are making tremendous strides in developing large language models for translating between English, Mandarin, German or French, but could there be a comparable data processing effort to do the same for minority languages? Genomic sequencing is increasingly cheap and done on a scale that was unimaginable ten years ago, but how can we ensure that comparable attention is devoted to collecting, mining and interpreting data about metabolism, development and morphology, thereby probing alternatives to genetic determinism? Investment in clinical data on specific pharmaceutical treatments drives medical advancements, but how can the development of a comparable data ecosystem to support research on lifestyle and social interventions, which may have an equal or better chance to improve individual health and wellbeing, be ensured? Making AI less opaque and more accountable includes interrogating the make-up, evolution and future directors of the data ecosystem, taking into account the multiple goals which AI – and the underpinning data resources – are meant to serve.

List of references

Bowker, Geoffrey C./Star, Susan Leigh (1999): Sorting Things Out: Classification and its Consequences, Cambridge, MA: The MIT Press.

Council of Canadian Academies (2022): Leaps and Boundaries. The Expert Panel on Artificial Intelligence for Science and Engineering, Council of Canadian Academies, Ottawa, ON: Council of Canadian Academies (https://www.cca-reports.ca/wp-content/uploads/2022/05/Leaps-and-Boundaries_FINAL-DIGITAL.pdf).

Elliott, Kevin C./Cheruvelil, Kendra S./Montgomery, Georgina M./Soranno, Patricia A. (2016): "Conceptions of Good Science in Our Data-Rich World." In: BioScience 66/10, pp. 880–889.

Floridi, Luciano/Illari, Phyllis (eds.) (2014): The Philosophy of Information Quality, Cham: Springer.

Kitchin, Rob (2014): The Data Revolution: Big Data, Open Data, Data Infrastructures and Their Consequences, London: SAGE.

Leonelli, Sabina (2016): Data-Centric Biology: A Philosophical Study, Chicago and London: Chicago University Press.

Leonelli, Sabina (2018a): La Ricerca Scientifica nell'Era dei Big Data, Milan: Meltemi Editore.

Leonelli, Sabina (2018b): "Re-Thinking Reproducibility as a Criterion for Research Quality." In: Research in the History of Economic Thought and Methodology 36B, pp. 129–146.

Leonelli, Sabina (2021): "Data Science in Times of Pan(dem)ic." In: Harvard Data Science Review 3/1, (https://doi.org/10.1162/99608f92.fbb1bdd6).

Leonelli, Sabina (2023): Philosophy of Open Science (Elements in the Philosophy of Science Series), Cambridge: Cambridge University Press.

Leonelli, Sabina (forthcoming): "Is Data Science Transforming Biomedical Research? Evidence, Expertise and Experiments in COVID-19 Science." In: Philosophy of Science.

Leonelli, Sabina/Lewandowsky, Stephan (2023): The Reproducibility of Research in Flanders: Fact finding and Recommendations – KVAB Thinkers' Report 2022, KVAB Standpunten 81, Brussels: Royal Flemish Academy of Belgium for Science and the Arts.

Leonelli, Sabina/Lovell, Rebecca/Wheeler, Benedict W./Fleming, Lora/Williams, Hywel (2021): "From FAIR Data to Fair Data Use: Methodological Data Fairness in Health-related Social Media Research" In: Big Data & Society 8/1 (https://doi.org/10.1177/20539517211010310).

Leonelli, Sabina/Tempini, Niccolò (eds) (2020): Data Journeys in the Sciences, Cham: Springer.

Leonelli, Sabina/Williamson, Hugh F. (2023): "Artificial Intelligence in Plant and Agricultural Research." In: Alok Choudhary/Geoffrey Fox/Tony Hey (eds.), Artificial Intelligence for Science. A Deep Learning Revolution, New Jersey et al.: World Scientific Publishers, pp. 319–333.

Mackenzie, Adrian (2017): Machine Learners: Archaeology of a Data Practice. Cambridge, MA: The MIT Press.

The Royal Society (2019): The AI Revolution in Scientific Research, London: The Royal Society.

Zook, Matthew/Barocas, Solon/boyd, danah/Crawford, Kate/Keller, Emily/Gangadharan, Seeta Peña/ Goodman, Alyssa/et al. (2017): "Ten Simple Rules for Responsible Big Data Research." In: PLoS Computational Biology 13/3, e1005399.

AI in mathematics

On guided intuition and the new environment of calculation

Gérard Biau in conversation with Anna Echterhölter, January 28, 2023

Echterhölter: As a mathematician by training, as a professor at the Probability, Statistics, and Modeling Laboratory (LPSM), and as the director of the Sorbonne Center for Artificial Intelligence (SCAI), you are in a unique position to observe current changes within various sciences in response to AI. Given that research topics and new instruments arrive and fade in all of the sciences, would you say AI is currently changing mathematics?

Biau: There are different ways of answering this question. About 20 years ago, mathematics was a kind of a solo science, that you did alone in a library. Today, we are very strongly influenced by AI tools (of course this does not only concern mathematicians) which are very much having effects on mathematics in the sense that now mathematicians use Google, they communicate with each other via email, we have recommendation systems to find papers on the web, etc. Mathematicians have an aptitude and an openness for using AI tools in their research. That is the first important point.

The second point that must be underlined, is that machine learning is a real game changer for mathematicians because it is an experimental science. In mathematics, when we wrote papers 20 years ago, there were only one or two authors, it was a kind of confidential science.

Today, in machine learning papers, mathematics is part of the paper and there are five, six, seven authors, because it's experimental. The mathematician here is often part of an interdisciplinary team. This is very interesting, because, as a result, mathematicians come to play a new role.

We cannot ignore the impact of machine learning on mathematics. It can reinvent the field with new vocabulary and tools, and the new generation is totally free with this new system.

The third point is that it is an auspicious moment for mathematics. There are new fields of mathematics which have been created because of AI and machine learning. For example, until 20 years ago, high dimension was totally absent from the statistics world, or even from the world at large. Everything changed around the year 2000, when genetic data arrived. With genetic data, you have millions of dimensions that are much larger than the sample size. So new tools were invented, new seminars were created, new papers were written on this and new methods, of course, were invented for high-dimensional phenomena.

To continue, I could also cite optimization. These neural nets we are talking about need optimization to find the right parameters to make the right decision. When you were an optimizer in the 90s, it was kind of hard to find an academic position. But today, if you are an optimizer, you have offers from universities and IT companies. AI is also a game changer from this point of view. It is the same with topological data analysis, which is the field that analyses the geometrical properties of the cloud points (its density, the number of components…). Topological data analysis is something new that did not exist 15 years ago.

This is an entirely new field which is newly created in mathematics and which is an emerging area for the understanding of the properties of big data sets today. I could also talk about the so-called "physics-informed learning", a totally new field that merges data science with scientific computing. It is a mixture of differential equations, evolutionary phenomena and machine learning. These two fields are merging because now we have data, we have algorithms and we have a new point of view on these topics. New areas of mathematics have been created due to AI and this is an important lesson.

Fourthly and finally, something is emerging today that is not so present in the mathematics community, but that is there. I am speaking about the use of AI to prove new theorems or to guide intuition. This is a field that is very, very important and highly interesting. I am not certain if it will transform mathematics, but what I'm certain about, is that it will help mathematicians develop new tools.

Echterhölter: For non-mathematicians it can come as a surprise how much talk there is about intuition within this most exact of all disciplines. You have brought up the important category of "guided intuition", which describes quite a fundamental change of a mathematical research practice. Hitherto, finding new proofs has been associated with pen and paper, walks in the woods, sustained periods of concentration in solitary silence and maybe a chalkboard. Then came digital tools and especially AI. What happens to this type of problem solving? Does it give way to new modes, settings and tools of finding proof in mathematics entirely? In particular, you have mentioned applications that assist in mathematics like "Minerva". Could you elaborate on the role that this new tool, or mathematical AI, plays for mathematicians in "guiding their intuition"?

Biau: There is this tool, Minerva, which is able to solve very simple problems, say, provide proofs and answer elementary questions (Dyer/Gur-Ari 2022). It can be defined as a language model that is capable of solving mathematical problems and scientific questions using step by step reasoning. There is no tool today that can really solve very complex problem, but Minerva is an interesting step in this direction, and who knows what will happen in the next couple of years.

Beyond Minerva, today we have companies such as DeepMind that are interested in using machine learning for guiding intuition. In this regard, there was an important paper in *Nature* (Davies et al. 2021), in which they used machine learning to propose new relations in pure mathematics, thus allowing the mathematicians to verify relations suggested by the computer. The authors use data to discover potential patterns and relations between mathematical objects and use these observations to guide intuition and propose conjectures. This is a new type of collaboration between AI and mathematics.

In October 2022, just one weekend before the Paris conference, there was another breakthrough, another DeepMind paper in *Nature* (Fawzi et al. 2022). This time, the computer used machine learning to find a new way to multiply matrices. It is crucial to stress that multiplying matrices is a very important concept for machine learning, as it is full of matrices! It is very difficult to have efficient and clever ways to multiply matrices in order to save time and space. Now, DeepMind's algorithm was able to find a new way of multiplying two 4x4 matrices, which is already something, suggesting new algorithms. This is an important step.

You asked me if my discipline changed because of this? The answer is no, at this point. But maybe one day, a computer will probably help mathematicians. Peter Scholze, a German mathematician and a Fields Medalist in 2018, a great man, is working at the interface between algebraic geometry and topology. One of his recent proofs was verified by a computer and also presented in *Nature* in 2021 (Castlevecchi 2021). That was a big achievement, because it was the first time that the full proof, a very complicated proof, was certified by a computer. So, the computer is helping, even if it's not really AI-based.

All in all, the real moment will come when machines will propose a new proof, or a proof of a theorem, that has not yet been proven. That would be a real breakthrough for me. For now, what do algorithms actually do? They just look at many, many papers on the web and, without understanding, imitate what they find in them. This is already something, there is clearly intelligence at play here. The big breakthrough will happen when the machines will, as I said, suggest a new proof, make a connection between two areas of mathematics, or suggest a new way of looking at a problem. These things are typically human. I haven't seen this in a machine before.

I assume this is the same with literature, with art... When the machines will propose something that we have never seen before, then we, the mathematicians, will be in danger somehow (laughs), but for now I'm not worried.

Echterhölter: How are these new AI applications different from once successful software like Mathematica and the computer as a numerical tool?

Biau: Software such as Mathematica or MAPLE, which have been used for a long time by mathematicians, are very different in that they perform complicated calculations and operations as directed by the operator, i.e. the mathematician. They are therefore very valuable tools to help mathematicians perform difficult calculations and simplify results. However, they work very differently than the algorithms that I mentioned above, which use data to propose new results to mathematicians. Eventually, of course, all these tools will converge.

Echterhölter: Would you say that some groups within mathematics are more open towards using these new tools and turning to the guidance of machines?

Biau: The question of how AI is changing the way we think is very interesting. One way of looking at this would be to observe how students, the young generation, behave. I have seen a major change with my students in mathematics, graduate and PhD students. The way they do mathematics today is entirely different from the way my students did it 10, 15 years ago. Now, they are fully integrating and utilizing new tools, for example to compute a series, or to prove that a function has a given property... We do not even attempt to prove it with mathematical arguments, we just trust the computer. It is a new way of learning mathematics, which fully integrates the machine as part of the process.

Echterhölter: You served as the president of the French Statistical Society (Société française de statistique) from 2015 to 2018. This was a time during which programming libraries for AI multiplied and the public got an idea about what was going on in the aftermath of Alpha Go's win over a human at the traditional board game. Deep learning produced its first staggering results, although it had been around much longer. This success entailed a shift in the underlying statistical approaches. One general transformation seems to have been from Markovian models to convolutional neural networks and mass data approaches. Given that AI has a statistical anatomy, how did the statistics community react to this new heyday of AI after its 30 years of winter? Did AI have immediate adversaries among statisticians?

Biau: Statistics today finds itself in a rather paradoxical situation. On the one hand, it is indispensable for the understanding, analysis and implementation of modern machine learning methods, which are all based on data and therefore on techniques involving the science of randomness. On the other hand, the application conditions of statistics within machine learning are very different from its usual perimeter, since statistics is now confronted with models of gigantic dimensions and ever larger sample sizes. It is therefore a real challenge for statisticians today to be able to answer all these new questions! To do this, they have to adapt their tools, devise new methods and develop concepts, some of which have not changed for several centuries! But rest assured, statisticians are adapting perfectly to this new world and I am impressed by the speed at which the discipline is evolving. The younger generations of statisticians have perfectly understood the issues at stake and I have no worries about the future of the discipline.

Echterhölter: What is the specific relation of statistics to data, in comparison to mathematics, and does this specific relation change at all just because of AI? In the 19th century statistical societies in many countries produced and collected data, and did not just develop stochastics. One precursor to the French Statistical Society is a good testament to this rule: a founder of the "Société de statistique de Paris", Louis-René Villermé, was among the first to formulate the social question from 1860 onwards, and did so by backing up his claims about the health of workers with detailed numbers and data. Historically speaking, this means that statistical societies were as much about observing, describing and criticizing society through numbers, as they were about developing mathematical methods. During the 20th century this clearly changed, but does the discipline of statistics have to maintain a specific relationship to data and databases?

Biau: This is an interesting question. While data is at the heart of statistics (indeed, etymologically, the word statistics comes from the German word "Staatenkunde" (the knowledge or science of the state), the latter has tended to evolve, around the 90s of the 20th century, somewhat away from reality, towards what is known as mathematical statistics, which encompasses the abstract study of models of inference and prediction. Interesting as it is, mathematical statistics does not really touch reality and remains in the ideal world of mathematics. But all that is changing today with the need to implement concrete and efficient methods for dealing with astronomical amounts of data. In some ways, this is a return to the roots for statistics, which must focus on its original raw material, namely data! In a way, statistics can thank AI.

Echterhölter: What does risk assessment for this new technology of AI look like in your research community? For instance, is this new technology a threat to some fields of mathematics? Do topics within statistics go extinct because of it? And to look beyond the ivory tower, how are hazards beyond mathematics discussed in your community?

Biau: Of course, we can talk about the amazing progress of machine learning in computer science and mathematics, but we could also talk about the progress of GTP-3 and other tools such as DALL-E… and how they are changing science.

Behind all this, however, there are some very important issues that need to be addressed. Ethics of course, but also sustainability and environmental issues. Consider, for example, that the training phase of GPT-3-based versions of ChatGPT emits tons of CO_2. This is something we should be aware of when we use these tools. The amount of energy needed for this type of algorithm is just crazy, and I'm not talking about all the energy used in the data centers!
Moreover, there is also the very important question of social acceptance of AI. We are increasingly becoming slaves to algorithms, not only in our science, but in some ways in what we eat, how we drive, how we meet, the internet, etc. Is

this really what humanity wants? We have a lot of social problems in the world today and while I'm obviously no expert in sociology, I can't help but think that behind some of these problems is a widespread fear of a world that is becoming increasingly dehumanized by technology and AI. I think this is something fundamental that we need to think about seriously.

List of references

Castelvecchi, Davide (2021): "Mathematicians Welcome Computer-assisted Proof. Proof-assistant Program Handles an Abstract Concept, Revealing a Bigger Role for Software in Mathematics." In: Nature 595, pp. 18–19.

Davies, Alex/Veličković, Petar/Buesing, Lars/Blackwell, Sam/Zheng, Daniel/Tomašev, Nenad/Tanburn, Richard/et al. (2021): "Advancing Mathematics by Guiding Human Intuition With AI." In: Nature 600, pp. 70–74.

Dyer, Ethan/Gur-Ari, Guy (2022): "Minerva: Solving Quantitative Reasoning Problems with Language Models." June 30, 2022 (https://ai.googleblog.com/2022/06/minerva-solving-quantitative-reasoning.html).

Fawzi, Alhussein/Balog, Matej/Huang, Aja/Hubert, Thomas/Romera-Paredes, Bernardino/Barekatain, Mohammadamin/Novikov, Alexander/et al. (2022): "Discovering Faster Matrix Multiplication Algorithms With Reinforcement Learning." In: Nature 610, pp. 47–53.

Artificial Intelligence as a cultural technique

Sybille Krämer in conversation with Jens Schröter, March 5, 2023

Schröter: You published a volume called "Mind, Brain, Artificial Intelligence" back in 1994. How has your view of so-called 'artificial intelligence' changed since then?

Krämer: I was – and still remain – convinced of the culturally shaped exteriority of the human mind: Having a brain is a necessary, but by no means the sufficient condition of our cognition. To think it is not a purely mental process in the head but is characterized by three other aspects: (I) the use of language and tools, (II) the social interaction with others, and (III) our corporeality and metabolism-based embeddedness in the ecosystem of our planet. This is the horizon in the 80s/90s when Artificial Intelligence (AI) aroused both fascination and criticism in me.

The fascination was based on the fact that rule-based symbol processing in the form of 'symbolic machines', which was practiced as a human intellectual technique long before the invention of the computer – for example in written calculation or logical deduction – always characterized a subarea of human problem-solving. To see how far machines with this paradigm of symbol processing can be developed – in the 80s these were the Expert Systems as a spearhead – does not reveal how human-like these machines work, but vice versa how machine-like humans have organized and still organize some domains of their cognition. So the remarkable fact for me about the then prevailing form of AI was not at all that computers can model the brain (according to the formula brain and mind like hardware and software) but that they adapt or simulate a cultural-technical practice, namely the handling of written symbols. It is not by chance that Alan Turing (1950) explicitly makes the human calculator, which enters, rearranges, and deletes symbols on checkered paper, the model of his mathematical-technical concept of the Turing machine. The difference is that the checkered paper has now become an endless tape.

On the other hand, my criticism was directed towards the myth of 'disembodied intelligence', associated with the symbol-processing approach of AI, as soon as this is generalized as a model of human thinking and our being-in-the-world. This was one of the critical arguments of Hubert Dreyfus (1972), thereby going back to Heidegger. By virtue of our bodily situatedness, we have a primordial relation to the world that is independent of explicit symbol processing, a pre-symbolic intuitive understanding that implicitly structures our practices. Then, something came to the fore that marked the limits of the symbol-processing paradigm.

This was roughly the tableau of my initial involvement with AI at the end of the last century.

However, with the mass data made possible by the Internet, social platforms, and ubiquitous computing – used to train artificial neural networks, especially in Deep Learning – the role of AI in society has fundamentally changed. Here are some symptoms of this change:

I. Artificial Intelligence has arrived in everyday life (search engines, face recognition, spam filters, navigation, chatbots, etc.) – in other words, it is not only used as a selective expert system. In everyday applications, it mostly remains hidden from users, often – though not always – operating below the threshold of perception. This is changing with Large Language Model-based chatbots, which respond to colloquial prompts and thus advance to an everyday technique.
II. The increased generative potential is conspicuous since both images and texts can be generated today with natural language instructions, each of which is unique, not plagiarized.
III. Since learning systems are no longer explicitly instructed via programming, but are primarily trained by sample data and error feedback, the internal models formed in the process remain opaque: The area of non-knowledge in systems, that nevertheless function well, is growing.
IV. Early AI was seen as a tool to uncover functional processes of the human mind (= symbol processing) or brain (= connectionism). Now, learning algorithms teach us about the discriminations implicit in our social practices that condense into training data. By practically executing biases represented in training data, algorithms at the same time bring them into the open.
V. Statistical language analysis and language generation have superseded attempts to model semantics, meaning, and comprehension. The ap-

proaches of the Large Language Models, especially the 'family' of ChatGPTs, show: What the machine generates is not based on understanding, but on the statistical combination of elementary tokens (small groups of letters below the level of meaning) according to the most probable linkages. Thus, the astonishment, in how many respects ChatGPTs can produce plausible texts, corresponds to the insight that precisely no intelligence is required for this. What is necessary, however, is combinatorial access to billions of texts – which is not feasible for humans – in order to create products whose reference to reality is fictional – i.e. without any claim to truth. Does quantity – the unsurpassable large training data volumes – turn into quality here? Or has the demarcation line between quantity and quality become questionable in general?

Schröter: How would you classify the development of so-called 'artificial intelligence' in the history of formalization that you have studied in detail? Today's dominant machine learning methods belong to a rather statistical paradigm – does this belong to the history of formalization or rather not?

Krämer: Formalization does not mean calculating with numbers, but manipulating graphic signs according to given rules. The philosopher Leibniz first articulated this distinction (Krämer 2016). In written reckoning, the eye, hand, and brain work together and create a 'machine room of intelligence' that consists of formal pattern manipulation and is independent of using a real physical machine. The signs can represent numbers, but they do not have to. The procedure itself is an interpretation-independent operation of forming and transforming strings of signs. In memory of handwritten calculating: If a table with one and one, one minus one, one times one, one divided by one is available, then elementary arithmetics can be carried out with paper and pencil, without having to know at all that numbers are processed. This, at least, is the sense of formality that emerged with the development of mathematical and logical calculi in the modern era. Of course, formalization has no end in itself: If a consistent object domain is discovered as a reference domain of a calculus, domain-specific problems can be solved formally and new insights can be gained.

This being said, any operation with numbers, regardless of how the calculation is performed and whether probability and statistics play a role in it, is necessarily formal. How formality and statistics are related is exposed when the sentence is correctly understood that in 2021 each woman in Germany had 1.58 children.

But we had to add another dimension with regard to the relationship between machine learning/statistics and formalization. It is the transition from problem-solving to predictive algorithms, which is crucial for contemporary digitization. Problem-solving algorithms determine a result in a stereotypical mechanical way: By applying the rule of calculation correctly, the result will be correct too. You can 'trust' the algorithm. Predictive algorithms, on the other hand, refer to the future and predict the probability that a possible event will perhaps occur. Already in the case of problem-solving algorithms, the 'knowing-that' splits from the 'knowing-how' in the application: The knowing how to do something becomes transparent, teachable, and learnable; the knowing why it works remains hidden and is at best transparent to mathematicians, but not to the calculators.

In contrast, in predictive algorithms, the machine acquires a knowing-how in the form of an internal model, i.e. the functional competence to make an input correspond to an output. The 'knowledge' implicit in this internal model usually cannot be inferred from the output and remains opaque; apart from that, these internal models change with every use and in innumerable permutations. Moreover, with predictive algorithms, the social and political importance of the presupposed labelling grows, i.e. the mostly human selection and marking of training data as well as the social scaling of thresholds in the internal model building.

We see: Every algorithmization implements and embodies a specific relationship of knowledge and non-knowledge, of transparency and opacity; but in predictive algorithms, the domains of non-knowledge and uncertainty radically increase.

In view of this situation, doesn't the idea of 'Explainable AI' also create an illusion? Do we perhaps have to radically change our attitude and perspective with regard to the relation between knowing and not knowing? Is it not rather a matter of reopening the fundamental questions of knowledge/non-knowledge, of acting under uncertainty, and all this in the opposite direction too: A medical doctor interpreting an X-ray is much more likely to act under the sword of Damocles of uncertainty than a system trained to make these diagnoses with thousands of analyzed X-rays. Are common terms like 'knowledge society' emphasizing enough that every new knowledge creates new not-knowing, that we cannot always eliminate uncertainty but have to learn how to deal with it? And that human action cannot escape this ambivalence?

Schröter: How would you relate to the development of so-called 'artificial intelligence' in contrast to the somewhat fuzzy discourse of 'digitalization'? How would you relate to the assumption, that at least neural networks are rather analog technologies, again because of the finely graded weighting of the activity of artificial neurons, and because of their parallelism (cf. Sudmann 2018)?

Krämer: The digital exists – this may come as a surprise – before and independently of the computer. By digitization, I mean a process in which a continuum is broken down into basic elements and discretized so that they can be coded and combined with each other in a more or less arbitrary way. A prototype for digitization is the alphabet. Although the flow of oral speech knows breaks, they do not correspond at all to the blank spaces between words and sentences in alphabetic writing. With the finite repertoire of alphabetic characters, an unlimited number of combinations can be produced in the two-dimensionality of a surface. This non-linear 'nature' of writing is revealed for example by the phenomenon of the crossword puzzle which exists only as a two-dimensional, graphic medium illustrating the novel configurations that spatial writings open up in comparison to temporal speech. Moreover, alphabetically ordered lists sort large amounts of information, think of the traditional telephone directories, which allow casual access to amounts of data that cannot be surveyed by humans. A 'database principle avant la lettre' developed in social practice is already being applied: the abandonment of narration in favor of formal sorting and addressing of pieces of information that are independent of each other. This database principle gave rise to the academic flagship projects of print-oriented modernity in the form of dictionaries, encyclopedias, and lexicons.

Let us summarize. Two things are important with regard to my concept of digitization:

(1) There is an embryonic digitality already connected to alphanumeric literacy. This does not only apply to the European alphabetization: The hexameters of the Chinese Book of Changes "I Ching", for example, are written with dual code, which can be translated without constraint into machine-processable Unicode. The digital is to be understood independently of computer use.
(2) The relationship between analog and digital is relative in so far as we understand it in terms of the continuous-into-discrete transformation. The transition from fluid speech to discrete writing is a transition from an analog

> to a digital medium. But if the transformation from a printed text to a machine-readable and -analyzable document encoded in TEI, is considered, then the printed typeface is in the role of the analog and only the encoding instantiates the process of digitization.

In a significant way, the connection between digitality and Artificial Intelligence is clarified by their latest development: The already mentioned contemporary chatbots in the context of Large Language Models (GPT-4, Bard etc.) operate on the basis of small, meaningless groups of letters, the 'tokens'. Here, too, we are dealing with the decomposition of something continuous into smaller meaningless units. Hardly anything can better illustrate how 'deeply' the techniques of Artificial Intelligence are allied with the digital, understood as a process of dynamic discretization.

It should be recalled that linguistics characterizes human language by its 'double articulation'. From a limited repertoire of meaningless elements such as phonemes or letters, an unlimited number of meaningful words and sentences can be formed. The question arises if a digital principle is already nested in spoken language – at least implicitly. However, there are good reasons to assume that the phoneme is the result and product of the grapheme, the smallest written unit. In fact, only the emergence of phonetic writing has split and divided communication in its totality of prosody, mimic, gesture, deixis, and verbality and crystallized the phonetic dimension as an independent communicative strand and condensed it to an object like perceivable 'language'. If this is true, it would be the writing that puts the grid of digitizing over human language.

And one last remark: If your question aims at a possible return of the analog by artificial neural networks, I am skeptical about any neuromorphic diction and rhetoric. Bird flight also inspired human flight experiments, without airplanes imitating the natural model. Is it not the same in relation to natural and artificial neural networks? Everything that matters in contemporary Artificial Intelligence, is mostly not programmed but trained by huge databases, and what can explain its technical power is something that finds no role model in nature. The procedure of error feedback, for example, which has an analog in the social practice of teaching when corrected dictations are returned, finds no parallel in neurophysiology. Or with regard to the architecture of the hidden layers – a central component of the Deep Learning process: If each layer analyzes selected aspects of the input with different weighting, or if these computational processes take place in the layers one after the other – all this also has no analog in our brain. Not to mention, by the way, the energy efficiency that is so typical for our brain.

Schröter: What role do you think methods of so-called 'artificial intelligence' could play in the field of digital humanities? How could machine learning be used in the cultural sciences and humanities, and even in philosophy?

Krämer: In this context, I'd like to talk about the 'sting of the digital'. What I would like to express here is that the debate about the Digital Humanities and their acceptance by the traditional humanities can provide impulses for a self-correction of the humanities' self-image. This self-correction refers to the absolutization of hermeneutics and interpretation as the royal road and definiens of the humanities (Krämer 2023). Furthermore, using 'sting' as a metaphor refers to criticizing the belief that the humanities have nothing to do with empiricism or with material and quantifiable things and processes. Incidentally, both of these biases have already been subject to erosion in the late last century, even independently of the emergence of Digital Humanities.

The humanities' disciplines encompass not only the traditional fields from history to linguistics, literature, music, and art studies, but also archaeology, ethnology, and even regional and cultural studies. They have always worked with materials, that is, with things, documents, and artifacts of all kinds, which are to be collected, dated, classified, annotated, compared, archived, and so on. In this ecosystem of scholarly work in the humanities, empirical questions – and thus numbers and counting – always had a certain status. But the traditional humanities with their hypostasizing of interpretation as key methodology, have long remained blind to the materiality of their research objects and consequently to the importance of numbers and countability in many subfields of their research.

Nevertheless, it is precisely here that research questions open up that can be meaningfully addressed by the Digital Humanities under the conditions of contemporary digitization. This is always the case when large data corpora, which relate to lifeworld and/or cultural-historical contexts and can no longer be surveyed, let alone examined, by human eyes and hands, can now be analyzed with data-driven, computer-based methods. However, this is only possible through the subtle, difficult, never-ending interaction between researchers and computer-generated, data-driven procedures. It goes without saying that interpretation on the part of human actors is constantly involved: no number – and no data – interprets itself.

In prosaic terms, the question of sense and nonsense of the Digital Humanities could be transformed into the question of what role empirical questions play in the respective discipline. Against this background, it is not surprising that datafication and digitization first took hold of the natural sciences and, in the 20th century, also of economics and the social sciences, before it has now arrived in the 21st century humanities. Perhaps the discussion about the legitimacy of Digital Humanities serves as a proxy function for the less exciting question of when and how the empirical can or should gain a birthright in the humanities.

We must not make the mistake of reestablishing C.P. Snow's two-culture difference (Snow 1959), which is unacceptable today, within the humanities. Even the traditional humanities have always been dependent on dealing with numbers and data, think of concordances that have existed since the 13th century, catalogs of works or historical dating, etc., just as, conversely, the Digital Humanities always have to interpret their results in the light of their research questions. There is no such thing as interpretation-free empirics.

In the opposite direction, however, I also find problematic contemporary attempts to identify and ennoble computational procedures themselves as hermeneutic procedures, as Dobson did in 2019, for example, in order to provide the Digital Humanities with legitimacy in the Humanities. As already emphasized, I am more inclined to weaken the hermeneutic paradigm as a unique selling point of the Humanities by recognizing that their academic practices include a plethora of activities in the preparation of their research objects that precede and prepare the ground for interpretation in the first place.

However, there is an interesting and revealing addition to this statement. Computers are forensic machines (Kirschenbaum 2012), like microscopes and telescopes directed toward the data universe to find patterns that mostly escape human attention. Of course, the optical analogy is limp insofar as it ignores the generative aspect of processing and synthesizing music, images, and text. However, what is at stake in explaining the forensic function is the dimensions of the culturally unconscious. What people miss in their practices, a machine can register.

This can be explained by the computer-philological example of author attribution. If styles of individual authors become identifiable by means of a ranked list of the 'incidental' functional words used – how often are words like 'and', 'nevertheless', 'however', etc. being used? – then the machine is able to identify an author by attributes of his or her use of language that is not at all part of the stylistic devices intentionally employed, but rather is subverted in writing and occurs unconsciously in the performance of written articulation. It is not about something that is hidden behind what is written, but that is given in what is written down. It is implicit in the surface of the text and can therefore be taken from it.

What emerges here within the dimension of author attribution is generalizable: Despite the use of terms such as 'Deep Learning', information processing technologies – also in the form of Artificial Intelligence algorithms – are a surface technology for the identification, analysis, and production of patterns. What is true for numbers and data is also true for patterns: Whether patterns have meaning, sense, and relevance, be it for life or for a research question is up to humans to decide, applying the pattern discovery capacity of the machine for their specific purposes.

It has hardly been registered so far that 'close' and 'distant reading' converge in this question. The cultural scientist Carlo Ginzburg (1983) – as a micro-historian, he was an advocate of close reading – saw a 'circumstantial' or 'indication paradigm' emerging as a methodological dispositive of the humanities in the transition from the 18th to the 19th century. The inventor of the detective Sherlock Holmes, the author Arthur Conan Doyle, the art historian Giovanni Morelli, and the psychoanalyst Sigmund Freud developed their insights by studying unnoticed details at crime scenes, in faked paintings, and in traumatized souls. In this way, Ginzburg was able to show why Doyle's detective novel became the most successful crime novel series: because readers are involved in the process of finding clues. The propagandist of distant reading, Franco Moretti (2013), in turn, by comparing all detective novels in Doyle's epoch (a fact Ginzburg could not have had an overview of), comes to a very similar conclusion, namely that of the exceptional position Doyle's "Sherlock Holmes" novels had.

The micro perspective of close reading and the macro perspective of distant reading are not opposing perspectives but can complement each other. Furthermore, something else becomes clear here: Statistical methods are often reproached by the humanities because they only represent the average and are therefore an instrument for the enforcement of mediocrity and the renunciation of creativity. However, statistically operating computational methods do not only calculate average and mean values, but by virtue of this computational capacity they can also uncover the knitting pattern of the individual from a most unusual perspective, just as forensics can uncover a singular course of events or author attribution can uncover author identities. However, this always works only probabilistically, i.e., by a probability statement. In short: Statistics is not the enemy of casuistry and of the individual case, but – used sensibly – can be precisely its aid.

Schröter: Can so-called 'artificial intelligence' be described as a 'cultural technique'? Or does it rather presuppose certain cultural techniques?

Krämer: Every technology is socially constituted and thus a cultural phenomenon. And yet a distinction must be made between 'technology' and 'cultural techniques'. In the context of the in 1999 started Helmholtz Center for Cultural Techniques in Berlin – I was a member of the eight-member founding group – the term 'cultural technique' aimed to orient research in the humanities more strongly towards the materiality, mediality, and technicality of their research objects. In this Helmholtz group, cultural techniques were regarded as routinized everyday procedures for dealing with symbolic and technical artifacts that are sedimented in everyday practices, the mastery of which provides a basis for social participation, but also for social differentiation. Cultural techniques are crucial resources of scientific and artistic practices and also underlie higher-level cognitions.

We are familiar with the fact that writing, reading, and calculating are cultural techniques of the era of printing. From this point of view, it is obvious that digital literacy implies a decisive development of those cultural techniques that have been typical for alphanumeric literacy in the 'Gutenberg Galaxy'... The elementary handling of keyboards, smartphone use, the ability to communicate by email, and, above all, to search for information on the Internet are decisive aspects of contemporary digital cultural techniques, without which participation in social life is hardly conceivable. At the same time, these are practices at whose mastery or non-mastery fault the lines of contemporary society emerge, both socially, but also generationally. But does this also include the processes of Artificial Intelligence?

For the era of Expert Systems – i.e., in 'woodcut' terms: the AI of the last century – I would have answered this firmly in the negative. But precisely because contemporary Artificial Intelligence has seeped into our everyday behavior in many different forms, the situation has changed. Without streaming, navigating, searching the net, online banking, spam filters, etc., contemporary participation in everyday life seems almost impossible to realize – although in principle this remains possible, just as illiterate people can lead a special existence in literal cultures. This dependence on the cultural techniques of Artificial

Intelligence also applies to complex mental work: Without computer-generated visualization, medical diagnoses and operations are hardly feasible anymore, stock market trading thrives on real-time analyses, driving assistants in cars have become standard, and fitness watches control training and mobility. A significant step in everyday usability of AI is the software trained with large data corpora, allowing users to instruct image and text generation with natural language – and its colloquial character is important.

However – and this also seems to be a novelty in the degree of the associated dangers – Artificial Intelligence procedures often run as background processes that are hardly registerable for users, let alone recognizable and accessible. In a harmless dimension, when taking photos with a smartphone or in the use of auto-correction functions, but more problematically in the creation of personal data profiles as 'waste products' of Internet navigation.

Artificial Intelligence nowadays is implemented into the use of apps, objects, and procedures. The cultural technique consists in being able to deal with virtual objects in a functionally and factually appropriate way without having to understand how this use of data can be exploited in a functionally and factually non-intended, but commercialized way. What I have characterized as the dispositive of technology use – i.e., being able to control and use without having to understand – acquires an ethical-political signature here. Can we conclude from this that the cultural technique of Artificial Intelligence also consists in learning how to preserve data sovereignty? Or is this idea of sovereignty, rooted in the European Enlightenment with its maxim of 'thinking for oneself', an illusion – and perhaps was from the very beginning? For it is precisely the suitability of these everyday applications which become smarter with each use, that is in turn restricted, if not hindered, by mechanisms of data protection: Who isn't annoyed by the popping up of the cookie consent form, which degrades data sovereignty to check-marking? How much more helpful could digitization be in Germany if patient data or even the data available in administrations were merged? A dilemma is emerging between smart everyday usability and responsible handling of Artificial Intelligence's 'background cultural technology'. 'Dilemma' is understood here as a conflict situation and a predicament that cannot simply be transformed into a positive solution.

Schröter: Would you see the use of machine learning in different sciences as a kind of upheaval – or rather as a continuation of the increasing role of computers in the sciences (e.g., in the form of computer simulation)?

Krämer: Wherever the dynamics of media innovations are concerned, they are always to be understood in the tension between continuity and breakup, between tradition and disruption.

To give a distant example: The absence of book religion in ancient Greece allowed written texts to advance into a non-canonical discursive space debating the pros and cons of truth claims. What was previously known only from the oral practices of court proceedings in Greece, was now transposed into a written medium. Thus a type of text emerged, often in dialogue form as in Plato, which insisted on arguing about truth – and this became a relevant starting point for the Western type of philosophizing. This change is often called the transition from orality to literality, a highly problematic thesis, in whose garb mostly the Eurocentric assumption of the superiority of alphabetic writing was transported. Of course, orality is not replaced and made obsolete by literacy. Rather, writing opens up a symbolic space in which new ways of using and dealing with language become possible. And the oral also takes on new signatures, for example in the genre of the scientific lecture.

But back to the digital: Undoubtedly, the computer is currently becoming a universal tool in the sciences, from simple word processing to computer simulation. I use the word 'computer' here as a chiffre for the ecosystem of scientific information processing based on ubiquitous datafication. To stay with computer simulation, it is not simply that computer simulation now joins experiment and theory as a third research pillar in the sciences. Rather, this simulation opens up a new kind of mediation between analytical theory and empirical experiment: Experimenting with theories becomes possible (Gramelsberger 2008) and gives rise to a 'theory laboratory'. Computer simulation opens up a space in which traditional instruments of knowledge such as theories and experiments gain a new profile, combined with new options for knowledge.

Under the conditions of extensive datafication on the one hand and 'learning algorithms' on the other, this new profile is that computers can work with mass data in ways unattainable by human power. The forensic capability of computers, familiar with criminalistic use, can now be extended to many areas of scientific research, where it can be used to uncover patterns that are beyond human perception.

If the computer acts like a microscope and telescope on datafied worlds in data-driven research methods, then data corpora reveal and uncover what remains invisible to limited human perception. These computer-processable traces are mostly statistical, hence numerical constellations. And since neither traces nor data and certainly not numbers are self-interpreting, it is clear that only the research motivation, creativity, and synthesis of human interpreters can produce meaning and content from these traces, data, and numbers. Humans combine computer-generated results with theses, theories, and narratives and thus turn data processing into knowledge production.

Therefore, the question of the relationship between continuity and upheaval, between continuation and innovation in the scientific use of computers must be answered with a 'both/and' – as is usually the case with disjunctive questions.

The continuity of the development is unmistakable: It is well known that machine learning and the imitation of the human nervous system played a role already in 1956 at the conference at Dartmouth College, where McCarthy introduced the term 'artificial intelligence'. Turing had already raised these questions in the 1940s. In 1957, Frank Rosenblatt conceived the first artificial neural network with the Perzeptron; in 1966, Joseph Weizenbaum created the first chatbot with Eliza – and shook up the humanities scholars at the latest as a result of the illusion evoked by users of Eliza that an empathetic human was speaking here. Over the years, many other stations were added: Expert Systems in medicine, oral speech synthesis, winning chess, Go and quiz programs, chatbots such as Siri and Alexa, and finally, the image and text-generating artificial neural networks based on Deep Learning methods, training, and testing: Artificial Intelligence – regardless of its many slumps and crashes in the public consciousness and the seasonal metaphors like 'winter of artificial intelligence' that are readily used for this purpose – forms Ariadne's thread in the history of technology and science of the last decades.

Nevertheless, there is also an innovative, disruptive dynamic – and its symptom is the cultural-technical embedding of Artificial Intelligence in everyday practices. This cannot be monocausal traced back, for example to the use of Deep Learning processes from around 2012, but includes at least two other indispensable components: the datafication, doubling our world into the shadow image of a computer-processable data universe, and the extremely increased computing power of the hardware. The Deep Learning procedures become better and better with each increase in the amount of data – which was not true of machine learning in the early days of Artificial Intelligence – and increased amounts of data, in turn, require increased computational power, and so on. From the swirling dance of these three conditions with each other, has now entered the family of Large Language Models to the public; this has already been interpreted as the 'iPhone moment' of Artificial Intelligence. It is also significant for interpreting Artificial Intelligence now becoming a cultural technique, that it was OpenAI enabling the download of ChatGPT for all interested people (100 million users after only two months). All big players in this field will go to the market with their own versions, and Microsoft already announced to incorporate Large Language Models into its Outlook and Office programs. Search engines – but they were that before, ergo: continuation and break!

Schröter: How can the already so-called 'artificial intelligence' be placed in the history of the 'exteriority of the mind', which you have been investigating for quite some time?

Krämer: We are familiar with understanding humans as meaning-giving and symbol-oriented living beings who constantly interpret their world. Who would and could contradict this? But does looking for meaning and interpretation take it all? Civilizations develop by increasing the areas structured in a way that is independent of interpretation, reflection, and understanding. This is true not only for formal operations in the context of intellectual work but also for ritualized everyday practices. We celebrate Christmas even without a Christian message, drive cars without an understanding of technology, cook without an awareness of chemical interactions, and successfully apply computational algorithms. Alfred Whitehead (1911) remarked laconically at the beginning of the 20th century that the level of development of a civilization is shown by how many of its important operations can be performed without thinking about them.

Let us note: The dispositive of the use of technology consists in being able to apply and control without having to understand. And exactly this technical dispositive is transmittable to subareas of mental work too.

In addition, there is the collective character of the mind: Humans do not simply have natural intelligence but participate in different degrees in the socially shaped and distributed mind, acquired, passed on, and handed down in the collective. Our cognitive capacity can only be reconstructed as social epistemology. It already starts with an almost trivial fact: 85 percent of what we know, we cannot verify and justify on our own, but we acquire this knowledge through words, writings, and images from others. And trust is that very bond, the 'glue' that turns received information into knowledge for us. Here, with the knowledge machines of AI, an important moral problem emerges: How far can we trust the apparatus and the algorithms? Not at all in the case of the ChatGPTs, which generate their plausible-sounding texts as purely fictional products without any reference to reality, without any internal truth check (work is being done to change this). These machines have no mind, and no understanding, but calculate the probabilities of small tokens and word patterns.

Back to the question of the extended human mind: Without the exteriority of auxiliary means, starting with spoken language, including the manifold forms of visual signs, up to ornaments, pictures, graphs, diagrams and maps, scientific cultures and other functional areas in society would be unthinkable. To paraphrase Ludwig Wittgenstein: Why do we say that our thinking is located in the head and why do we not say that the speaking mouth or the writing hand is thinking too? We do not think on paper, but with paper.

In the context of the human mind's evolution in the interplay of eye, hand and brain, the cultural technique of flattening plays a central role. Here, 'flattening' is not meant in a pejorative way, but rather in a sense that inscribed and illustrated surfaces embody an irreplaceable, often creative, potential as a workspace for designing, as a thought laboratory, or as a workshop for composition and combinatorics. Just as we use geographic maps to orient and move in unfamiliar terrain, the diagrams and graphs of science provide a cartographic impulse for orientating and operating in conceptual spaces of knowledge: invisible entities, and non-spatial abstractions become representable and processable in two-dimensional spatiality. Our conception of time is also rooted in this potential for spatialization; we need only to think of the historian's timeline or the measuring of time by clocks. The inscribed or illustrated surface as a medium in between temporal one-dimensionality and spatial three-dimensionality is a translation manual from time into space and vice versa. To avoid misunderstanding: There are no flat corporal objects empirically, yet we treat inscribed and illustrated surfaces as if they are two-dimensional. Given the diagrammatic practices of knowledge, we realize how strongly the computer and the digital are linked to the exteriority of artificial flatness.

This is not only true for computer programs, which have to be written down before they can be used as machine instructions; it is also true for the model of the Turing machine, which works with a tape that can move back and forth, or is true for the multiplication of surfaces, which is typical for the architecture of the 'hidden layers' in Convolutional Neural Networks, and it is not least true for all the visualizations that are necessary to transfer computer-generated outputs into a form that can be understood by humans. And this applies basally already to encoding in TEI: Implicit reading conventions that we master as tacit knowledge by distinguishing and recognizing headings, footnotes, paragraphs, and proper names from one another in a text, must be made explicit line by line when encoding into a computer-processable script. The computer is a surface technology; therein lies its power and its limitations. As a microscope and telescope into the data universe it is unsurpassed – but also only within the data universe. What is not in this universe, does not exist for the computer.

Schröter: In 1998, you published the beautiful volume "Medien, Computer, Realität" (Media, Computers, Reality). The subtitle was "Concepts of Reality and New Media". What 'conceptions of reality' are associated with so-called 'artificial intelligence'?

Krämer: The idea that explaining our brain is to think along the lines of computerized operations, i.e., that phantasm (of the beginnings of Artificial Intelligence) to assume that the computer is the appropriate model for the human mind, is taken ad absurdum precisely because the latest chatbots are based on Large Language Models. The fascinating range of text genres produced by chatbots is – as we all know – free of all understanding on the part of the machine. The machine does what it does best after being fed huge corpora of Anglo-Saxon training data to calculate probabilities of letter tokens and word combinations.

The idea that technical apparatuses and processes displace and substitute people is problematic too. What AI actually demonstrates, is that we have to understand the relationship between humans and technology as co-performance – as a shared activity and interaction. Could we go so far as to think of human/machine interaction under the precinct of contemporary digitization, according to the model of alternating moves that are performed in a game?

Therefore, the talk of so-called 'self-learning programs' is distorting. Even when a computer defeated the four best poker players in the world, the winning program Libratus still had to be trained at night during the competition by its creators on the basis of game data. Rainer Mühlhoff (2019) elaborated on the socially distributed nature of Artificial Intelligence by pointing to the work armies of cheap click workers whose job it is to label the training data. In processes like CAPTCHA, where we have to read distorted strings or to name image objects to prove and identify ourselves as human, we fill the pool for training data of learning algorithms in involuntary pandering.

Schröter: It has become a standard argument to criticize so-called 'artificial intelligence' on the one hand because of the 'bias' of the data sets and on the other hand because of the lack of 'explainability'. In your opinion, are there other important criticisms of 'artificial intelligence'?

Krämer: First, the short answer. There are at least 3 points of view:

I. The resource problem: Artificial Intelligence in the mode of artificial neural networks or so-called 'self-adaptive algorithms' require immense data corpora. Our data universe is not infinite. This is not only an ecological problem of high power consumption. It is also about the fact that the algorithms' appetite for data multiplies the options in terms of data abuse and raises questions about data protection, copyright, etc.
II. The history of Artificial Intelligence – starting with its name – is also the history of the use of distorting terms such as the talk of 'self-learning systems'. The degree of self-sufficiency that this term evokes does not exist. All algorithms made efficient by training owe their potential to depend on interactions with humans, whether in labelling data, in graduating parameters, in deciding when output is considered 'efficient enough', etc. Or this talk of data as 'raw material', which also resonates in the phrase 'data mining': Data are artifacts, even if the data are based on measurements of the real. They are human-made: manufactured, not found. Here, an alarming proximity to the idea of 'nature as raw material' comes into play and thus, to a worldview oriented towards the exploitation of nature, the limits of which we are now – sometimes dramatically – confronted with.
III. Finally, it is important to mention the dominance not only of the English language, but of the Anglo-Saxon cultural asset and heritage that goes into the huge training bases of contemporary Large Language Models. The queries and instructions possible in the national languages, as well as what the system provides colloquially, are based on (machine) translation.

Let's keep in mind: Mistakes of today's AI are the technical advances of tomorrow! For example, the metamorphosis into a racist led to the removal of chatbot Tay (released by Microsoft 2016) from the network, and this metamorphosis became an instructive topic of debate; similarly, BlenderBot (released by Meta 2022) mutated into a supporter of conspiracy theories. Learning algorithms mirror the practices on the basis of which they learn, as if through a magnifying glass: It is up to us to learn how to use the computer as an instrument of self-recognition – and not only in the form of the fitness bracelet. We should address Artificial Intelligence less from the perspective of modeling and technical projection of mind and intelligence, but more as a virtual mirror of human communication. Elena Esposito (2022) has convincingly argued that not Artificial Intelligence, but artificial communication, is the operational basis of current computer use.

A further, even more complex answer to the question of criticism suggests itself to me: Is the gesture of critique itself, which founded academic modernity, perhaps reaching its limits at present, as Rita Felski (2015) suspects?

The gesture of 'critique' is deeply anchored in the humanities' self-image of scholarly work. Unfairly shortened to the formula: Saying 'no' is always possible, saying 'yes' is under suspicion of apology. But are we as humanistic scholars really 'by profession' in the position of a meta-position towards that which we criticize? With the consequence that we are entitled to actually judge and evaluate from the superior standpoint of a knowledge that has both an affinity for technology and at the same time looks ahead to the future? Wasn't it precisely a concern of the convinced hermeneut Hans-Georg Gadamer (1975 [1960]) that humanities scholars should be regarded not in the bird's-eye perspective as observers, but in the participants' perspective as players in the events of the world, entangled in prejudices? Perhaps this is the reason why I do not focus on the critique of AI, but want to shake up the prejudices in which the humanities are caught when they take a stand on digitalization and Artificial Intelligence. To enlighten about technology means first to understand technology to some extent and second, to free its use from myths.

What is critical, is not so much AI itself as a technical endeavor, because we need technology to solve the problems of this planet in a way that can be both accepted and welcomed by the people whose behavior needs to change. Rather, what is critical, is our use and abuse of technical potential and the myths and ideologies surrounding it.

In fact, critical humanists like to focus on the ideologizations and mythicizations, apocalyptic and apologetic interpretations of Artificial Intelligence – and then often pass this off as a critique of AI itself or misinterpret it as such. I, therefore, argue for a kind of 'sobriety' in the discussion of AI. It is still about an – albeit interactive – 'toolbox', whose fields of application are growing by the hour, not to say proliferating.

It is not the intelligence and rationality of machines that we have to fear, but the irrationality of people.

List of references

Dobson, James. E (2019): Critical Digital Humanities. The Search for a Methodology, Champaign: University of Illinois Press.

Dreyfus, Hubert (1972): What Computers Can't Do. The Limits of Artificial Intelligence, New York: Harper & Row.

Esposito, Elena (2022): Artificial Communication. How Algorithms Produce Social Intelligence, Cambridge, MA: The MIT Press.

Felski, Rita (2015): The Limits of Critique, Chicago and London: The University of Chicago Press.

Gadamer, Hans Georg (1975 [1960]): Wahrheit und Methode. Grundzüge einer philosophischen Hermeneutik, 4th ed., Tübingen: J.C.B. Mohr.

Ginzburg, Carlo (1983): Spurensicherungen. Über verborgene Geschichte, Kunst und soziales Gedächtnis, Berlin: Wagenbach.

Gramelsberger, Gabriele (2008): "Computersimulationen – Neue Instrumente der Wissensproduktion." In: Renate Mayntz/Friedhelm Neidhardt/Peter Weingart/Ulrich Wengenroth (eds.), Wissensproduktion und Wissenstransfer: Wissen im Spannungsfeld von Wissenschaft, Politik und Öffentlichkeit, Bielefeld: transcript, pp. 75–96.

Kirschenbaum, Matthew (2012): Mechanisms: New Media and the Forensic Imagination. Cambridge, MA: The MIT Press.

Krämer, Sybille (2016): Leibniz on Symbolism as a Cognitive Instrument. In: Philosophy of Emerging Media ed. Juliet Floyd and James E. Katz, Oxford: Oxford University Press, pp. 307–318.

Krämer, Sybille (2023): "Should We Really 'Hermeneutise' the Digital Humanities? A Plea for the Epistemic Productivity of a 'Cultural Technique of Flattening' in the Humanities." In: Journal of Cultural Analytics 7/4 (https://doi.org/10.22148/001c.55592).

Krämer, Sybille (ed.) (1994): Geist, Gehirn, Künstliche Intelligenz: Zeitgenössische Modelle des Denkens. Ringvorlesung an der Freien Universität Berlin, Berlin and New York: De Gruyter.

Krämer, Sybille (ed.) (1998): Medien, Computer, Realität. Wirklichkeitsvorstellungen und Neue Medien, Frankfurt a.M.: Suhrkamp.

Moretti, Franco (2013): Distant Reading, London: Verso.

Mühlhoff, Rainer (2019): "Menschengestützte Künstliche Intelligenz. Über die soziotechnischen Voraussetzungen von 'deep learning'." In: Zeitschrift für Medienwissenschaft 11/2, pp. 56–64.

Snow, Charles Percy (1959): Two Cultures. The Rede Lecture: Cambridge: Cambridge University Press (https://www.rbkc.gov.uk/pdf/Rede-lecture-2-cultures.pdf).

Sudmann, Andreas (2018): "Szenarien des Postdigitalen: Deep Learning als MedienRevolution." In: Christoph Engemann/Andreas Sudmann (eds.), Machine Learning – Medien, Infrastrukturen und Technologien der Künstlichen Intelligenz, Bielefeld: transcript, pp. 55–74.

Turing, Alan (1950): "Computing Machinery and Intelligence." In: Mind 59/236, pp. 433–460.

Whitehead, Alfred N. (1911): An Introduction to Mathematics, London: Williams & Norgate.

List of contributors

Clemens Apprich is head of the Department of Media Theory as well as the Peter Weibel Research Institute for Digital Cultures at the University of Applied Arts in Vienna, where he holds the Professorship for Media Theory and History since 2021. He is guest researcher at the Centre for Digital Cultures at Leuphana University of Lüneburg, as well as an affiliated member of the Digital Democracies Institute at Simon Fraser University, of the Global Emergent Media Lab at Concordia University, and of the Research Centre for Media Studies and Journalism at the University of Groningen. His current research deals with filter algorithms and their application in data analysis as well as machine learning methods. Apprich is the author of "Technotopia: A Media Genealogy of Net Cultures" (Rowman & Littlefield International, 2017), and, together with Wendy Hui Kyong Chun, Hito Steyerl, and Florian Cramer, co-authored "Pattern Discrimination" (University of Minnesota Press/meson press, 2019). Currently, he is working on a new book about Animated Intelligence (Amsterdam University Press, forthcoming).

Gérard Biau is a full professor at the Probability, Statistics, and Modeling Laboratory (LPSM) of Sorbonne University, Paris. His research is mainly focused in developing new methodologies and rigorous mathematical theory in statistical learning and artificial intelligence, whilst trying to find connections between statistics and algorithms. He was a member of the Institut Universitaire de France from 2012 to 2017 and served from 2015 to 2018 as the president of the French Statistical Society. In 2018, he was awarded the Michel Monpetit – Inria prize by the French Academy of Sciences. He is currently director of Sorbonne Center for Artificial Intelligence (SCAI).

Isabelle Bloch graduated from the Ecole des Mines de Paris, Paris, France, in 1986, and received the master degree from the University Paris 12, Paris, in 1987, the Ph.D. degree from the Ecole Nationale Supérieure des Télécommunications (Télécom Paris), Paris, in 1990, and the Habilitation degree from University Paris 5 (Université Paris Cité), Paris, in 1995. She has been a Professor at Télécom Paris until 2020 and is now a Professor at Sorbonne Université. Her current research interests include symbolic and hybrid artificial intelligence, mathematical morphology, spatial reasoning, fuzzy set theory, 3D image understanding, structural, graph-based, and knowledge-based object recognition, and medical imaging.

Johannes Breuer is a senior researcher in the team Survey Data Augmentation at GESIS – Leibniz Institutes for the Social Sciences and (co-) leads the team Research Data & Methods at the Center for Advanced Internet Studies (CAIS). His research interests include the use and effects of digital media, computational methods, and open science. Further information: https://www.johannesbreuer.com/

Anna Echterhölter is professor for the history of science at the University of Vienna. She habilitated in 2017 at the Humboldt University Berlin. In addition to fellowships at the Max Planck Institute for the History of Science in Berlin and the GHI in Washington, she held interim professorships at the Humboldt and Technical University Berlin. Her research interests include the history of data and quantification, epistemic decolonization, German Pacific colonies.

Engineer and philosopher by training, **Jean-Gabriel Ganascia** is currently Professor of Computer Science at the Sorbonne University. He continues his research in artificial intelligence at LIP6 (Laboratory of Computer Science of Sorbonne University), where he leads the ACASA team. A specialist in data mining and machine learning, his current research activities focus on the literary aspects of digital humanities, computational ethics, and the ethics of digital technologies. He is a Fellow of EurAI (European Association for Artificial Intelligence) and a member of the CPEN-CCNE, the Subcommittee on Digital Ethics of the CCNE (French National Ethics Committee). He is also Chairman of the Steering Committee of the CHEC (Cycle des Hautes Études de la Culture), Chairman of the Ethics Committee of "Pôle Emploi", the French National Employment Agencies, and Chairman of the AFAS ("Association Française pour l'Avancement des Sciences"). Finally, he was the Chairman of the Ethics Committee of the CNRS (COMETS) between 2016 and 2021.

Matthieu Komorowski is a Clinical Senior Lecturer at Imperial College London and a consultant in intensive care and anaesthetics at Charing Cross Hospital in London. He was previously a visiting scholar at MIT, and an associate of Harvard University where he taught on machine learning in healthcare. Matthieu holds additional qualifications in space, mountain, diving and hyperbaric medicine. In 2022, he reached the final stages of the European Space Agency astronaut selection. His research group at Imperial College London has secured around £1.3 million in funding to develop and test artificial intelligence-based tools for sepsis in the NHS.

Sybille Krämer was Professor of Philosophy at the Free University of Berlin until her retirement in April 2018 and has been Senior Professor ('Guest Researcher') at the Leuphana University of Lüneburg since March 2019. The University of Linköping, Sweden, awarded her an honorary doctorate (Dr. h.c.) in 2016.

Giacomo Landeschi is Associate Professor of Archaeology and researcher at Lund University. He obtained his master's degree from the University of Pisa in 2005 and received his PhD from the IMT Institute for Advanced Studies in Lucca. As of 2013, Landeschi is employed by Lund University where he is one of the founding members of the Digital Archaeology Lab and research engineer in the Humanities Lab. His research interests include digital archaeology, archaeological theory, and landscape archaeology. Landeschi has been teaching BA and MA-level courses in several institutions including University of Pisa, Lund University, Umeå University, University of Copenhagen.

Sabina Leonelli is Professor of Philosophy and History of Science at the University of Exeter; Director of the Centre for the Study of the Life Sciences (Egenis); lead of the "Data Governance, Ethics and Openness" strand of the Exeter Institute for Data Science and Artificial Intelligence; Fellow of the Alan Turing Institute; Editor-in-Chief of History and Philosophy of the Life Sciences and Associate Editor of the Harvard Data Science Review; and recipient of the Lakatos Award 2018 and the Patrick Suppes Prize 2022 for research on the epistemology of data-intensive science. She has been awarded many competitive grants including two from the European Research Council, "The Philosophy of

Data-Intensive Science" (2014–2019, www.datastudies.eu) and "A Philosophy of Open Science for Diverse Research Environments" (2021–2026, www.opensciencestudies.eu). She is an alumna of the Global Young Academy and regularly engages with a variety of national and international initiatives in science policy, especially concerning the governance and use of large data infrastructures.

Matteo Pasquinelli is Professor in Media Philosophy at the University of Arts and Design Karlsruhe where he is coordinating the research group on Artificial Intelligence and Media Philosophy KIM. His research focuses the intersection of philosophy of mind, political economy and the automation of knowledge and cultural production. He edited the anthology "Alleys of Your Mind: Augmented Intelligence and Its Traumas" (Meson Press, 2015) and wrote, with Vladan Joler, the visual essay 'The Nooscope Manifested: AI as Instrument of Knowledge Extractivism' (AI & Society, 2022; also online as nooscope.ai). For Verso Books he wrote the monograph "The Eye of the Master: A Social History of Artificial Intelligence" (October 2023). Among others, he published essays for the Journal of Interdisciplinary History of Ideas, Qui Parle, Radical Philosophy, Les Mondes du Travail, South Atlantic Quarterly, Parrhesia, Theory Culture & Society, Multitudes, and e-flux.

Evangelos Pournaras is Associate Professor at University of Leeds, UK, where he leads the Distributed Intelligent Social Computing (DISC) lab. He is also a UKRI Future Leaders Fellow, an Alan Turing Fellow and a research associate at the UCL Center of Blockchain Technologies. He has a significant track record of high-profile publications and international research experience at ETH Zurich, EPFL, Delft University of Technology (PhD) and IBM T.J. Watson Research Center. Evangelos has won the Augmented Democracy Prize, the 1st prize at ETH Policy Challenge, including 5 paper awards and honors. Two of his projects are in the UNESCO IRCAI Global Top-100 list of AI projects that tackle sustainable development goals.

Markus Elias Ramsauer is a PhD candidate in the Department of History at the University of Vienna as part of the Volkswagen Foundation funded research project "How is Artificial Intelligence Changing Science? Research in the Era of Learning Algorithms". He has completed MA degrees in Cultural and Social Anthropology as well as in the History and Philosophy of Science at the University of Vienna. His current research interests are centered around questions of world and large-scale modeling in the second half of the 20th century.

Fabian Retkowski is a PhD candidate associated with the Interactive Systems Lab (ISL) at the Karlsruhe Institute of Technology (KIT). He holds a master's degree in computer science from KIT, earned in 2020. Currently, Mr. Retkowski is actively involved in the research project "How is Artificial Intelligence Changing Science?" funded by Volkswagen Foundation. His research focuses on the fields of artificial intelligence and natural language processing, particularly on topical segmentation and text summarization.

Gabriele Schabacher is Professor of Media and Culture Studies at the Institute for Film, Theatre, Media, and Cultural Studies (FTMK) at Johannes Gutenberg University Mainz. Since 2021 she is deputy spokesperson of the CRC 1482 "Studies in Human Categorization" and principal investigator of the project "Urban Control Regimes. Railway Stations as Infrastructure of Human Categorization". Her research areas include the media history of traffic, mobility, and infrastructures, digital technologies of surveillance, the cultural techniques of repair, the media history of seriality, and the theory of autobiography. Among her recent publications is a monograph on infrastructure work (2022), co-edited volumes on the cultures of repair (2018, together with Stefan Krebs and Heike Weber) and on the practices of workarounds (2017, together with Holger Brohm, Sebastian Gießmann, and Sandra Schramke), as well as the articles "In Control of Algorithms: Video Analytics and Human–Machine Relations at the Train Station" (forthcoming) and "Time and Technology. The Temporalities of Care" (2021).

Jens Schröter, Prof. Dr., is chair for media studies at the University of Bonn since 2015. Since 4/2018 director (together with Anja Stöffler, Mainz) of the DFG-research project "Van Gogh TV. Critical Edition, Multimedia-documentation and analysis of their Estate" (3 years). Since 10/2018 speaker of the research project (Volkswagen Foundation; together with Prof. Dr. Gabriele Gramelsberger; Dr. Stefan Meretz; Dr. Hanno Pahl and Dr. Manuel Scholz-Wäckerle) "Society after Money – A Simulation" (4 years). Director (together with Prof. Dr. Anna Echterhölter; PD Dr. Sudmann and Prof. Dr. Alexander Waibel) of the Volkswagen Foundation main grant "How is Artificial Intelligence Changing Science?" (Start: 1.8.2022, 4 years); April/May 2014: "John von Neumann"-fellowship at the University of Szeged, Hungary. September

2014: Guest Professor, Guangdong University of Foreign Studies, Guangzhou, People's Republic of China. Winter 2014/15: Senior-fellowship at the research group "Media Cultures of Computer Simulation". Summer 2017: Senior-fellowship IFK Vienna, Austria. Winter 2018: Senior-fellowship IKKM Weimar. Winter 2021/22: Fellowship, Center of Advanced Internet Studies. Recent publications: Medien und Ökonomie, Wiesbaden: Springer 2019; (together with Christoph Ernst): Media Futures. Theory and Aesthetics, Basingstoke: Palgrave 2021; (together with Julia Eckel, Christoph Ernst, eds.): Tech /Demos (Navgationen 1, 23). Visit www.medienkulturwissenschaft-bonn.de / www.theorie-der-medien.de / www.fanhsiu-kadesch.de

Urvi Sonawane is a final year medical student studying at Imperial College London. She completed her Remote Medicine intercalated BSc with 1st class honours and received the National Lung and Heart Institute Outstanding Achievement Prize in Remote Medicine for her final project. This project involved examining the impact of shear forces from facemasks on the skin, and testing the effects of an intervention on comfort, erythema and inflammation. Aside from this, she is an active volunteer for Streetdoctors, which is a national charity that aims to provide first-aid skills to young people. She had the role of co-team lead for the West London branch, during which 520 young people were trained. She has a keen interest in acute care and anaesthetics and is fascinated by the ways in which artificial intelligence may change healthcare in the future.

Andreas Sudmann teaches media studies at the University of Bochum and currently serves as the scientific coordinator and principal investigator for the research group "How is Artificial Intelligence Changing Science? Research in the Era of Learning Algorithms" (funded by the Volkswagen Foundation, started in 2019/2022) at the University of Bonn. Current research interests include: digital media cultures and in particular artificial intelligence, digital methods, theory and history of media, media critique, science and technology studies.

Alexander Waibel is a Professor of Computer Science at Carnegie Mellon University (USA), and at the Karlsruhe Institute of Technology (Germany). He is the director of the International Center for Advanced Communication Technologies (interACT). Prof. Waibel is known for his work on AI, Machine Learning, Multimodal Interfaces and Speech Translation Systems. He proposed early Neural Network based Speech and Language systems, including the TDNN (the first shift-invariant "Convolutional" Neural Network), Modular Neural Networks neural transfer learning and sequential learning models. Combining advances in ML with work on multimodal interfaces, Waibel and his team developed pioneering solutions to cross-lingual communication, including the first consecutive (1991) & simultaneous (2005) translation systems, mobile speech translators, multimodal smart rooms and human-robot interfaces. Waibel published extensively in the field (>800 papers, >30.000 citations) and received many patents and awards. He is a member of the National Academy of Sciences of Germany and a Fellow of the IEEE. He received his BS, MS and PhD degrees from MIT and CMU, respectively.

Sabine Wirth is Junior Professor for Digital Cultures at the Department of Media Studies at Bauhaus-Universität Weimar. Her research interests are the history and media theory of interfaces within personal, mobile and ubiquitous computing environments, digital image cultures, and the popularization of AI technologies. Current research project: "Curating the Feed: Interdisciplinary Perspectives on Digital Image Feeds and Their Curatorial Assemblages".

Editorial

Since Kant, critique has been defined as the effort to examine the way things work with respect to the underlying conditions of their possibility; in addition, since Foucault it references a thinking about "the art of not being governed like that and at that cost." In this spirit, **KI-Kritik / AI Critique** publishes recent explorations of the (historical) developments of machine learning and artificial intelligence as significant agencies of our technological times, drawing on contributions from within cultural and media studies as well as other social sciences.
The series is edited by Anna Tuschling, Andreas Sudmann and Bernhard J. Dotzler.

Andreas Sudmann (PD Dr.) is a media scholar at the universities of Bochum and Bonn in Germany. His research interests include AI, digital cultures, media theory, history of media, and media critique.
Anna Echterhölter (Prof. Dr.) is professor of history of science at Universität Wien. Her main research areas are the history of data and German colonialism.
Markus Ramsauer is PhD candidate in history of science at the Department of History at Universität Wien.
Fabian Retkowski is PhD candidate in computer science at the Institute of Anthropomatics at Karlsruhe Institute of Technology.
Jens Schröter (Prof. Dr.) holds the Chair of Media Studies at Rheinische Friedrich-Wilhelms-Universität Bonn. His main research area is the theory and history of digital media.
Alexander Waibel (Prof. Dr.) works at the Institute of Anthropomatics at Karlsruhe Institute of Technology. His main research areas are artificial intelligence, machine learning, automatic speech recognition & translation, multimodal and perceptual user interfaces as well as neural networks.